ETHNOBOTANY OF INDIA

Volume 2

Western Ghats and West Coast of Peninsular India

ETHNOBOTANY OF INDIA

Volume 2

Western Ghats and West Coast of Peninsular India

Edited by

T. Pullaiah, PhD

K. V. Krishnamurthy, PhD

Bir Bahadur, PhD

APPLE ACADEMIC PRESS

Apple Academic Press Inc. | Apple Academic Press Inc.
3333 Mistwell Crescent | 9 Spinnaker Way
Oakville, ON L6L 0A2 | Waretown, NJ 08758
Canada | USA

© 2017 by Apple Academic Press, Inc.
First issued in paperback 2021
Exclusive worldwide distribution by CRC Press, a member of Taylor & Francis Group
No claim to original U.S. Government works

ISBN-13: 978-1-77463-120-1 (pbk)
ISBN-13: 978-1-77188-404-4 (hbk)

Library and Archives Canada Cataloguing in Publication

Ethnobotany of India / edited by T. Pullaiah, PhD, K. V. Krishnamurthy, PhD, Bir Bahadur, PhD.

Includes bibliographical references and indexes.
Contents: Volume 2. Western Ghats and West Coast of Peninsular India.
Issued in print and electronic formats.
ISBN 978-1-77188-404-4 (v. 2 : hardcover).--ISBN 978-1-77188-405-1 (v. 2 : pdf)

1. Ethnobotany--India. I. Pullaiah, T., author, editor II. Krishnamurthy, K. V., author, editor III. Bahadur, Bir, author, editor

GN635.I4E85 2016   581.6'30954   C2016-902513-6   C2016-902514-4

Library of Congress Cataloging-in-Publication Data

Names: Pullaiah, T., editor.
Title: Ethnobotany of India. Volume 1, Eastern Ghats and Deccan / editors: T. Pullaiah, K. V. Krishnamurthy, Bir Bahadur.
Other titles: Eastern Ghats and Deccan
Description: Oakville, ON ; Waretown, NJ : Apple Academic Press, [2016] |
Includes bibliographical references and index.
Identifiers: LCCN 2016017369 (print) | LCCN 2016021535 (ebook) | ISBN 9781771883382 (hardcover : alk. paper) | ISBN 9781771883399 ()
Subjects: LCSH: Ethnobotany--India--Eastern Ghats. | Ethnobotany--India--Deccan.
Classification: LCC GN476.73 .E857 2016 (print) | LCC GN476.73 (ebook) | DDC 581.6/309548--dc23
LC record available at https://lccn.loc.gov/2016017369

Apple Academic Press also publishes its books in a variety of electronic formats. Some content that appears in print may not be available in electronic format. For information about Apple Academic Press products, visit our website at **www.appleacademicpress.com** and the CRC Press website at **www.crcpress.com**

Ethnobotany of India 5-volume series

Editors: T. Pullaiah, PhD, K. V. Krishnamurthy, PhD, and Bir Bahadur, PhD

Volume 1: Eastern Ghats and Deccan

Volume 2: Western Ghats and West Coast of Peninsular India

Volume 3: North-East India and Andaman and Nicobar Islands

Volume 4: Western and Central Himalaya

Volume 5: Indo-Gangetic Region and Central India

ABOUT THE EDITORS

T. Pullaiah, PhD

Former Professor, Department of Botany, Sri Krishnadevaraya University, Anantapur, Andhra Pradesh, India

T. Pullaiah, PhD, is a former Professor at the Department of Botany at Sri Krishnadevaraya University in Andhra Pradesh, India, where he has taught for more than 35 years. He has held several positions at the university, including Dean, Faculty of Biosciences, Head of the Department of Botany, Head of the Department of Biotechnology, and Member, Academic Senate. He was President of the Indian Botanical Society (2014), President of the Indian Association for Angiosperm Taxonomy (2013), and Fellow of the Andhra Pradesh Akademi of Sciences. He was awarded the Panchanan Maheswari Gold Medal, the Dr. G. Panigrahi Memorial Lecture Award of the Indian Botanical Society, the Prof. Y. D. Tyagi Gold Medal of the Indian Association for Angiosperm Taxonomy, and a Best Teacher Award from Government of Andhra Pradesh. Under his guidance 53 students obtained their doctoral degrees. He has authored 45 books, edited 15 books, and published over 300 research papers, including reviews and book chapters. His books include *Flora of Eastern Ghats* (4 volumes), *Flora of Andhra Pradesh* (5 volumes), *Flora of Telangana* (3 volumes), *Encyclopedia of World Medicinal Plants* (5 volumes), and *Encyclopedia of Herbal Antioxidants* (3 volumes). He was also a member of Species Survival Commission of the International Union for Conservation of Nature (IUCN). Professor Pullaiah received his PhD from Andhra University, India, attended Moscow State University, Russia, and worked as postdoctoral Fellow during 1976-78.

K. V. Krishnamurthy, PhD

Former Professor, Department of Plant Sciences, Bharathidasan University, Tiruchirappalli, Tamill Nadu, India

K. V. Krishnamurthy, PhD, is a former Professor and Head of Department, Plant Sciences at Bharathidasan University in Tiruchirappalli, India, an adjunct faculty at the Institute of Ayurveda and Integrative Medicine, Bangalore and is at present Consultant, Sami Labs Ltd, Bangalore. He obtained his PhD degree from Madras University, India, and has taught many undergraduate, postgraduate, MPhil, and PhD students. He has over 50 years of teaching and research experience, and his major research areas include plant morphology and morphogenesis, biodiversity, floristic and reproductive ecology, and cytochemistry. He has published more than 185 research papers and 28 books, operated 16 major research projects funded by various agencies, and guided 32 PhD and more than 50 MPhil scholars. His important books include *Methods in Cell Wall Cytochemistry* (CRC Press, USA), *Textbook of Biodiversity* (Science Publishers, USA), Bioresources of Eastern Ghats (Bishen Singh and Mahendra Paul Singh, Dehra Dun, India), Plant Biology and Biotechnology (2 volumes, Springer India, New Delhi) and *From Flower to Fruit* (Tata McGraw-Hill, New Delhi). One of his important research projects pertains to a detailed study of Shervaroys, which form a major hill region in the southern Eastern Ghats, and seven of his PhD scholars have done research work on various aspects of Eastern Ghats. He has won several awards and honors that include the Hira Lal Chakravarthy Award (1984) from the Indian Science Congress; Fulbright Visiting Professorship at the University of Colorado, USA (1993); Best Environmental Scientist Award of Tamil Nadu state (1998); the V. V. Sivarajan Award of the Indian Association for Angiosperm Taxonomy (1998); and the Prof. V. Puri Award from the Indian Botanical Society (2006). He is a fellow of the Linnaean Society, London; National Academy of Sciences, India; and Indian Association of Angiosperm Taxonomy.

Bir Bahadur, PhD

Former Professor, Department of Botany, Kakatiya University, Warangal, Telangana, India

Bir Bahadur, PhD, was Chairman and Head of the Department, and Dean of the Faculty of Science at Kakatiya University in Warangal, India, and has also taught at Osmania University in Hyderabad, India. During his long academic career, he was honored with the Best Teacher Award by Andhra Pradesh State Government for mentoring thousands of graduates and postgraduate students, including 30 PhDs, most of whom went on to occupy high positions at various universities and research organizations in India and abroad. Dr. Bahadur has been the recipient of many awards and honors, including the Vishwambhar Puri Medal from the Indian Botanical Society for his research contributions in various aspects of plant Sciences. He has published over 260 research papers and reviews and has authored or edited dozen books, including *Plant Biology and Biotechnology* and *Jatropha, Challenges for New Energy Crop,* both published in two volumes each by Springer Publishers. Dr. Bahadur is listed as an Eminent Botanist of India, the Bharath Jyoti Award, New Delhi, for his sustained academic and research career at New Delhi and elsewhere. Long active in his field, he was a member of over dozen professional bodies in India and abroad, including Fellow of the Linnean Society (London); Chartered Biologist Fellow of the Institute of Biology (London); Member of the New York Academy of Sciences; and a Royal Society Bursar. He was also honored with an Honorary Fellowship of Birmingham University (UK). Presently, he is an Independent Director of Sri Biotech Laboratories India Ltd., Hyderabad, India.

for his research contributions in various aspects of plant Sciences. He has published over 260 research papers and reviews and has authored or edited dozen books, including Plant Biology and Biotechnology and Vetiver Chemistry for New Energy Crop, both published in two volumes each by Springer Publishers. Dr. Bahadur is listed as an Eminent Botanist of India by the Bharath Jyoti Award, New Delhi, for his sustained academic and research career at New Delhi and elsewhere. Long active in his field, he was a member of over dozen professional bodies in India and abroad, including Fellow of the Linnean Society (London), Chartered Biologist Fellow of the Institute of Biology (London), Member of the New York Academy of Sciences, and a Royal Society Bursar. He was also honored with an Honorary Fellowship of Birmingham University (UK). Presently, he is an Independent Director of Sri Biotech Laboratories India Ltd., Hyderabad, India.

CONTENTS

LIST OF CONTRIBUTORS

S. John Adams
Department of Pharmacognosy, R&D, The Himalaya Drug Company, Makali, Bangalore, India

Afroz Alam
Department of Bioscience and Biotechnology, Banasthali University, Rajasthan–304022, India. E-mail: afrozalamsafvi@gmail.com

Bir Bahadur
Department of Botany, Kakatiya University, Warangal 506009, Telangana, India. E-mail: birbahadur5april@gmail.com

S. Noorunnisa Begum
Centre of Repository of Medicinal Resources, School of Conservation of Natural Resources, Foundation for Revitalization of Local Health and Traditions, 74/2, Jarakabande Kaval, Attur PO, Via Yelahanka, Bangalore 560–106, India. E-mail: noorunnisa.begum@frlht.org

G. Hariramamurthi
Trans-Disciplinary University (TDU), School of Health Sciences, FRLHT, 74/2, Jarakabande Kaval, Attur post, Via Yelahanka, Bangalore – 560064, Karnataka, India. E-mail: hariram_01@yahoo.com

Namera C. Karun
Department of Biosciences, Mangalore University, Mangalagangotri, Mangalore – 574199, Karnataka, India, E-mail: karunchinnappa@gmail.com

S. Karuppusamy
Department of Botany, The Madura College (Autonomous), Madurai – 625011, Tamilnadu, India. E-mail: ksamytaxonomy@gmail.com

K. V. Krishnamurthy
Consultant, R&D, Sami Labs, Peenya Industrial Area, Bangalore – 560058, Karnataka, India. E-mail: kvkbdu@yahoo.co.in

K. Ravi Kumar
Centre of Repository of Medicinal Resources, School of Conservation of Natural Resources, Foundation for Revitalization of Local Health and Traditions, 74/2, Jarakabande Kaval, Attur PO, Via Yelahanka, Bangalore 560–106, India. E-mail: k.ravikumar@frlht.org

M. N. B. Nair
Trans-disciplinary University, Veterinary Ayurveda Group, School of Health Sciences, 74/2, Jarakabandekaval, Attur Post, Yelahanka, Bangalore, India, E-mail: nair.mnb@frlht.org

B. N. Prakash
Trans-Disciplinary University (TDU), School of Health Sciences, FRLHT, 74/2, Jarakabande Kaval, Attur post, Via Yelahanka, Bangalore – 560064, Karnataka, India. E-mail: vaidya_bnprakash@yahoo.co.in

T. Pullaiah
Department of Botany, Sri Krishnadevaraya University, Anantapur – 515003, Andhra Pradesh, India. E-mail: pullaiah.thammineni@gmail.com

N. Punniamurthy
Veterinary University Training and Research Centre, Pillayarpatty, Thanjavur–Trichirapally National Highway (Vallam Post), Near RTO's Office, Thanjavur – 613403, India, E-mail: thanjavurvutrc@tanu-vas.org.in

B. S. Somashekhar
Fulbright Fellow, Associated with: FRLHT and IINDICUS (Institute for Indigenous Cultures and Studies), 74/2, Jarakabande Kaval, Attur P.O., via Yelahanka, Bengaluru, 560106, India. E-mail: bssomashek-har@hotmail.com, cultureindicus@yahoo.in

Kandikere R. Sridhar
Department of Biosciences, Mangalore University, Mangalagangotri, Mangalore – 574199, Karnataka, India, E-mail: kandikere@gmail.com

P. M. Unnikrishnan
United Nations University – Institute for the Advanced Study of Sustainability, Tokyo, Japan, E-mail: unnipm@gmail.com

D. K. Ved
Centre of Repository of Medicinal Resources, School of Conservation of Natural Resources, Foundation for Revitalization of Local Health and Traditions, 74/2, Jarakabande Kaval, Attur PO, Via Yelahanka, Bangalore 560–106, India. E-mail: dk.ved@frlht.org

LIST OF ABBREVIATIONS

AICRPE	All India Coordinated Research Project on Ethnobiology
AMR	antimicrobial resistance
ANI	Ancient North Indian
ASI	Ancestral South Indian
ASI	Anthropological Survey of India
CAMP	conservation assessment and management prioritization
CE	common era
EIC	English East India Company
FPIC	free, prior informed consent
FRLHT	Foundation for Revitalization of Local Health Traditions
IK	indigenous knowledge
LEK	local ecological knowledge
MPCAs	medicinal plants conservation areas
NBR	Nilgiri Biosphere Reserve
NWFP	non-wood forest produce
PURSE	Promotion of University Research and Scientific Excellence
RALHT	Rapid Assessment of Local Health Traditions
RBG	Royal Botanic Gardens
RP	reverse pharmacology
SGs	sacred groves
TAMP	Threat Assessment and Management planning
TBGRI	Tropical Botanical Garden and Research Institute
TDR	trans-disciplinary research
TEK	traditional ecological knowledge
VOC	*Verenigde Oost-Indische Compagnie*
ZOO	Zoo Outreach Organization

PREFACE

Humans are dependent on plants for their food, medicines, clothes, fuel and several other needs. Although the bond between plants and humans is very intense in several 'primitive' cultures throughout the world, one should not come to the sudden and wrong conclusion that post-industrial modern societies have broken this intimate bond and interrelationship between plants and people. Rather than, plants being dominant as in the 'primitive' societies, man has become more and more dominant over plants, leading to over-exploitation of the latter, and resulting in a maladapted ecological relationship between the two. Hence a study of the relationships between plants and people—ethnobotany and, thus, between plant sciences and social sciences, is central to correctly place humanity in the earth's environment. Because ethnobotany rightly bridges both of these perspectives, it is always held as a synthetic discipline.

Most people tend to think that ethnobotany, a word introduced by Harshberger in 1896, is a study of plants used by 'primitive' cultures in 'exotic' locations of the world, far removed from the mainstream people. People also think wrongly that ethnobotany deals only with non-industrialized, non-urbanized and 'non-cultured' societies of the world. Ethnobotany, in fact, studies plant-human interrelationships among all peoples and among all. However, since indigenous non-western societies form the vast majority of people now as well as in the past, a study of their interrelationships with people becomes important. More than 10,000 human cultures have existed in the past and a number of them persist even today. They contain the knowledge system and wisdom about the adaptations with diverse nature, particularly with plants, for their successful sustenance. Thus, ethnobotanical information is vital for the successful continuance of human life on this planet.

Ethnobotany is of instant use in two very important respects: (i) indigenous ecological knowledge, and (ii) source for economically useful plants. The first will help us to find solutions to the increasing environmental degradation and the consequent threat to our biodiversity. In indigenous societies biodiversity is related to cultural diversity and hence any threat to biodiversity would lead to erosion in cultural diversity. Indigenous cultures are not only repositories of past experiences and knowledge but also form the frameworks for future adaptations. Ethnic sources of economically useful plants have resulted in serious studies on bioprospection for newer sources of food, nutraceuticals, medicines and other novel materials of human use. Bioprospecting has resulted in intense research on reverse pharmacology and pharmacognosy. This has resulted in attendant problems relating to intellectual property rights, patenting and the sharing of the benefits with the traditional societies who owned the knowledge. This has also resulted in serious documentation of traditional knowledge of the different cultures of the world and to formalize the methods and terms of sharing this traditional knowledge. It has also made us to know not only *what* plants people in different cultures use and *how* they use them, but also *why* they use them. In addition it helps us to know the biological, sociological, cultural roles of plants important in human adaptations to particular environmental conditions in the past, present and future.

This series of the five edited volumes on ethnobotany of different regions of India tries to bring together all the available ethnobotanical knowledge in one place. India is one of the most important regions of the old world, which has some of the very ancient and culturally rich diverse knowledge systems in the world. Competent authors have been selected to summarize information on the various aspects of ethnobotany of India, such as ethnoecology, traditional agriculture, cognitive ethnobotany, material sources, traditional pharmacognosy, ethnoconservation strategies, bioprospection of ethnodirected knowledge, and documentation and protection of ethnobotanical knowledge.

The first volume was on Eastern Ghats and adjacent Deccan region of Peninsular India, while the second is on Western Ghats and Western Peninsular India, a region of very great socio-cultural history of not only India but also the whole Indian Ocean region. Published information is summarized on different aspects. Our intention is that this may lead to discovery of many drugs, nutraceuticals and other useful products for the benefit of mankind.

Since it is a voluminous subject we might have not covered the entire gamut, but we have tried to put together as much information as possible. Readers are requested to give their suggestions for improvement of the remaining volumes.

ACKNOWLEDGMENTS

We wish to express our grateful thanks to all the authors who contributed their research/review articles. We thank them for their cooperation and erudition. We also thank several colleagues for their help in many ways and for their suggestions from time to time during the evolution of this attractive and readable volume.

We wish to express our appreciation and help rendered by Ms. Sandra Sickels, Rakesh Kumar, and the staff of Apple Academic Press. Above all, their professionalism has made this book a reality and is greatly appreciated.

We thank Mr. John Adams, Senior Research Fellow of Prof. K.V. Krishnamurthy, for his help in many ways.

We wish to express our grateful thanks to our respective family members for their cooperation.

We hope that this book will help our fellow teachers and researchers who enter the world of the fascinating subject of ethnobotany in India with confidence, as we perceived and planned.

CHAPTER 1

INTRODUCTION

K. V. KRISHNAMURTHY,[1] BIR BAHADUR,[2] and T. PULLAIAH[3]

[1]Consultant, R&D, Sami Labs, Peenya Industrial Area, Bangalore–560058, Karnataka, India

[2]Department of Botany, Kakatiya University, Warangal–506009, Telangana State, India

[3]Department of Botany, Sri Krishnadevaraya University, Anantapur–515003, India

CONTENTS

ABSTRACT

This chapter deals with an introduction to the subsequent chapters covered in this volume. It describes the physical and biological features of the West Coast and Western Ghats of Peninsular India. It also deals with an introduction to the study areas and their ethnic diversity and plants of importance in medicine, food and other requirements of tribal people. An introduction to conservation measures involving sacred groves is also given. The ethnobotany of mangroves and bryophytes of these regions is also included.

1.1   THE WEST COAST AND THE WESTERN GHATS

The Indian subcontinent consists of the Himalayan mountains as the northern border, the almost flat expanse of the Indo-Gangetic plains in the middle and the triangular peninsular India with uplands and plateaus in the south bordered on both sides by narrow coastal plains along the seaboards (Bay of Bengal on the east, Arabian Sea on the west and the Indian Ocean on the South) (Valdiya, 2010). Each of these regions has not only distinct structural and lithographic features and physiography but also different geologic evolutionary histories. Peninsular India is 2,200 km long in the N-S direction and 1,400 km broad (in the broadest region) in the E-N direction, with its apex terminating at Kanyakumari at the extreme south. The western upland of peninsular India forms the Sahyadri Range (08° 19′ 18″–21° 16′ 24″ N and 72° 56′ 24″–78° 19′ 40′E). This range extends 1,600 km southwards from the Tapti river valley in southern Gujarat to Kanyakumari in southernmost Tamil Nadu. This range is conveniently divided into three zones (Valdiya, 2010): (i) Northern Sahyadri in Gujarat and Maharashtra, which is made of Late Cretaceous basaltic lavas; (ii) Central Sahyadri in Goa and Karnataka, which is made of Archaean gneisses and high-grade metamorphic rocks; and (iii) Southern Sahyadri in Kerala and Tamil Nadu which is made of Late proterozoic Charnockites and Khondalites. According to Valdiya (2010) "the NNW-SSE-trending fractures and faults, defining the ranges forming linear blocks make the Sahyadri a horst mountain of sorts. Its west-facing steep to a near vertical flank is characterized by a multiplicity of precipitous encarpments disposed en echelon and alternating with very narrow irregular terraces. These features have given rise to a 'landing stair' known as the Western Ghats." Thus, the mountain range is the Sahyadri and the escarpment is the Western Ghats (W. Ghats). The Sahyadri hill range meets the Eastern hill ranges (E. Ghats) at Nilgiris area. The Sahyadri hill range is located about 15–100 km inland from the west coast, depending on its location. The total estimated area of W. Ghats is 1,64,280 km^2, which is about 5% of the total area of India (Nayar et al., 2014).

On the northern Sahyadri and on a larger part of the central Sahyadri the eastern flank slopes gently eastwards; it drains the Godavari, Krishna, Tungabhadra, and Kaveri rivers, all of which flow eastward over long distances to discharge their water into the Bay of Bengal. There are also west-flowing shorter rivers, such as Ulhas, Vaitarni, Kalinadi, Gangavali, Sharavati, Netravati, Mandovi, Payaswini, and Valapattan that arise in the hill range; these rivers often flow deep vertically down (sometimes as waterfalls, such as Jog falls). From the southern Sahyadri arise the rivers like Noyyal and Vaigai that flow eastward, and Ponnani, Periyar, Pamba, Achankovil and Kakkad that flow westward.

The average elevation of the Sahyadri is 1,000 to 1,200 m. The high peaks are Salher (1,567 m), Harishchandragarh (1,424 m), Mahabaleshwar (1,438 m), Kalsubai (1,714 m), Thadiannamalai (1,745 m), Doddabetta (2,637 m), Kolaribetta (2629 m), Mukurti (2554 m), Anaimudi (2695 m), Vavulmala (2339 m), Kodaikanal (2133 m), Chembra Peak (2100 m), Elivaimala (2088 m), Banasura (2073 m), Kottamala (2019 m), Meesapulimala (2640 m), Elaimalai (2,670 m), Vandaravu peak (2553 m), Kattumala (2552 m), and Anginda (2383 m). In general. southern Sahyadri has taller peaks than the central or northern Sahyadri. Between the Nilgiris and Anaimalais is the Palghat gap (25–30 km wide) (Valdiya, 2010). There is also a minor Sengotta gap at the extreme south. The rainfall in the hill range ranges from less than 1,000 mm to over 7,450 mm, greater rainfall being seen in southern Sahyadri. The rain-shadow regions (on eastern slopes) get only 500 mm rainfall on an average. The average temperature ranges between 15–24°C.

The West Coast (W. Coast) is characterized by an array of near-shore terrestrial cliff faces, dunes, sandy shores and urban, village, agricultural and industrial landscapes. There are also near-shore islands (for example the Anjdiv Island off Goa). In some places there are estuaries, coves, deltas, lagoons, embayments, backwaters, mangrove vegetation, salt marshes, mud flats and salt panes. The Konkan Coast is 8–24 km wide and is a rocky shore of cliffs, bays, coves and small beaches. The Kanara Coast (in Karnataka) is 30–50 km wide, becoming 70 km wide near Mangalore. The Malabar Coast (in Kerala) is 20–100 km wide. A 80 km long and 5–10 km wide lagoon is barred by a 55 km long sand spit and this gives rise to the Vembanad Lake. There are a few more barred lagoons in the Malabar Coast. The mangrove systems of W. Coast are strikingly very small and patchy here and there when compared to those of the E. Coast. The area covered by brackish water in W. Coast is around 3,30,000 ha, while that of Kerala alone is around 500 km^2.; the Vembanad brackish water area alone has an area of 200 km^2. The W. Coast of peninsular India is affected by very prominent NNW-SSE faults, cut and locally displaced by ESE-WNW to E-W oriented shear zones. One of the consequences of continuing fault reactivation is the evolution of the spectacular escarpment referred to as W. Ghats that sharply defines the western flank of the Sahyadri range (Valdiya, 2010).

Phytogeographically, the W. Ghats can be divided into four regions (Abraham, 1985): (i) Region between Tapti River and Goa; (ii) Region between Kalinadi and Coorg; (iii) The Nilgiris; and (iv) The Anamalai, Cardomomum and Palani hills. To these can be added the southernmost W. Ghats, the Agasthiamalai, which forms a unique phytogeographic zone. The W. Coast region can be added as the sixth phytogeographic region. The main vegetation types of W. Ghats are scrub savanna, semi-deciduous forests, dry-deciduous forests, moist deciduous forests, wet evergreen forests, montane forests, grasslands, shola forests, wetland vegetation, marshes (particularly *Myristica* marshes), etc. In addition to these, there are mangrove forests. The W. Ghats form one of the 34 hotspots of the world with a high degree of endemic taxa of plants, animals, fungi, lichens and microbes. It is a UNESCO World Heritage Site and is one of the eight "hottest hotspots" of biological diversity in the world (Myers et al., 2000). More than 60 genera and 1,500 species of plants are reported to be endemic to W. Ghats, although according to Nayar et al. (2014) there are only 1,270 endemic flowering plant species in this region. Around 27–35% of India's plants are reported to exist in W. Ghats and the adjacent coastal region. According to Nayar et al. (2014) there are around 8080 specific and subspecific flowering taxa, of which 7,402 species, 593 subspecies and varieties are confirmed, while the status of 85 taxa need to be verified beyond doubt. The W. Ghats is also home to 145 wild plant species related to cultivated taxa: it also contains more than 30% of the country's mammal, bird, reptile, amphibian and fish species (Bawa et al., 2007). Many animals are endemic to W. Ghats (Gunawardene et al., 2007). W. Ghats is recognized as a World Heritage Site with 39 of its regions included as very sensitive areas.

1.2   ETHNIC DIVERSITY AND KNOWLEDGE SYSTEMS AND FACTORS THAT IMPACTED THEM

The generally accepted model of human evolution emphasizes that the modern human species originated about 200,000 years ago in East Africa and then started to migrate to different regions of the world around 70,000 -50,000 years ago. The origin and settlement of Indian people are still matters that are debated. Although the initial migrations to India might be accidental or by chance, the subsequent migrations were essentially due to attraction to India's biological wealth. Humans appear to have spread to many parts of India by the middle of Palaeolithic period (around 50,000 to 20,000 years ago) (Misra, 2001). It is generally agreed that the earliest widespread occupants of major part of N. India were from the Dravidian base, but with the arrival of Indo-European language speakers they were pushed in more and more numbers to peninsular India; initially peninsular India had a much scarcer populations of Dravidians (Basu et al., 2003; Kanthimathi et al., 2008). Most, if not all, migrations after the 16th century were due to pulls from their destinations and pushes in their homeland (Gadgil et al., 1998).

There are 461 tribal communities in India (Singh, 1992) who speak about 750 dialects, which can be classified into four language groups: Austro-Asian, Dravidian, Sino-Tibetan and Indo-European. The W. Ghats/W. Coast region of India first came under human influence during the Palaeolithic Age around 20,000 years ago (Gadgil and Thapar, 1990), although some consider that it happened around 15,000 years ago (Subash Chandran, 1997). The latter date is supported by the discovery of stone tools from some of the river valleys of W. Ghats, while artifacts of this Age were discovered in certain other areas of W. Ghats. These evidences indicate the hunter-gatherer mode of subsistence of these early occupants. However, in the Mesolithic Age (between 12,000 and 5,000 years ago) most of these hunter-gatherer communities got transformed into food cultivators, particularly as *podu* cultivators in the hill valleys. In Chapter 2 of this volume Krishnamurthy et al. give a detailed account on the most important ethnic communities of W. Ghats and W. Coast of peninsular India. They are mostly responsible for the development of ethnic knowledge on plants of this region as well as for the domestication of some useful plants that had originated in this region, such as pepper, *Garcinia* species, etc. They were also responsible for cultivating some of the exotic plants that were introduced into this region long back in history. These include taxa like cloves, *Myristica*, cardamomum, *Areca catechu*, etc.

The west coast of peninsular India is one region in the whole world in general and particularly in the Indian Ocean- Arabian Sea region that had the greatest of impacts in its ethnic diversity and ethnobotany. Right from third millennium BCE this region was a trade zone and people from Mediterranean, Rome, Greek, Arab, East African and S.E. Asian regions were visiting the west coast for trade on several items from and to India. Trade had a great impact on the ethnic societies all along the west coast; the traditional life style, profession, culture and social life of the some of the tribes were changed totally. They were made to involve themselves in trade either directly or indirectly. Trade also forced some ethnic societies to resort to cultivation of certain plant species, which they were previously collecting from the wild. Sustained use of plant resources gave rise to a system where excess collected/cultivated were made available for trade. There was a change in religion, particularly Christianization or Islamization of originally 'Hindu' tribal members, either by force or volition that resulted in a great genetic mix-up. Because of the impact of Arabs, E. Africans, Dutch, Portuguese and English people, there was also creolization and production of pidgins (mixture of languages), which again changed the prevailing socio-cultural environment. Added to trade and religious effects, polity also had its effects on the ethnic diversity and ethnobotany of this region. Details on all the above impacts are detailed in the third chapter of this book by Krishnamurthy.

The west coast of peninsular India underwent a lot of changes after the 15th century due to the arrival of colonial powers of Europe, such as the Dutch, Portuguese and English for exploitation of the resources and to trade them. The Portuguese and Dutch were greatly interested in the traditional medicinal knowledge and the raw drugs, and medicinal formulations of the local ethnic communities so that they can be used by the Europeans at home. The Dutch exploited the Ezhava community's medical knowledge and this resulted in the Dutch Governor at Cochin to compile the 12-volume book Hortus Malabaricus. The Portuguese doctor Garcia da Orta who got settled down at Goa exploited the traditional medical knowledge available to him in the west coast and Deccan and came out with a classical medical book. These two major works along with two or three other works brought to light the superiority of Indian traditional medical wisdom, which was not based on Hippocratic approach. Krishnamurthy and Pullaiah have dealt with the European contribution to Western Indian Enthobotany in the period between 16th and 18th centuries CE in the fourth chapter of this book.

The world has known a great variety of cultures each with its own knowledge, belief and value systems. Religion, formal or informal, primitive or modern has always played a very important role in determining these knowledge and value systems. However, there are certain basic parallels and common aspects in the different traditional knowledge and belief systems that are in vogue even today in many parts of the world. One such common feature is conceiving life of people in terms of three common interrelated and inseparable domains: spiritual, human and natural domains. Traditional Knowledge in the nature domain includes thematic fields related to food and health practices and their biotic sources, and that in the social domain included knowledge about local organization, leadership, management of natural resources, conflict resolution, gender relations, art and language. The spiritual domain includes knowledge and belief systems about the invisible divine world, spiritual forces, ancestors and about how these systems translate this knowledge into values and ritualistic practices. None of these three domains remains and operates in isolation and hence all the three are inseparable and are highly integrated. Somasekhar has dealt in detail with the traditional knowledge, worldviews and belief systems of the various ethnic groups of W. Ghats and their relevance in integrating their natural, spiritual and practical worlds in Chapter 5.

1.3 UTILITARIAN ASPECTS OF ETHNIC COMMUNITIES OF WESTERN PENINSULAR INDIA

Of the three approaches available to study and document ethnobotanicals, the utilitarian or economic approach (the other two being cognitive and cultural) primarily motivated research in bioprospecting of plants used as sources of food, medicine and other utilitarian values to human beings. This ethnodirected approach has enabled the discovery several useful ethnic plant resources and has introduced newer nutraceuticals, pharmaceuticals and molecules of great importance for human consumption. India is one of the major areas of the world known for the diversity of its ethnic societies and ethnic knowledge on plants. As is to be mentioned in Chapter 2 of this volume, the W. Ghats and W. Coast of peninsular India have several ancient tribal communities with very rich knowledge on ethnobotanicals.

In 1990, Mike Balick emphasized the value of ethnobotany in the identification of therapeutic agents from the rainforests and that only a few of them have been studied for their potential medicinal uses (Balick, 1990). His work also indicated at the wealth of unstudied plants, particularly those whose medicinal efficacy has been proved by ethnic knowledge of many ancient tribals. Since then there have been numerous scientific articles on the use of traditional biodiversity knowledge in bioprospecting of medicinal plants. Indian scientists also took up this lead and documented traditional ethno-medicinal knowledge held by various tribal communities. Noorunnisa Begum et al. have summarized all information relating to the ethnomedicinal knowledge of the tribals of western peninsular India in Chapter 6 of this volume. They have not only traced briefly the history of ethno-medico-botany of this region but also have enlisted the plant taxa used by different ethnic communities for treating various ailments.

Many Indian ancient tribes are of the pastoral type and have been instrumental in domestication of wild cattle breeds as well as in maintaining the domesticated and native breeds for future breeding programs. Pastoralists mainly depend upon livestock keeping for their livelihood; they allow their cattle to graze on common property resources. The pastorals are usually nomadic, but in recent times have become settled and look after their animals often out a 'God-given' duty. The life of the *Toda* tribe would be unimaginable without the daily rituals associated with buffaloes, which are very dear to them. They were quite aware of the problems of breeding and maintaining these cattle varieties; particularly they were aware of the diseases that affect the cattle. Hence, they had, by trial and error, identified potential ethnoveterinary as well as fodder plant taxa that are found around them respectively for curing the various ailments and feeding the cattle they had domesticated and were using for getting milk, manure and fuel, plowing and carting. A number of ethnic communities of western peninsular India are pastorals, at least in the past; as examples we can mention the *Kurubas* of Karnataka and *Todas* of Nilgiris. Most of the ethnoveterinary knowledge of these communities are transmitted orally from one generation to another through a family lineage and because of breakage of this link in many cases, such knowledge is greatly lost. Some community-based approaches and efforts of some NGOs like BAIF, ANTHRA, SEVA and FRLHT this knowledge is being retrieved and documented. In Chapter 7 of this volume Nair and Punyamurthy have dealt with in detail about the ethnoveterinary plant taxa. Their article highlights the various plants used to cure common and rare diseases of cattle (amounting to at least 20) by the various ethnic communities. Their paper also deals with documentation of local ethnoveterinary practices and assesses these practices and knowledge for their efficacy and safety, recommendation of positively assessed practices for immediate implementation and the need to revitalize healthy ethnoveterinary traditions. They have also addressed the problem of the antimicrobial resistance and the relevance of traditional plants to redress this problem.

Around 3 million people of the world, particularly those who live in the tropical and subtropical belt, are at the risk getting malarial fever caused by the protozoan species of the genus *Plasmodium* and on an average around 200 million malarial cases are known in the globe (WHO, 2014). The problem of malarial fever is unique that it has to be tackled: at the level of the parasite, *Plasmodium* and at the level of its spreading-vector, the mosquito. Unless both these are taken care of, malaria cannot be easily eradicated.

Traditional ethnic communities that live in tropical/subtropical regions of the world, particularly in India and Africa, have been living with the malarial incidence for several centuries and have come up with simple plant-based remedies for both prevention and cure of the disease and also for combating the vector mosquitoes (preventive measures) and the parasitic *Plasmodium* (curative measures) once infected. In Chapter 8 of this volume Prakash et al. have dealt with the medicinal flora of W. Ghats that are used against malaria and the endogenous development aspects related to it.

India has some extremely unusual vegetables and fruits as well as an unparalleled variety of them. Although the nutritional information regarding many of them are available, we do not have much information of those used by the ancient ethnic communities, particularly of western peninsular India. Unlike the modern food systems, those of ethnic communities cover the full spectrum of life. In the last three to four centuries there had happened a great disconnection between traditional people and the food they used to take, particularly due to globalization and homogenization which have replaced traditional and local food cultures. High yield crops that require high-put farming technologies and monoculture agriculture have largely degraded ethnic ecosystems and have harmed traditional agro-ecological zones and practices. Moreover, the modern food technologies and industries have caused diet-related chronic diseases and different forms of malnutrition and have distorted greatly local food security and safety (Burlingame, 2011). Since food is a human right and since modern food systems have affected traditional food systems there is an urgent need to protect the heritage of indigenous people and the health of their culturally determined foods and food systems from being lost. Traditional foods are not only foods but also medicines as they have both nutraceutical and therapeutic values. Many of the tribal-used food sources have amazing therapeutic effects on many common ailments of humans. If these tribal food plants/nutraceuticals are properly exploited, India can potentially contribute in a significant way to the world fruit and vegetable market not only in terms of food value but also in terms of therapeutic value. In Chapter 9 of this volume Sridhar and Karun discuss in detail the food plants of W. Ghats that have both food and therapeutic values. These plants should not only be conserved and protected but also should be subjected to detailed research with reference to their chemistry, energy value, nutrient value and therapeutic effects, especially in comparison to food plants obtained from modern agriculture (Erasmus, 2009).

Species of plants provide a vast array of products used by humankind worldwide. Certain plants have been exploited directly from the wild, while a number of them are from cultivated sources. In spite of vast overall development, plant resources largely remain poorly understood, underexploited and poorly documented as most of them are under the domain of ancient traditional ethnic societies (Krishnamurthy, 2003). Knowledge of plants use from these traditional societies has not been translated into greater and wider use largely because of poor documentation of ethnic knowledge on plants. Besides serving as foods and medicines many plants are known from ethnic sources to serve as ornamentals, timbers, fibers, dyes, fuel/renewable energy and for a host of other products used in industry and commerce. The W. Ghats of India particularly is a good source of non-food and non-medical plants of great value, a knowledge about which was largely provided by the different ethnic communities living there. Karuppusamy and Pullaiah have given a detailed account of such plants in Chapter 10 of this volume.

Mangroves are intertidal vegetations that are characteristically located in littoral, sheltered and low-lying tropical and subtropical coasts, although they dominate river delta, lagoons and estuarine complexes developed from terrigenous sediments (Tomlinson, 1986). In India, although mangroves dominate the eastern coastal line, the western coast does have mangrove patches whenever west-flowing rivers join the Arabian sea. Mangrove resources have been traditionally exploited by the coast-living ethnic communities of the western peninsular coastal India for various purposes. Pullaiah et al. have dealt with in detail the ethnobotany of mangroves in Chapter 11 of this volume. The uses range from food and medicine to other uses.

The modern scientifically oriented and techno savvy society often mistakenly considers that religion and belief systems are not interested in conserving various biodiversity elements. On the other hand, religious values very often help to conserve and sustainably manage biodiversity (Ramanujan, 2004). The long-prevailing practice of conserving patches of various types of forest vegetation with an in-built temple or idol/totem of deity/spirit was in vogue in India (and a few other places of the world) (Hughes and Subash Chandran, 1998; Hughes and Swan, 1986). This is particularly true of the W. Ghats. Sacred groves are the best institution that were designed and established by traditional communities of this region not only to conserve the ecosystem/vegetation around them but also to sustainably obtain their basic requirements in a non-destructive manner. In Chapter 12 of this volume, the efficacy of sacred groves as the most effective conservation method for protecting the biodiversity of W. Ghats has been detailed by Krishnamurthy, citing concrete living examples of sacred spaces.

In ethnobotanical literature attention is often given to flowering plants as they are the most dominant and obvious group of plants. Lower plants, such as Algae, Bryophytes and Pteridophytes do not find adequate attention from ethnobotanists. India, particularly the W. Ghats, is a region rich in Bryophyte flora and various ethnic communities have often come across them and tried to utilize them for various purposes. In Chapter 13, Afroz Alam discusses the various aspects of Ethnobryology. This chapter indicates to us that focused research in future should be done on the ethnobotany of these neglected groups of plants in diverse localities of India and elsewhere in the world.

KEYWORDS

- **Ethnic Diversity**
- **Ethnoveterinary Plants**
- **Peninsular India**
- **Sacred Groves**
- **Traditional Food**
- **Traditional Medicines**
- **West Coast**
- **Western Ghats**

REFERENCES

Balick, M.J. (1990). Ethnobotany and the identification of therapeutic agents from the rainforest. In: Chadwick, D.J. & Marsh, J. (Eds.). Bioactive Compounds from Plants. Wiley, Chichester, pp. 22–32.

Basu, A. Mukherjee, N., Roy, S., Sengupta, S., Banerjee, S., Chakraborty, M., Dey, B., Roy, M., Roy, B., Hattacharyya, N.P., Roychoudhury, S. & Majumder, P.P. (2003). Ethnic India: A genomic view with special reference to peopling and structure. *Genomic Res. 13,* 2277–2290.

Bawa, K.S., Das, A., Krishnamurthy, J. Karanth, J.U., Kumar, N.S. & Rao, M. (2007). Ecosystem a Profile: Western Ghats and Sri Lanka Biodiversity Hotspot-Western Ghats Region. Critical Ecosystem partnership Fund, Virginia, USA.

Burlingame, B. (2009). Preface. pp. V–VI. In: Kuhnlein, H.V., Erasmus, B. & Spigelski, D. (Eds.). Indigenous Peoples' Food Systems. FAO, Rome.

Chandran, M.D.S. (1997). On the ecological history of the Western Ghats. *Curr. Sci. 73,* 146–155.

Erasmus, B. (2009). Foreword. pp. ix–x. In: Kuhnlein, H.V., Erasmus, B. & Spigelski, D. (Eds.). Indigenous Peoples' Food Systems. FAO, Rome.

Gadgil, M., Joshi, N.V., Manoharan, S., Patil, S. & Shambu Prasad, U.V. (1998). Peopling of India. In: Balasubramanin, D. & Appaji Rao, N. (Eds.). The Indian Human Heritage. Universities Press, Hyderabad, India. pp. 100–129.

Gadgil, M. & Thapar, R. (1990). Human Ecology in India. Some Historical Perspective. *Interdisciplinary Sci. Rev. 15,* 209–223.

Gunawardene, N.R., Daniels, A.E.D., Gunatilleke, I.A.U.N., Gunatilleke, C.V.S., Karunakaran, P.V., Nayak, K.G., Prasad, S., Puyravaud, P., Ramesh, B.R., Subramanian, K.A., Hughes, G.J.D. & Subash Chandran, M.D. (1998). Sacred groves around the earth: an overview. In: Ramakrishnan, P.S. (Ed.). Conserving the sacred for biodiversity management. Oxford & IBH publishing Co. Pvt. Ltd., New Delhi, pp. 69–85.

Hughes, J.D. & Swan, J. (1986). How much of the earth is sacred space? *Environ. Rev. 10,* 247–259.

Kanthimathhi, S., Vijaya, M. & Ramesh, A. (2008). Genetic study of Dravidian castes of Tamil Nadu. *J. Genetics. 87,* 175–179.

Krishnamurthy, K.V. (2003). Textbook on Biodiversity. Science Publishers, New Hampshire, USA.

Misra, V.N. (2001). Prehistoric human civilization of India. *J. Biol. Sci. (Suppl.) 26,* 421–431.

Nayar, T.S., Rasiya Beegam, A. & Sibi, M. (2014). Flowering Plants of the Western Ghats. 2 Vols. Jawaharlal Nehru Tropical Botanic Garden and research Institute, Thiruvananthapuram, India.

Ramanujan, M.P. (2004). Sacred Groves—A Dying Wisdom? In: Muthuchelian, K. (Ed.). Biodiversity Resource Management and Sustainable Use. Madurai Kamaraj University, Madurai, India, pp. 180–190.

Singh, K.S. (1993). Peoples of India (1985–92). *Curr. Sci. 64,* 1–10.

Tomlinson, P.B. (1986). The Botany of Mangroves. Cambridge University Press, New York.

Valdiya, K.S. (2010). The making of India: Geodynamic Evolution. Macmillan Publications Ltd., New Delhi.

WHO (2014). World Malaria Report. World Health Organization. Geneva.

CHAPTER 2

ETHNIC DIVERSITY

K. V. KRISHNAMURTHY,[1] BIR BAHADUR,[2] and S. JOHN ADAMS[3]

[1]*Consultant, R&D, Sami Labs, Peenya Industrial Area, Bangalore–560058, Karnataka, India*

[2]*Department of Botany, Kakatiya University, Warangal–506009, India*

[3]*Department of Pharmacognosy, R&D, The Himalaya Drug Company, Makali, Bangalore, India*

CONTENTS

ABSTRACT

This chapter summarizes the basic information related to the most important ethnic/tribal communities of Western peninsular India. The history of peopling of this region is provided in the introductory part, followed by a description of the most important tribes. The focus is on their population size, way of life, ethnic knowledge on plants and socio-cultural characteristics. The importance of this study is given at the end.

2.1   INTRODUCTION

There are two conflicting models on the origin, evolution and spread of modern human species. According to the multiregional model, archaic humans originated in the African continent, migrated throughout the Old World over a million years ago and then evolved into the modern humans many times in different parts. Hence, according to this model, the East Asians, Australians, Europeans and Africans have had separate ancestries, although with subsequent enough gene flows between them to prevent speciation (Thorne and Wolpoff, 1992). According to the very widely supported recent model of human evolution, all modern humans had a relatively recent African origin (about 200 thousand years ago), followed by a subsequent dispersal throughout the Old World, that completely replaced the existing archaic human population (Disotell, 1999). A single migration out of Africa, however, received less support initially based on fossil and genetic studies and, at least, two geographical routes were suggested to have been taken by the early migrants out of Africa (Cavalli-Sforza et al., 1984; Lahr and Foley, 1994). The first route led up to Nile valley, across the Sinai Peninsula and north into the Levant (Middle East); these migrants might have died out quickly in the Levant and did not proceed further anywhere from Levant. The second route was from North Africa (i.e., the mouth of Red Sea between the Horn of Africa and Arabia) to Arabia by using primitive boats; from Arabia the humans migrated to Central Asia from where they then dispersed to various parts of the World. This migration, through the second route, took place around 70,000 to 50,000 years ago.

The Indian subcontinent possesses the greatest of ethnic diversities of the World, probably because of its strategic position at the trijunction of African, north Eurasian and Oriental realms. All the discussions on Indian Ethnodiversity so far made have been on three basic approaches: racial, linguistic and religious. The existing confusion and controversies have largely been due to mixing up of the discussion involving more than one of the above three approaches. The racial groups involved are Caucasoid, Negrito, Australoid, Mangolioid and Negroid, the linguistic groups involved are Austro-Asian, Indo-European, Dravidian and Sino-Tibetan, and the religious groups involved are Hindus, Muslims, Christians, Buddhists, Jains, Sikhs, Parsis and Jews. It is most likely that the first entry of modern human species into India must have been around 70,000 to 50,000 years before the present. According to Gadgil et al. (1998) the likely major migrations of modern humans into India include the following linguistic groups: (i) Austro-Asian language speakers around 65,000 years ago, probably from the North-East. These were the earliest to arrive in India prior to the beginning of agriculture. Many hunter-gatherer tribes of present-day India are descendants of these people. These people also have the most diverse vocabulary; (ii) Dravidian language speaking in several waves around 6,000 years ago from the Middle East. The bulk of present-day Indian mainland populations is constituted by the descendants of these people. They came with knowledge of crop cultivation and animal husbandry. A number of tribes of India belong to the Dravidian group. As an example, we may cite the *Kanis* of southernmost Western Ghats (W. Ghats, hereafter); (iii) Indo-European speakers in several waves around 4,000 years ago. These people are mostly nomadic. As an example, we may cite the *Pardhis*. These people have the least diverse vocabulary; and (iv) Sino-Tibetan speakers in several waves around 6,000 years ago. Fairly recently, Reich et al. (2009) and Thangaraj (2011), based on a detailed analysis of different ethnic populations of India with reference to their evolutionary genetic markers, such as mtDNA and Y chromosome, have shown that small groups of ancestors founded most Indian groups, which then remained largely isolated with limited cross gene-flow for long periods of time. According to Thangaraj (2011), there are two main ancestral groups for the present Indian population, an "Ancient North Indian (ANI)" and an "Ancestral South Indian (ASI)" groups. The former is directly related to the Middle East, Central Asia and Europe, while the latter is not related to groups outside India. Thangaraj and his group, however, have suggested only three early major migrations from Africa [in contrast to the four suggested earlier by Gadgil et al. (1998)] into India: (i) via sea to Andamans; (ii) via land to S. India through W. Coast (ASI populations); and, (ii) via land into N. India (ANI populations). From ANI and ASI groups, the populations of all parts of India were then derived. It is very likely that some present-day South Indian populations were derived from Middle East People, who came to S. India (and Sri Lanka) via the Arabian Sea around 6,000 years back or even earlier. However, there are evidences of a genetic mixing of these two ancestral groups mentioned above, which was reported to have been initiated around 4,200 years ago, when the Indus Valley Civilization was waning, followed by huge migrations within India. This mix-up stopped to a great extent around 2,000 years ago, probably due to *Manusmriti* (written around 100 BCE) resulting in the evolution of caste system. There are then many still-unanswered questions pertaining to how the Indian subcontinent was peopled.

2.2   PEOPLING OF WESTERN GHATS AND WEST COAST OF PENINSULAR INDIA

It is very difficult to correctly explain how the W. Coast and W. Ghats were peopled in the historic period after the first entry of modern human species into India around 70,000 to 50,000 years ago. It is most likely that this peopling was effected in waves at different times of past history of human civilization into different parts of India. The Austro-Asian and Sino-Tibetan language speakers are absent in the W. coast region, while the descendants of other two groups (Indo-European and Dravidian) are widely present in this region (Gadgil et al., 1998). The Indo-Europeans are predominantly present in the north and the Dravidians in the middle and south of western peninsular India. Peopling of this part of India would not have happened before 20,000 years from the present as is at this period that the monsoons became distinctly weak at the height of glaciation in the northern latitudes (Gadgil and Thapar, 1990). Initially, peopling must have been sporadic and thin in different parts of this region but the density and area of occupation gradually increased in the subsequent historical periods.

   Archaeological data of 23,050 ± 200 BCE on human presence are available from late prehistoric sites in Maharashtra region (Sali, 1989). Around 10,000 BCE there was an increase in the number of sites and occupation in a range of diverse ecological sites by communities using microlithic implements (Chakrabarti, 1999). Along the Konkan Coast these groups settled in hill terraces and rocky outcrops near the sea coast, further suggesting the exploitation of marine resources by the settled communities. The Indo-European descendants must have become hunter-gatherer and nomadic tribes/ethnic communities of northern W. Ghats, while the Dravidians must have become cultivators of plants and cattle breeders, although there might have been exceptions of some Dravidians becoming hunter-gatherers, in southern W. Ghats. Although it has been claimed that in northern W. Ghats cultivations must have started around 4,000–3,000 years ago and in Southern W. Ghats around 2,000 years ago (Gadgil et al., 1998), Tamil literary, archaeobotanical and epigraphic evidences indicate simultaneous cultivation efforts throughout southern W. Ghats around 4,000 years ago (Krishnamurthy, 2006).

Increased aridity in the climate of Indus Valley, salinity rise in the lakes and the drying up of the Saraswati river during the fourth millennium before the present may have caused a migration of the agropastoral Harappan people in large numbers towards the W. Coast and W. Ghats due to latter's strategic position, the relatively humid climate (Subash Chandran, 1997) as well as the availability great biodiversity and bioresources. This larger migration happened in the megalithic period. Harappan civilization experts have shown that the Late Harappan culture had extended so far south as the Krishna river valley of the Belgaum region. The suspected objects of Harappan worships like the Mother Goddess, the prototype of Lord Shiva, the humped bull, the serpent, etc. are also seen in the northern and middle W. Ghats region. Many sacred groves of W. Ghats have a rich deposit of ancient terracotta figures characteristic of the Harappan civilization (Subash Chandran, 1997). The partially wild land race of rice of Late Harappan farmers might have given rise to the salt-tolerant rice land race of W. Coast region around 3,500 years before the present. From around 2000–1000 BCE there were temporary or permanent settlements of trading people from many countries (see details in Chapter 3 of this Volume).

2.3 MAJOR ETHNIC COMMUNITIES

'People of India Project' of the Anthropological Survey of India has recognized 4,635 communities with about 50 to 60 thousand endogamous groups in India (Gadgil et al., 1998). This project also enumerated 461 tribal ethnic communities of which 174 have identified subgroups. More than 40 indigenous tribes occur in W. Ghats. Some of them occur in more than one state of W. Ghats region. The most important tribal ethnic communities are described here.

2.3.1 KANIS

Also known as *Kanikkars*, the *Kanis* form an important tribal community of southernmost W. Ghats, which occurs both in Tamil Nadu (Bhagat Singh, 2015) and Kerala (Figure 2.1). *Kanis* occur in families of 5 to 56 in this hill range. The total number of individuals range from 1,500 to 2,000. Some of them are hunter-gatherers (gathering mainly wild edibles and honey), cultivators of crops (particularly manioc in recent years) or are involved in inland fishing. According to Viswanathan (2010) they use 352 medicinal plants, although Ayyanar and Ignacimuthu (2011) report their using only 90 plant species. In Kulathupuzha of Kerala they use around 120 plant species for medicinal purposes. The most important among them is *Trichopus zeylanicus* ssp. *travancoricus*, which is considered to be equivalent to Ginseng.

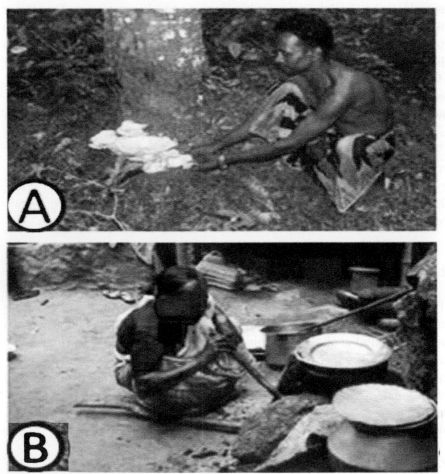

FIGURE 2.1   (a) *Kani* man collecting mushrooms; (b) *Kani* women cooking. (Source: Sargunam et al., 2012; used with permission from S. S. Davidson, India.)

2.3.2 KURICHIYAS

Also known as *Kurichans*, the *kurichiyas* live in Waynad and adjacent Kannur forests (Figure 2.2). They form 9% of the tribal population of Kerala. Traditionally specialists in archery (as the name indicates) and martial arts they were used by the king to combat British colonists, but now have become settled agriculturalists. They live in clusters of houses called *Mittam* or *Tharavadu* and their head is *Karanavar* or *Pittan*. The *kurichiyas* are a matrilineal society, deeply religious (worshipping village deities, particularly *Kaali*) (Ayappan, 1965) bury their dead with a grand funeral function and speak *kurichiya* language which is related to Malayalam. They retain their worldviews very strongly and are not affected much by modern trends. Their population as per 2001 census is 32,746. *Kurichiyas* are reported to use a number of ethnomedicinal plants including 70 species for skin problems alone and 40 species for ethnoveterinary purposes. They also have knowledge about 72 wild varieties of food plants.

FIGURE 2.2 *Kurichiyas* tribal women (Source: http://www.ayurvedayogaretreat.com/tribes-wayanad/).

2.3.3 PANIYAS

The *Paniya* tribe mainly inhabits Waynad area of Kerala and Nilgiris of Tamil Nadu. It is also seen in Kannur and Malappuram. *Paniyas* are short, dark-skinned, broad-nosed, curly-haired people. Some consider them as African in origin but it is contested by many. Traditionally the people of this tribe have been bonded labors and were once sold along with the plantations by landlords. Being agricultural laborers as their name indicates the *Paniyas* have great knowledge on ethnoagricultural practices. They live in bamboo huts covered with mud and charcoal powder mixture. *Paniyas* constitute 22.4% of the tribal population of Kerala and are a poor and depressed people. Now-a-days, they are involved in the cultivation of cardamomum, pepper, ginger and coffee. They also collect tubers, potherbs and mushrooms. They speak corrupted Malayalam-Tamil mixture. Women wear munda cloth and wear ear rings in big ear-holes tucked with palm leaf rolls. Their original religion is animism but worship *Bagawati* or '*Kuli*' (a god of neither sex). Their priests are *Nolumbukaran*. They do not have rituals connected to birth, puberty or marriage. They bury the dead (Figure 2.3).

FIGURE 2.3   Men and Women of *Paniyas* tribes (Source: https://en.wikipedia.org/wiki/Paniya_people).

2.3.4   PULAYAS

Also known as *Cherumas, Pulayas* are one of the earliest inhabitants of Kerala region (Krishna Iyer, 1967; Anantha Krishna Iyer, 1909). Their language is a primitive form of Tamil mixed with Malayalam words. *Pulayas* are considered as Protoaustraloids, and are black, short and with flat nose and long mandibles. They are believed to have come and settled in Kerala during the Middle Ages and were originally hunter-gatherers; they were living in Kerala from *c* 4,000 BCE onwards. Soon they became land-owners and, in fact, were the original land owners of this region. They claim themselves to be descendants of the Chera kings, the Tamil rulers of the Chera kingdom (now Kerala), as their name *Cherumas* indicated. Their supremacy declined with the arrival of Aryan peoples, were made to forego the lands they owned and were forced to be agrestic slaves to work as agricultural labors. *Pulayas* have excellent knowledge on ethnic agricultural practices and crops. They worship in *Kavus* (=sacred groves) and their deities include *Kaali* and *Chattan*, besides ancestral spirits.

2.3.5 EZHAVAS

The *Ezhavas* form a very important ethnic community of not only southern Kerala but also of bordering Tamil Nadu, historically called *Nanjil nadu*. In fact, many consider *Ezhavas* of Kerala as migrants from Tamil Nadu; many of them speak both Tamil and Malayalam. They have very rich knowledge of traditional system of medicine, perhaps the ancient Siddha system of medicine, as well as on traditional medicinal plants. Achuden, the chief architect of *Hortus Malabaricus* (see details in a later chapter of this volume) is an *Ezhava*. *Ezhavas* were traditionally toddy collectors. They are matrilineal and are organized under five clans. They are exogamous and non-vegetarians.

2.3.6 CHOLANAIKKARS

The *Cholanaikkars* or *Malainaikkars* are one of the most primitive tribes. They number only around 360. Their name indicates that they inhabit in the interior forest 'cholas' or 'sholas' (=valley); the post-fix '*naikkars*' means kings. It is believed that they migrated from adjacent Karnataka forests to the forests of Silent Valley in Kerala. They are one of the last-remaining hunter-gatherers of South India. They speak the *Cholanaikkan* language with a mixture of Malayalam words. They refrain themselves from contacting other communities. They are usually of a short stature but are well-built and dark-complexioned. They live in rock shelters called "*kallulai*." Each group of *cholanaikkars* is called *chemmam*. They strictly follow the traditional rules fixed by their ancestors. Now-a-days they gather minor forest produce for their subsistence.

2.3.7   KADARS

Kadars are found not only in the Anamalais of Tamil Nadu (Bhagat Singh, 2015) but also in the adjacent W. Ghats of Kerala and Karnataka (Figure 2.4). *Kadars* are considered as belonging to the negrito tribal category of the African descent and are one of the most ancient tribals of not only W. Ghats but of entire India. However, no folk story is available with them about their origin. They live deep inside forests as their name *'Kadars'* indicate. *Kadars* are also known as *Kadins*, *Kadans* and *Kadane*. They are hunter-gatherers and practice this mode of life even till recently and some families still practice it and ardently follow their traditional culture. As per 2011 census there were 650 people (male 325 and female 325). They speak Tamil mixed with Malayalam or Kannada. *Kadars* are classified by UNESCO as one of the PTG (particularly vulnerable tribal groups) category of the tribals. It is on the verge of extinction. The huts of Kadars are located in hill slopes. Each village has about 15 huts built of bamboos and with straw roofs. These huts are built under huge trees. The villages are very much separated from one another and the members of the community meet only in weekly markets or *sandhis*. They use vessels made of bamboos, especially to store honey, water and oil. They make many more items out of bamboo. *Kadars* are hunters and gather many forest products like honey, wax, cardomum, and shikakkai. In recent years they have started cultivating millets, sorghum and ragi and also have started to work as laborers. The village of *Kadars* is known as *Mooppan*. The society of *Kadars* is patriarchal and endogamous and marriages are arranged. During betrothal function the bridegroom should give as gift a comb made of bamboo to the bride and it is a very important symbolic gesture. The *thaali* or sacred thread on the neck of bride is black-beaded. Divorced women cannot seek control of their children with their earlier husbands. *Kadars* follow a number of rituals, most of which are society-controlled. Mother and the new-born child should take coconut oil bath and then the child is named. The dead are buried in the north-east direction from its village. *Kadars* follow animism and they worship trees and some animals; certain trees are considered as male gods and certain others as female. They worship them as *Muni* and *Batrakali*, often worshipping with coconut and jaggery. In recent years they have started worshipping mainstream Gods like Lord Muruga, Shiva and Amman.

FIGURE 2.4 Kadar Tribes: (a) Young boy using the hunting weapon bow and arrow; (b) An age old couple of *Kadar* tribes; (c) Married women of *Kadar* tribes (Photo courtesy Dr. Bhagath Singh).

2.3.8 TODAS

The *Todas* form a very important tribal community among the foremost six communities of the Nilgiris. They once owned major part of Nilgiris. UNESCO classifies them as a PTG tribal group and it is on the verge of extinction (Bhagat Singh, 2014). *Todas* are also found in the adjacent Karnataka W. Ghats. They are unique and distinct from other tribal groups and are known for their appearance, dress, customs, rituals, worship, pastoral life, buffalo sacrifice, marriage and arts and crafts. Todas are also known as *Toduvar, Todavar, Todar* (from the Kannada root *Tuda*, which means high altitude, e.g., living in hills). '*Toda*' is also said to be derived from '*tur*' or '*tode,*' the most sacred tree of *Todas*. The origin of *Todas* is not clear and records about them are available only from the 11th century CE. Even their folk-tales and songs deal with more recent historical incidents. They claim themselves as descendants of *Pandavas*. Since their color and nose features indicate their European descent, they are considered by some not of Dravidian origin. They are also considered as descendants of ancient Romans, nomadics of ancient Jews or related to Sumerians. Their language *Toduvam* sheds some light on their origin and indicated its affinity to Dravidian group of languages. The traditional *Toda* huts are rainbow-shaped and are constructed with great artistry; they have a thatched roof and a low front door and hence avoids cold air entering inside and keep the inside warm. Their village is called *'mund'* or *'othaikkal mundu'* (=single–stone *mundu*) (it is the root for *'ootacamund,'* the main town in Nilgiris). There are around 75 munds in Nilgiris occupied by about 2004 people (males 959; females 1045) as per the recent census. They are pure vegetarians and eat rice and millets. Buffalo-rearing was their original profession. They rear buffaloes for two purposes: one set of buffaloes are dedicated to the 'Milk Temples' while the other serve for domestic purposes. Now Todas are engaged as laborers and as cultivators. There are two sub-categories of *Todas*: *Tardarol* moiety, considered as having a higher social status, and *Taviliol* moiety, considered to be of inferior status. They together have 15 clans. Each *mund* has only one clan. Marriage is outside the clans and is usually arranged by negotiation. 'Bow-arrow ritual' marks the seventh month of pregnancy of the married woman and a marriage is recognized as valid only after this ritual. *Todas* have a patriarchal society with a prevalence of polyandry; women enjoy sexual/social privileges. The death ritual of *Todas* is elaborate and is marked by buffalo sacrifice. The family deity is *Tekkisy*, who is believed to have created the *Todas* and their buffaloes. They also worship Lord Muruga and a few female deities. They usually avoid non-vegetarian food. They are specialists in the art of embroidery and the *shal* they make is called *Puthukkuli*; they also make embroidered clothes for wearing (Figure 2.5).

FIGURE 2.5   (a) Portrait image of *Toda* tribal old women; (b) Young man lifting the heavy weight stone as to achieve the task to get married (Photo courtesy Dr. Bhagath Singh).

2.3.9 KOTAS

The *Kotas* are an important Dravidian tribe. '*Kotas'* means those who live in hills. They form a very small PTG tribe and live at high altitude regions of Nilgiri hills, particularly in Thiruchhigodi area. They also live in some parts of southern Karnataka W. Ghats. As per 2011 census their number is only 308 (male 153; female 155). The *kota* village is called '*kokkal'* '*Keri'* and their huts are called '*pai.'* They speak the '*kota'* or *kovemanth* language, which has no script (mixture of Tamil and Kannada). *Kotas* are known for their colorful folk dances and are very good musicians and artisans (Bhagat Singh, 2015). They play their instruments and sing in festivals and funerals. Some of them are also potters, smiths and carpenters. Both men and women wear shals. *Kotas* have elaborate taboos that distance them from other tribes of Nilgiris. They worship *Kambattarayan* and *Kabatteswari* as well as *Kaali* and their temples are under the control of a *poojary* (=priest) (Figure 2.6).

FIGURE 2.6 Kota tribes (a) Singing song and playing musical during their ritual functions; B. Musical instrumental of Kota tribes (Photos courtesy Dr. Bhagath Singh).

2.3.10 BADAGAS

Like *Kotas*, the *Badagas* also live in Nilgiris. They are an agricultural community, dwelling in higher plateaus. They had settled down in Nilgiris around 16th century CE from Mysore region. The total number of *Badagas* is estimated to be around 3,50,000. They were once cultivating traditional hill crops under jum cultivation but now are involved mainly in cultivating potato, cabbages and tea. They speak Tamil-Kannada mixture. Badagas have a rich tradition of folk tales, songs and poems. They are Hindus and worship various village gods as well as Shiva. *Badagas* are vegetarians. They are exogamous and couples are usually from different clans. They are patrilineal (Hockings, 2012) (Figure 2.7).

FIGURE 2.7   *Badagas* tribal community of Nilgiris (Source: https://en.wikipedia.org/wiki/ Badagas).

2.3.11   KURUMBAS

Also called *Kurubas*, *Kurumas* or *Kurerba Gowdas*, these tribals are believed to be descendants of the Pallavas (Figure 2.8). They live as 5 or 6 families in small villages called *motta* or *kombai* in the valleys and forests of Nilgiris and Wyanad and their hunts. They are known for their black magic, witch craft and sorcery in the past. They were originally hunter-gatherers, shepherds and Yadavas but now are laborers in plantations. They speak the Kurumba language. They celebrate most Hindu festivals. The *Shola Nayakkars, Mullus, Uralis, Alus* or *Palu Kurumbas, Jenu Kurubas, Betta Kurubas*, etc. are all considered as subgroups or clans of *Kurumbas* (see Satishkumar, 2008). Many *Kurubas* have acted as priests in the past for their tribe as well as for other tribes. They use traditional medicinal plants in sorcery. Their huts are scattered on steep wooded slopes and are mainly made of bamboos with roofs fortified with mud, cow dung and grass. They use vessels made of bamboos and leaves of forests plants as utensils. They gather honey and cultivate small patches of land with ragi, samai and other millets. They have rich musical tradition and use bamboo pipes and mono- and bi-faced drums and plastic beads; women wear *thaali* after marriage. Soapnut was once used for bath. A few parallel-lines of dots are tattooed on their foreheads by women. Men wear loin clothes only while women wear a piece of cloth wrapped under their arms and up to knees. A *muthali* or *ejaman* administers the village with the help of three assistants respectively specialized in agricultural issues, marriage issues and as spoke person/messenger. Usually a *Urkoottam* (meeting of villagers) is organized to sort out all issues. During marriage-fixing function a token of betel leaves and a sum of Rs. 1.25 (now Rs. 101.25) is given by the bride groom to the bride's father. *Kurumbas* are a patrilineal society. *Kuruma* act is very famous in which four colors are traditionally used: red (through red soil), white (Bodhi soil), black (charcoal of tree) and green (leaf extract of *Cassia auriculata*). Kurumbas are of Dravidian descent.

FIGURE 2.8 *Kurumbas* tribes (a) Sacred groove; (b) *Kurumbas* tribal children; (c) drawing representation the worshiping activities of *Kurumbas* tribes; and (d) drawing representing the honey comb hunting (Photos courtesy Dr. Bhagath Singh).

2.3.12 HAVIKS

The *Haviks* or *Havyaks* form an important tribe of Karnataka; they are also present in Northern Kerala. They are Brahmins specializing in spice cultivation, especially pepper, cardamomum and betel nut. They possess excellent knowledge on all aspects of ethnoagriculture of plantation crops. These Brahmins are believed to have been brought to this region by the Kadamba king, Mayur Varma, for performing sacrificial rituals (Gauniyal et al., 2010). They speak Kannada. Hegde, Habbar and Bhat are the surnames used by them.

2.3.13   MUKRIS

The *Mukris* are one of the important tribes of Karnataka. They are tradition-ally hunter-gatherers. There are about 9,000 individuals of this community at present. Although now considered as a scheduled tribe, *Mukris* are ge-netically related to the *Haviks* and might be derived from the same ancestral stock.

2.3.14   HALAKKI AND KARE VAKKALS

Halakki Vakkals reside in North Karnataka. Their main job is to weave mats out of Pandanus leaves. They are largely concentrated in the coastal belt. The *Kare Vakkals* however, are concentrated in the hilly taluks of Yallapur and Haliyal taluks. They are exogamous and are essentially agrestal in occupa-tion. There are 75,000 *Halakkis* living in Koppas under direct control of their community heads. They were one of the earliest settlers in North Karnataka and are believed to have come from Tirupati area of Andhra Pradesh. There are both exogamous and endogamous clans within the *Halakkis*. They speak '*Halakki kannada.*' They have a rich folklore. *Halakkis* have a good knowl-edge on traditional agriculture and traditional crop plants.

2.3.15   SIDDHIS

The *Siddhis* of North Karnataka are a scheduled tribal community. Considered to be traditionally nomadic, the *siddhis* are Muslims. *Siddhis* also live in Gujarat. Interestingly the *Siddhis* have genetic affinity with S. African popu-lations. This suggests their probable migration through sea route and their settling down in these places, rather than migrating through land/coastal route. *Siddhis* are reported to use 69 species of plants for medicinal purpos-es. They are generally short-distance nomads in the semi-evergreen forests of North Karnataka. Many consider the *Siddhis* as being brought as slaves by the Portuguese. In a few villages the reminiscence of the African clan system is still persisting (Gauniyal et al., 2010).

2.3.16 JORWEE

Jorwee tribe of Inamgaon of Maharashtra are one of the oldest tribes. They are known to have cultivated rice around 3,400–2,700 years BP.

2.3.17 GAVLIS

Gavlis (also spelled as *Gowlis*) of Maharashtra and North Karnataka live on the upper hill plateau of the semi evergreen forests of W. Ghats, maintain a herd of buffaloes and cattle and practice some small-scale shifting cultivation. Although traditionally nomadic, grazing over the entire terrain is largely available to *Gavli* animals. Because they consume a lot of buttermilk, *Gavlis* do little hunting of other animals. *Gavlis* exchange ghee from their livestock for surplus paddy grains obtained from *Kunbis* (see Section 2.3.20). The folk history of *Gavlis* recounts how they moved from wetter hill tracts, switching from keeping sheep to keeping buffaloes and then assumed a distinct identity.

2.3.18 PHASEPARDHIS AND OTHER NOMADIC TRIBES

The *Phasepardhis* of Maharashtra snare ungulates, quails and partridges. The nomadic *Vaidas* of Maharashtra trap mongooses, civets as also hunt crocodiles. The nomadic or semi-nomadic *Nandiwallas* of Maharashtra hunt porcupines, monitor lizards and wild pigs with the help of dogs; they move from place to place with their bulls (=*nandis*). They are non-pastoral. Their number is around 8,000 (Gadgil and Malhotra, 1979). *Nandiwallas* form two distinct sovereign groups: *Tirumal Nandiwallas*, numbering around 3,000 and *Fulmali Nandiwallas*, numbering around 5,000. Both have clearly demarcated territories of operation and both have migrated from neighboring former Andhra Pradesh. Some *Fulmali Nandiwallas* have a settled life. *Mumba Devi* is worshipped by all *Fulmali Nandiwallas*, especially by those settled from nomadic life.

2.3.19   MAHASKE DHANGARS

The people of this tribe lived in the Mula river valley of Pune district which lies at the crestline of W. Ghats. They are pastoralists and were mainly dependent on water buffaloes. Their deity was *Waghjai* which resides in the river valley. The *Dhangar* settlements are exclusively on the flat hill-top plateaus at an altitude of 1,000 to 1,200 m, away from all villages. They live in small settlements of 5–10 families, tend their buffaloes on grazing in the forest and fight off tigers. They do very little shifting cultivation and do little hunting-gathering. Their main sustenance is on butter milk and on millets or rice obtained through barter for butter sold in weekly shandies. They do not require the service of any other community (such as priests, medicine men or barbers) and they hardly interact with others. Ecologically, this is not even the optimum habitat either for them or for their water-buffaloes but still persist to live in this environment.

2.3.20   KUNBIS

Kunbis (or *kunabis*) are a tribal community of North Karnataka and S. Maharashtra. Traditionally nomadic, they occupy the valleys and grow paddy. It is supplemented by some shifting cultivation of the lower slopes. *Kunbis* keep very few animals and hunting of wild animals over the entire terrain is largely their monopoly. They exchange their rice to the milk products obtained from *Gavlis*. They are particularly seen in the Mula river valley of Pune district at the Crestline of W. Ghats. They speak corrupted Marathi. They are believed to have migrated from Goa a long time back. They use around 45 ethnomedicinal plants. *Kunbis* maintain sacred groves as a resource as well as for cultural purposes. Sacred ponds form a component of these groves.

2.3.21   KOKNAS

Also called *Kokanas, Kuknas* or *Koknis*, this is a tribe of the hilly regions of Maharashtra and S. Gujarat. They are particularly common in Dangs district's Waghai forest. They are believed to have migrated from Konkan region to these places. They use 46 ethnobotanical species for medicine and food.

2.3.22   THAKARS

Also known as Thakers, this tribe inhabits Pune, Thane, Raigad, Ahmadnagar
and Nasik districts of Maharashtra. This tribe was originally forest-dwelling
but are now scattered both in forests and plains. Now the *Thakars* are mainly
dependent on forest and agricultural produce; the forest products are gathered by them. They store their agricultural produce, especially grains, in
wicker baskets called *Kangas*, which is worshipped by them. They use about
140 plants species for various medicinal purposes. They are Indo-Aryans
and speak Marathi. Their population is around 5 lakhs. They have a rich
tradition of folk songs and dances. They have two major sub-divisions: *Ma-
Thakars* and *Ka-Thakars*. *Thakars* were once described as "butterflies of the
wood."

2.3.23   GHADASI KOLIS

This tribe lives in the Mula river valley located in Pune district of Maharashtra
at the crestline of W. Ghats. The people of this tribe are shifting cultivators
and hunter-gatherers (Roy Burman, 1996). Their deity is *Bapujiboa*. They
stay in the higher hill-slopes at altitudes of 800–1000 m in settlements of
25–30 families. This tribe is always in conflict with *Marathas*. The Koli villages of Khanu and Chandar are the most difficult to access from anywhere.
Kolis maintain a number of sacred groves and bamboos form an important
component of these groves (Roy Burman, 1996).

2.4   CONCLUSIONS

India is known for its rich human resource as it is one of the earliest occupied
countries of the World by the modern human species. Peopling has been
going on in various parts of India from around 70,000–50,000 years ago.
Humans have occupied different environmental regimes of India many of
which have posed various problems for them. One such region is W. Ghats
and its adjacent W. Coast of Peninsular India. All ethnic tribes that occupied
this region from ancient times have largely been responsible for identifying
and using various plants of cultural, social and utilitarian importance growing around them for meeting several of their requirements: foods, medicines
(human and veterinary), aromatics, fodder, construction materials and others. They were also responsible for domesticating and cultivating some of

these plants from their wild relatives. They also developed their own traditional ways of conserving these plants. Hence, a study of the ethnic diversity of the Western peninsular India, as presented above, is very much needed to understand the cultural, social and utilitarian plant resources.

KEYWORDS

- **Ethnic Tribes**
- **Hunter-Gatherers**
- **Nomads**
- **Peopling**
- **Shifting-Cultivators**
- **Western Coast**
- **Western Ghats**

REFERENCES

Aiyappan, A. (1965). Some Patterns of Tribals Leadership. *The Economic Weekly*, Jan. 9, 1965. pp. 55–58.

Anantha Krishna Iyer, L.K. (1909). The Cochin Tribes and Castes. Vol.1. Higginbothams, Madras.

Ayyanar, M. & Ignacimuthu, S. (2011). Ethnobotanical survey of Medicinal Plants commonly used by Kani tribals in Tirunelveli hills of Western Ghats, India. *J. Ethnopharmacol. 134,* 851–864.

Bhagat Singh, A. (2014). Vazhum Moodhathaiyarkal-2 (=Living Ancestors-2). Thodargal (=Todas). *Kaadu 1*(2), 24–31.

Bhagat Singh, A. (2015). Vazhum Moodhathaiyarkal-3 (=Living Ancestors-3). Kotthars (=Kotas). *Kaadu, 1*(3), 22–29.

Bhagat Singh, A. (2015). Vazhum Moodhathaiyarkal-5 (=Living Ancestors-6), Anamalai Kaadargal (=*Kadars*). *Kaadu 1*(5), 38–45.

Bhagat Singh, A. (2015). Vazhum Moodhathaiyarkal-6 (=Living Ancestors-6) *Kaanigal* (=*Kanis*). Kaanigal (=*Kanis*). *Kaadu 1*(6), 26–34.

Cavalli-Sforza, L.L., Menozzi, P. & Piazza, A. (1994). The History and Geography of Human Genes. Princeton, University Press, Princeton, New Jersey.

Chakrabarti, D.K. (1999). India: An Archaeological History. Oxford Univ. Press, New Delhi.

Disotell, T.R. (1999). Origins of modern humans still look recent. *Curr. Biol. 9,* R647–R650.

Gadgil, M. & Malhotra, K.C. (1979). Role of Deities in Symbolizing conflicts of Dispersing Human Groups. *Indian Anthropol. 9,* 83–92.

Gadgil, M. & Thapar, R. (1990). Human Ecology in India. Some Historical perspectives. *Interdisciplinary Sci. Rev. 15,* 209–223.

Gadgil, M., Joshi, N.V., Manoharan, S., Patil, S. & Shambu Prasad, U.V. (1998). Peopling of India. In: Balasubramanian, D. & Appaji Rao, N. (Eds.). The Indian Human Heritage, Universities Press, Hyderabad, pp. 100–129.

Gauniyal, M., Pattanayak, I., Chahal, S.M.S. & Kshatriya, G.K. (2010). A genetic study of four population groups of Uttar Kannada district of Karnataka. *Anthropologist 12,* 205–210.

Hockings, P. (2012). The Badagas, sometimes refugees in a new Land. *DJ Dawn Jour. 1 (1),* 1–28.

Krishna Iyer, L.A. (1967). Kerala Charithram (Malayalam) Vol. I. National Book Stall, Kottayam.

Krishnamurthy, K.V. (2006). Tamilarum Thavaramum (in Tamil) (=The Tamils and Plants). Bharathidasan Univ., Tiruchirapalli, India.

Lahr, M.M. & Foley, R. (1994). Multiple dispersal and modern human origin. *Evol. Anthrop. 3,* 48–60.

Reich, D. Thangaraj, K. Patterson, N. Price, A.L. & Singh, L. (2009). Reconstructing Indian Population History. *Nature 461,* 489–495.

Roy Burman, J.J. (1996). A comparison of sacred groves among the Mahadeo Kolis and Kunbis of Maharashtra. *Ind. Anthropol. 26,* 37–45.

Sali, S.A. (1989). The Upper Palaeolithic and Mesolithic Cultures of Maharashtra, Pune.

Sargunam, S.D. Johnsy, G. Samuel, A.S. & Kaviyarasan, V. (2012). Mushrooms in the food culture of the Kanni tribe of Kanyakumari district. *Indian J. Trad. Knowl. 11*(1), 150–153.

Satishkumar (2008). Genetic profile of Jenu Kuruba, Betta Kuruba and Soliga tribes of Southern Karnataka and their Phytologenetic relationships. *Anthropologist 10,* 11–20.

Subash Chandran, M.D. (1997). On the ecological history of the Western Ghats. *Curr. Sci. 73,* 146–155.

Thangaraj, K. (2011). Evolution and migration of modern human: Inference from peopling of India. pp. 19–21. In: New Facets of evolutionary Biology (Abstracts). Madras Christian College, Chennai.

Thorne, A.G. & Wolpoff, M.H. (1992). The multiregional evolution of humans. *Sci. Am. 266,* 76–83.

Viswanathan, M.B. (2010). Scientific Screening of Kanis medicines from Kalakkad-Mundanthurai Tiger Reserve in Tamil Nadu of India. In: Golden Jubilee National Symposium on Plant Diversity Utilization and Management. Kerala Univ. Thiruvananthapuram, India, pp. 179.

CHAPTER 3

THE INFLUENCE OF TRADE, RELIGION AND POLITY ON THE ETHNIC DIVERSITY AND ETHNOBOTANY OF THE WESTERN PENINSULAR INDIA

K. V. KRISHNAMURTHY

Consultant, R&D, Sami Labs, Peenya Industrial Area,
Bangalore–560058, Karnataka, India

CONTENTS

ABSTRACT

This chapter deals with the impact of trade, religion and polity on the ethnic diversity and ethnobotany of Western peninsular India starting from third millennium BCE up to about the end of British rule in India. It explains how this region was subjected to the impacts of foreign traders, and external (as well as indigenous) religions, such as Christianity and Islam on culture, society and traditional botanical knowledge of local ethnic communities. These impacts caused the domestication of plants collected from wild and used (like pepper), introduction of exotic plants as plantation crops, changed the traditional profession of certain ethnic communities and also in the elimination of certain ancient tribes. These impacts were particularly great after the visit of Europeans in 16th century and the establishment of colonial British rule.

3.1   INTRODUCTION

It was indicated in the second chapter of this volume that peopling of West Coast (W. Coast) and Western Ghats (W. Ghats) of peninsular India should have happened at the earliest only around 20,000 years BP (Sali, 1989), as it was at this time that the monsoon became distinctly weak at the height of glaciations on the northern latitudes (Gadgil and Thapar 1990). Peopling at the beginning should have been very sporadic and thin but the density and area of occupation should have increased gradually in the subsequent historical periods. However, the major human influences in this region took place during the Old Stone Age over 12,000 years BP (Subash Chandran, 1997; Misra, 1989). Stone tools and artifacts belonging to this period were discovered from the following river valleys: Bharatapuzha (in Palakkad district), Beppur (Malappuram) Netravati (South Karnataka), Kibbanahalli (Mysore), Lingadahalli, Nidahalli and Kadur (Chikmagalur) and Honnali (Shimoga). Between 12,000 and 5,000 years BP (Mesolithic period) this region witnessed a gradual transition from hunter-gatherers to cultivators. However, fishing continued to be the occupation of those living in the coastal region. Many Mesolithic sites in places like Karwar, Ankola (both in N. Karnataka), Netravati valley (S. Karnataka), Nirmalagiri (Kannur), Chevayur (Kozhikode) and Thenmalai (Kollam) have been discovered (Subash Chandran, 1997). The presence of charcoal of 5,000 years BP in the last-mentioned site indicated that the Mesolithic people (of at least this region) burnt forests to initiate the slash- and –burn Swidden cultivation.

During Neolithic Age (5,000 to 3,000 years BP) this region saw primitive agriculture along with the continued pastoralism. Around 4,300 years BP there was animal domestication in Kodekal (Gulburga); in Hallur (Darwad) there was not only animal domestication (around 3,800 years BP) but also cultivation of millets and horsegram (around 3,500 years BP). The Jorwe tribe of Maharashtra cultivated rice around 3,400–2,700 years BP. The other Neolithic sites are Tambde Surla (Goa), Anmod (North Karnataka), Agumbe (Shimoga), hill slopes of Sita River (Uduppi), Kodagum and many sites in Kerala and extreme southern Tamil Nadu. Access to W. Coast from not only W. Ghats but also from the plains immediately east to the W. Ghats was also increasingly established in different parts along the entire length of western Peninsular India facilitating traffic of ethnic communities for various purposes including trade. For instance, the Nilaskal site in Agumbe gave the people easy access to the W. Coast. According to Sundara (1991) the Neolithic ethnic communities with their axes were able to descend from W. Ghats of S. Karnataka towards the coast during the end of the 4th millennium BCE in order to embark on shifting cultivation. Thus, it is suggested that shifting cultivation was likely to be older to the spread of iron tools in this region and was at the latest about 3,000 years BP. During 3,000–2000 years BP the coast and many areas in W. Ghats were intensively populated with people from different ethnic communities. This is evident from the presence of several Megalithic burial sites. Although climate change may be the reason for the beginning of cultivation (Caratini et al., 1991; Sukumar et al., 1993) man-made fire might have also been an important factor (Gadgil and Mehr-Homji, 1986).

3.2 PERIOD UP TO 5th CENTURY CE

The situation described above was the one prevailing in western peninsular India at or just before the initiation of deep impact of trade, religion and polity on its ethnic diversity and ethnobotany. It is at the period between third millennium BCE and fifth century CE that this region became a vital center of historical processes that molded the identity of peninsular India, within the wider Indian Ocean world in general and the Arabian Sea in particular. In many respects the Arabian Sea served as the historic core area of the Indian Ocean (Barendse, 2002) as it connected Middle East, Africa and Central, East and South East Asia. The historical processes, mentioned above, extended beyond agrarian expansion and the development of trading networks in the interior peninsular India and were concerned with the diverse ethnic communities within the spheres of polity, trade and religion that contributed to the cultural and social identity of the entire Indian Ocean world (Ray, 2003) as well as to that of the western peninsular India. In the words of Thapar (2000) this period is "a crucial period not only because it saw the initial pattern of Indian culture take shape, but also because it can provide clues to a more analytical understanding of the subsequent periods of Indian history," especially in peninsular India.

3.2.1   TRADE

All ancient people of India, including those in the western peninsular India, collected or produced only the required amount of material needed for their subsistence as well as, more significantly, to reinforce social ties and to repay obligations, both social and ritual. Trade changed this whole approach. Local trade, according to some, started around 10,000 years ago but maritime and long- distance trade that emerged and started flourishing in the period between 3rd millennium BCE and 5th century CE was not merely an elite activity and an offshoot of agricultural expansion but also caused a shift from pastoral and hunter-gatherer economy to a village economy based on plant cultivation and animal domestication. Trade and exchange of required materials are thus social products and form an internal component of most, if not all, societies that were evolving. Once agriculture was well underway the need for a variety of plants increased. The plants that had been used in the wild were gradually brought into cultivation. People also began to move useful plants from one area to another, and, very often, trade and religion were used as instruments to effect this movement. Thus, movement of plants, in a way, was responsible for the increased trading and for the spread of religion (as well as for an increase in warfare) (Prance, 2005). This resulted in the domination of territorial and commercial interests. This is particularly true for the western peninsular India from the 3rd millennium BCE onwards and here trade in spices was a major cause of territorial and commercial domination and aggression. This desire for spices from very early dates in history had a very great impact on the ethnic people and their ethnobotanical knowledge of the W. Coast of peninsular India. Local ethnic communities were forced to actively collect spices from the wild initially and subsequently to domesticate and cultivate them in order to meet the increasing demands of trade. Trade in spices and other ethnobotanicals from this region introduced a number of social, cultural and behavioral changes in the ethnic communities and in their attitude toward their regularly used ethnobotanicals. The cultural and social values placed on plants became more and more replaced by utilitarian approaches towards them. All the local communities started trading in a small way, at least in their excessively collected/produced plant products, although some local ethnic communities concentrated only on trading and exclusively became traders. As examples we can mention the *Banias* of Gujarat and the *Sarawat Brahmins* of Konkan region. The community of traders by sea was variously called: *Sagarapatoganas, Navikas, Mahanavikas,* etc. However, the participation of these local trading communities in the trade network of W. Coast of peninsular India was minimal, since it was the foreign demand for luxuries that largely triggered it (Figure 3.1).

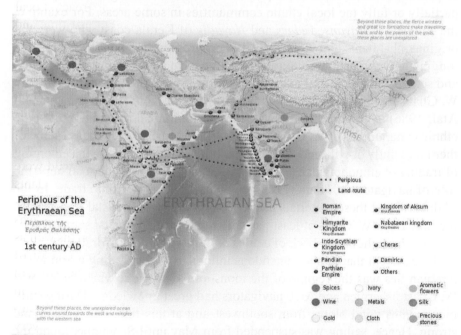

FIGURE 3.1   The Indian Coast of Erythraean Sea showing major ports as in the first century CE. (Source: https://en.wikipedia.org/wiki/Periplus_of_the_Erythraean_Sea#/media/File:Periplous_of_the_Erythraean_Sea.svg).

The 5th to 4th centuries BCE provided evidence of sailing between W. Coast of India with parts of the Indian Ocean region, both towards the west and east of India. The W. Coast of India is one of the two major trading systems that emerged to connect the many regional circuits, the other being the East Coast of India. This period may be said to be the starting time for Indo-Roman, Mediterranean and Indo-African trade. In the commerce that got started, the initiative usually came from the recipient countries or from Arabs and Axumites who acted as middle men. The foreign traders included Nabataeans, Sabaeans, Homerites, and Arabs. Some of the local ethnic communities also became middlemen in this trade network; some became financiers of trade. The initiative was more likely taken largely by merchants based in Egypt and the Eastern Mediterranean, especially in Alexandria. Some of the traders who were not originally from the W. Coast of peninsular India came to this region and started settling down replacing partially or fully the local ethnic communities in some areas. For example, the *Pulayas*, *Parayans*, *Vedars* and *Nayadis* of Kerala coast who were the original inhabitants before the arrival of progressive and dominant races belonging to Dravidian, Mediterranean, Polynesian and Aryan stocks invaded and settled in the W. Coast during this period, were either driven away to the W. Ghats or were enslaved to do manual work for them (Pushpangadan and Atal, 1986). Although fishing has been the traditional occupation of some ethnic communities of W. Coast (and continues to be so even today) some of them forcefully or willingly took to boat-building activity with the emerging of trading of ethnobotanicals. Some traditional fishermen had to assume the role of navigators of boats/vessels involved in local maritime trade. In the Malabar Coast these materials had to be brought to the coastal region from the hinterland/W. Ghats at least partially through small boats; some materials were brought on the back of cattles or carvans pulled by draft animals.

One system that had a tremendous influence on the trading world in the Arabian Sea was the regime of the monsoon wind about which only some W. Coast fishermen and Arab navigators had great knowledge. During summer months wind blows from southwest and at this time it is violent and strong. Hence, sailing was suspended from May until September along the Arabian Sea in the W. Coast of India. A reversal takes place around October and northeast monsoon dominates between November and April. The season for trade from Aden to Malabar is October to February. Hence sailing was seasonal and, for several months, there was no activity at ports and for those involved in maritime trade. They took to their traditional professions during these off-days. Sailing seasons determined the price and movement of commodities. In view of this, this trade between W. Coast and Mediterranean region is called Monsoon Exchange, a term coined by historian J.R. McNeill (as distinct from Columbian Exchange to denote European trade after 15th Century) (McNeill, 2000; Crosby, 1972).

The traded materials included spices, aromatics, raw drugs, dyes, food stuffs, woods, cotton and textiles, sesame seeds and oil, betel nut and many other plant (and animal) products. The *Periplus Maris Erythrae* (see Schroff, 1912; Casson, 1989) written in Koine Greek Language by an unknown sailor mentioned the following as the traded items during the first century CE from the ports of W. Coast of Peninsular India: frankincense, myrrh, Cassia, bdellium, a range of gum resins like *duaka*, *kankamon* and *mokrotu*, dates, ebony and sissoo woods, spikenard, *Cinnamon* leaves (*malabathrium*), textiles especially block-printed cotton textiles of ritual significance, food stuffs, pepper, etc. In his initial survey of 311 papers, referred to as *Geniza Documents*, relevant to the Indian trade, Goiten (1966) found that no trade orders were transferred directly from the Indian Ocean to the Mediterranean but these terminated at Cairo instead. The list of 77 items transferred from India (and Arabia and E. Africa) to Cairo included herbs (36 items), cotton (6 items), tropical fruits, such as coconut (5 items), timber (1 item) and dyes, such as Indigo. The Sindani/Sandan indigo mentioned in Geneva papers is identified with the Konkan coast. Archaeological evidences indicate trade in coconut, Job's tears, rice, etc. Mention must also be made about the *Vienna papyrus* which speaks about a trade agreement made at Muziris (or Mouziris) in the Malabar coast and a Greek merchant regarding the transportation of traded goods to be imported from India and bound for Alexandria. The standard traded items were Gangetic nard (spikenard), pepper, textiles and ivory. Mention must also be made about Indo-African trade both to and fro. The African millets were evident in India during this period. The archaeological record from W. India indicated the presence of sorghum and millets between 2000 and 1200 BCE. Finger millet, cowpea, watermelon and the lablab or hyacinth bean reached India during the second millennium BCE and pigeon pea, okra, and castor followed them (see Carney and Rosomoff, 2009). The African plants made their way past the Arabian Peninsula to India along two principal trade routes: one linked Ethiopian highland to the Horn of Africa and the other connecting East African highlands of Zanj (now the Swahali Coast). Recovered artifacts suggest that East Africa and India were trading as early as 3000 BCE (Lejju et al., 2006). Notable among South Asian plants that went to Africa through India were *taro* (or *cocoyam*) and banana which became a staple food plant of people of African humid tropics.

The major ports in the West Coast involved in maritime trade ethnobotanicals and other items were (from north to south) Souppara, Bassein,

Kalleina (Near Bombay) Semulla, Palaipatmai, Melizelgara, Erannoboas, Sesekreienai island, Kaineita island, Hog islands, Nauora, Tundis, Mouzris, Nelkinda, Bakare, Baltita, etc. (Figure 3.1).

Although bartering was still prevalent, at least at the inland trade circuits, several materials functioned as money: barley, pepper, cowries (mollusc shells), lead, copper, bronze, tin, silver and gold.

3.2.2   RELIGION

Religion played a major role in the transformation and social change among the ethnic communities of western peninsular India. It represented a synergy as crucial as social and economic integration. It also played a major role in the attitudinal changes in the use of ethnobotanical knowledge by ethnic communities from cultural, social and utilitarian perspectives. Most traditional ethnic communities of this region practiced before the common era (CE) the basic religions of animism or its slightly advanced version of totemism (many primitive communities follow these religions even now). Village or clan gods were largely worshipped. Formal religions like Hinduism, Buddhism and Jainism were beginning to be followed slowly in different parts of western peninsular India. For example, Buddhist monastic sites have been discovered in Maharashtra region during the Satavahana reign; about 80 sites with 1,200 rock-cut monastic centers are known from this region. Jain religious archaeological remnants and caves have been discovered in Karnataka and Tamil Nadu. The influence of religion should be seen at several levels. One facet of religion was legitimizing political authority, and rituals and ceremonies were very important factors in this. Religious functionaries often formed close links with caravans carrying ethnobotanicals and other items and trading groups whom they accompanied through forested tracts of W. Ghats and in sea voyages. This is especially true for Judaism and Christianity during this period of history in W. Coast.

An invaluable literary source for understanding of the Indian Ocean network and west coast of India is the Christian Topography written in the 6th century CE by an Egyptian monk, Cosmos, known as Indopleustes (McCrindle, 1897). In the early part of his life, he was a merchant and had traveled widely in the Arabian Sea. This work is significant in two respects: (i) it provides information on the spread of Christianity in the W. Coast and on conversion of some local ethnic people to Christianity, as well as on the settlement of Christians in this region, and (ii) it provides information on ethnobotanicals of this region, particularly that were traded and exchanged.

This work especially mentions pepper and coconut, the former as the native of Malabar. It also mentions the maritime trade involving seasame and cotton clothes from Khambat, Kalyan and Chaul on the Konkan coast and pepper from the Malabar coast. Cloves, aloes, silk and sandal wood came from the east to both Malabar and Konkan coasts. In the collection, handling and trade of ethnobotanicals local ethnic community people were involved. Regarding the spread of Christianity this work particularly mentions about Apostle Thomas (St. Thomas) who traveled throughout India in the first century CE. He finally came to Malabar coast during the realm of King Mazdai. It was here that he was finally condemned to death by the king and attained martyrdom. A set of six *Pahlavi* inscriptions written on Christian stone crosses is found in S. India and five of these were located in Kerala in the churches of Kottayam, Murrucira, Katamaram, and Alanga. The Christian community people (i.e., non-converts) started settling down in small numbers in the Malabar coast from the beginning of CE and were speaking either *Pahlavi* or Persian at the beginning; their number started increasing after 6th century CE.

3.2.3 POLITY

In peninsular India the earliest political dynasty was the *Satavahana* dynasty, which appeared around 1st century BCE in the Western Deccan. The subsequent dynasties included the *Abhiras* of the 3rd century CE. The Kerala coast during this period was under the *Chera* Tamil kings before whom Tamil chieftains ruled parts of the southernmost W. Coast of peninsular India. These chieftains and *Chera* kings themselves were belonging to ancient ethnic communities and the present-day *Cherumas* or *Pulayas* claim themselves to be descendants of the *Cheras* (see Chapter 2 of this volume). The Sangam Tamil literature and the Tamil epic *Chilappathikaram* speak a lot about the customs and culture of the ethnic communities of this region; they also speak in detail about the various plants used by these communities for cultural, social and utilitarian purposes (Krishnamurthy, 2006).

Thus, the ancient "ethnosphere" of the Arabian Sea involving the W. Coast of India during the period from 3rd millennium BCE to 5th century CE was a multiethnic and multipolar network of cultural and commercial exchange without any Centre or State dominating and was not a unipolar world system. "World" prices of traded commodities were still not fixed at one location, unlike the situation after 15th century CE (Barendse, 2002).

3.3   PERIOD UP TO 15ᵗʰ CENTURY CE

From the 15ᵗʰ century CE onwards there is evidence for rapid transformation in all three spheres (trade, religion and polity) in western peninsular India. In the political sphere the period was marked by the emergence of large regional kingdoms, such as the *Pallavas*, *Rastrakudas*, and the *Calukyas*. In the context of religion, Hindu temples rose as the centers for social, cultural and economic activity. In the arena of trading networks one can see the beginnings of merchant guilds, such as the *Manigrammam* and the *Ayyavoles*. The ancient (old) world system of trade in ethnobotanicals and socio-cultural developments in the western peninsular India did not simply disintegrate in the 15ᵗʰ century despite heavy strains in the system, but, persisted without drastic changes.

Chaudhuri (1985, 1990) had dealt with the economy and civilization of the Indian Ocean region (including western Peninsular India) from the rise of Islam in the 7ᵗʰ century CE. Wink (1990) also emphasized that this Ocean presented a unified Arabic-speaking world. Here the Muslims monopolized trade in ethnobotanicals and other commodities and imposed a unified currency based on the gold *Dinar* and the silver *Disham*. The Muslims participating in the trade with W. Coast of India were either Arabs or Persians and settlements of Muslim traders grew in the Konkan and Malabar Coasts. In fact, the Arabic Sea was called the Islamic Sea. In addition to foreign Muslims, there were *Banias* (in S. Gujarat coast), Goans, *Saraswat Brahmins* (Konkan and Malabar coasts) and Tamil merchants (for S.W. coast), and *Mappillas* (in Malabar Coast) of India, who were involved in trade. They were all dealing with items, such as Malabar teak, many spices and condiments, Canara rice, manufactured goods, medicinal plants and raw drugs, aromatics, dyes like Indigo, etc., in which many local ethnic communities were actively participating in procuring/cultivating, handling transporting and financing. Muslim trading networks moved Asian rice, citrus and sugarcane from W. Coast of peninsular India into Middle East, Mediterranean and E. African destinations. They also transported sorghum, pearl millet and other plant products from African continent to India (Carney and Rosomoff, 2009).

In spite of the above, ethnic, national or religious labels were of a limited use during this "ecumenical" age. Thus, the Arabian Sea was an " archipelago of towns" (or their surroundings) involved in trading on the coast, and populated by mariners, traders, middlemen, local navigators, fishermen, laborers, financiers, boat/ship builders, etc. living within various autonomous communities of ethnic peoples (Barendse, 2002).

3.4   PERIOD AFTER 15th CENTURY

A lot of details are available, although highly scattered, from the prodigious quantities of European archives regarding the ethnobotany and ethnic societies of western Peninsular India after 15th century. Local records, literature, books, etc. are also available.

3.4.1   TRADE

After the 16th century the Arabian Sea was being principally dominated by the Portuguese *Estato da India,* the Dutch *Verenigde Oost- Indische Compagnie* (VOC) and the English East India Company (EIC). The Danish role was small and was for a very brief period only. The most important trade ports/coastal trade centers of this period on the W. coast of peninsular India were the following: Konkan Coast—Surat, Daman, Bassein, Mumbai, Chaul, Revbandar, Dabhul, Rajapur, Vengurla, Panjim, Goa, etc.; Canara Coast—Karwar, Onore, Barcelore, etc.; Malabar coast—Calianpore, Kannur, Calicut, Ponnani, Trichur, Cochin, etc. (Figure 3.2). Some ports like Surat, Calicut or Cochin, although were trade centers too, were mainly agricultural market centers obtaining goods from their farming hinterland. The life of other ports depended entirely on the maritime trade, making them true 'brides of the sea" (Barendse, 2002). These ports and their immediate surroundings were populated by people belonging to diverse ethnic communities, both from abroad and from local areas. Those from abroad included the Portuguese, Dutch, Mozambique (who were slaves of Portuguese), Syrians, Coptians, Jews, Italians, Armenians, Turks, Arabs, Christians, and a few from S.E. Asia. A rough statistics of their population in different parts of W. Coast during this period is given by Barendse (2002). Those who were controlling the different ports and their surroundings at different times were also mentioned by him. For example, Daman was the main port of Portuguese, Bombay was initially under Portuguese but was then transferred to the English, Dabhul was controlled by the *Saraswat Brahmins,* Vengurla, and Ports in Malabar coast were under the control of Dutch, some under Portuguese and a few others under the control of English (at a later period) (see details in Barendse, 2002).

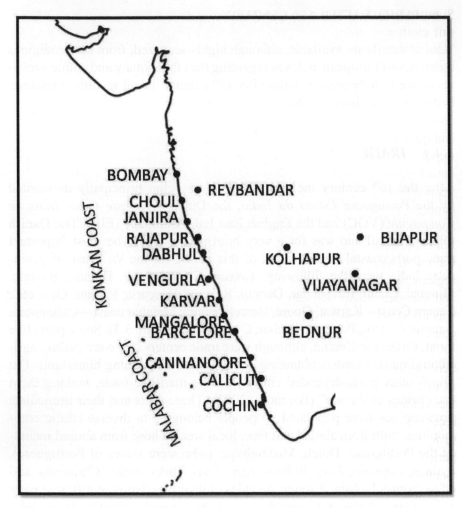

FIGURE 3.2   Indian West Coast as of 16th century, showing the major ports and trading centers (map reconstructed based on historical data. Cartographic limits are not accurate.)

Daman under Portuguese was handling trade in rice, coconut and sugar/sugarcane, the chief botanical items produced by local ethnic communities. The *Saraswath Brahmins* around the year 1600, were trading with Hormuz,

Suqutra (Socotra) and Yemen from Dabhul, mainly on textiles, such as sail cloth and cheap piece cloth woven in Belgaum and Kolhapur. From Vengurla the VOC mainly exported cotton yarn to Holland. Russia imported Indian textiles, indigo, and pepper. Indian textiles were significantly traded in Russia and many Indian textile merchants got settled in 17th century CE in parts of Russia, like Narva. The Indian merchants had special knowledge about piece cloth, which others lacked (Dale, 1994). As to the employment and quantities, but not to the investment, involved pepper trade played but a minor role in the commerce of Arabian Sea. Like other spices pepper was a luxury good. However, clearly trade from Malabar coast was dominated by pepper and as the VOC puts it: "it was the bride around whom everybody dances" (see Barendse, 2002). But to most ethnic inhabitants of Malabar coast the trade in several other products was at least as important as pepper. For example, rice was exported in substantial quantities from Cannanore and Calicut to Muscat; this rice was invariably brought to the coast from interior regions; arrack was another product of voluminous trade from Canara coast and this was mainly brought to the coast through women peddlers. There was also trade in coconut and coir to Gulf countries. Thus, rice, arrack and coir export far surpassed pepper trade.

The intensity of marine trade from W. Coast after the 15th century is evident not only from the number of ships that visited/left from various ports of this coast but also from the number of ship-wrecks so far reported from the Arabian Sea. In 1600 CE Lisbon received 3 or 4 ships annually, in 1700 received 2 ships annually, France 2 or 3, London 4 or 5 and Amsterdam 6 to 8. The number of ships entering ports of Aden and Mocha during the period 1616–1705 from different W. Coast ports is as follows: Vengurla-6; Dabhul-9; Chaul-7; Rajapur-7; Konkan-15; Calicut-8 and Malabar-31. Similarly, the number of ships arriving at Muscat in 1672, from Karwar was 1, from Konkan 14 and from Malabar coast 27 (Barendse 2002).

In the 17th century small towns, such as Karhad, Modul, Malwan and Kolhapur of Konkan coast were the major suppliers of the "Bhaleghatte cloth" to the Portuguese: cheap cotton *dhotis* and *dobradas* destined for the African market. They also produced *dogeri*, which was taken by *Khatri* merchants to Muscat and Persia. The involved people were engaged in weaving during monsoon and in agriculture during other seasons.

In the 17th and 18th centuries there was extension of trade linkages with the expanding modern world-system or the world economic order from the W. Coast of peninsular India: there was trade in ethnobotanicals and finished/ harvested plant products, such as medicine, raw drugs, textiles, coffee, cotton, sugar, black pepper, indigo, tobacco, rice and other grains: some of these were exports and some were imports. Unlike the previous periods, trade came almost under the control of Europeans. The collaboration and support of indigenous people of this region in this trade was vital for production/collection, transport and handling and supplying them to the European traders. However, the period between 1690 and 1720 was very critical as there was a collapse of the "bubble of companies" controlled by Dutch (VOC) and to a large extent by the Portuguese, and the EIC took a stronger hold from then onwards. The collapse started around 1690 and was almost complete by 1720 (Barendse, 2002).

After 15th century CE trade overland was dominated by the seasonal movements of the migratory *banjaras* (traders) and these movements were often linked to the migration of cattle in summer from pastures in Deccan to those in the Konkan coast. It was incidentally also related to a trade in draft cattle between the fringes of forests and the village communities, thus making inland trade an adjunct to such movements of nomadic ethnic communities and the cattle trade. For instance, the inhabitants of Bardes (in Goa) who owned 2,000 oxen plus 2,000 oxen of others monopolized the trade in arrack and coir ropes. The oxen drivers made four trips to and across the W. Ghats annually not only for allowing oxen to graze in summer and partly to procure merchandise (Barendse, 2002). Thus, the caravan merchants were mostly typical peddlers. Since the fortunes of Konkan ports were linked to these merchants who passed across W. Ghats and since most Konkan towns were surrounded by extensive grazing meadows for the cattle, overland trade was greatly promoted. A similar situation prevailed in Malabar coast, where tradable goods from inland were carried on pack animals, particularly near Alur. Cotton cloth export from Travancore, although minimal, had a large internal market; there were more than 4,200 looms between Thengapattinam to Cape Camorin catering to this need.

3.4.2   RELIGION

Religion played a very important role in the changes noticed in the ethnic diversity (and consequently on ethnobotany) of western peninsular India. Although the effect of Christianity was noticed from the first century onwards, the effect was substantial after the visits of Europeans to this region after 15th century CE. Special mention must be made about the conversion of local inhabitants to Catholicism promoted by agents of *Real Padroado Portugues do Oriente*. Writing in 1550 from Quilon in the Malabar coast, the Jesuit missionary Nicolao Lancilotto reported as follows: "since the inhabitants of these countries are very miserable, poor and cowardly, some were baptized through fear, others through worldly gains and others for filthy and disgusting reasons which I need not mention" (quoted in Boxer, 1963). It is not wrong to assume that this Christianization process "created local populations of 'Portuguese' cultural orientation" (Baxter, 1996). The effect of Islam started after 7th century CE, although Arabian traders came to this region much earlier than Christians. For instance, the Arabian merchants who got settled in Calianpore (the historic border between Kerala and Karnataka) since the 7th century got mixed with indigenous population, who then got converted to Islam. Their descendants differ from North Indian Muslims (often the descendants of Moghuls), both as to their language (former people speak Malayalam) and to their customs. These people were called *Mappillas*. They are also different from Arab and Persian Muslims, who controlled trade in Malabar Coast before the arrival of Portuguese and Dutch (Miller, 1976). In Cannanore there was a predominance of Muslims, both *deshi* (local) and *paradeshi* (outsiders), with more than 20,000 inhabitants even around 1660s. Ponnani, located south of Calicut, was also Muslims-dominated; in fact, Ponnani was the religious town of Muslims. Muslim *Kunjalis* dominated this town. Cochin was unique in that it had about 7,000 Indian Christian families, a larger number of Hindus and a substantial number of Jews. Among Christians there were orthodox Thomas Christians and Catholic Christians. Jews, who started coming there in large numbers from 4th century CE onwards, were around 6,000 families in the 17th century; they mainly lived in Matancheri area. Many Jews held powerful positions including heading ministries under the King of Cochin. By the 18th century there was a rise in Jewish merchants in Cochin and of Thomas Christians in Travancore resulting in significant changes in ethnic diversity. It should also be mentioned that there was great rivalry between the Christian fractions both in Kerala and in Goa, especially during 17th and 18th centuries.

Religion (as also trade) promoted creolization, a process by which new "languages" arise from intense contact between two or more languages in a reasonably small area. Linguistic admixture as evidenced in creolization is, in fact, a product of cultural (and religious) admixture in specific circumstances. When Portuguese dominated Malabar and Goa Portuguese-lexified creoles were formed in Malabar (Portuguese–Malayalam interaction) and in Goa (Portuguese–Konkani interaction) with a pidgin stage in between. There are evidences of creoles being spoken respectively in Cannanore, Tellicherry, Waynad, Matre, Calicut, Cranganore, Vypeen, Cochin, Alleppey (in Kerala) and Konkani coast and Goa. This creolization entailed inequality, social hierarchization, issues of domination and subalternity, mastery, and servitude, control and resistance, power and entanglement, etc. Such creoles declined rapidly by the 20th century, although there are pockets where it is still in existence (Cardoso, 2015).

A possible contribution of Africans is also relevant for these creoles in both Malabar and Konkani coasts, as there is historical evidence for the import of a significant number of Africans into the *Estado da India* as slaves, especially from S. E. Africa. A number of African words are formed in Indo-Portuguese texts, such as Garcia da Orta's book. Creolization made distinct cultural and social changes in the local ethnic communities of Malabar, Canara and Konkari coasts.

3.4.3 POLITY

Even by the late 16th century, lands throughout the western peninsular India were under the control of kings/empires, although no single State in the Arabian Sea area politically controlled maritime trade from the coastal ports. In the late 17th century, at Surat, the Omani trade *flottillas* has their own well-armed and manned customs. Oman had, in fact, created a system of fortified warehouses throughout the W. Coast, like the ones at Surat, Janjira on the Konkan coast and at Cannanore. The Omanis administered all these. Daman was given to Portuguese by the Sultan of Gujarat in 1578 on lease. Bombay was initially under Portuguese control but was again gained by the English. Revander port was under Portuguese control but the up-town Chaul was under the domination of Ahmadnagar king and later under Bijapur and Maratha rulers. Dabhul port was under *Saraswat Brahmins*, while Vengurla was under Dutch control although the former had control over it earlier. Goa and adjacent regions were under the Portuguese and gain control extended to south up to Cape Rama. The W. Ghats near Cape Rama was the historic border between Islamic and Hindu kingdoms, e.g., the Vijayanagar empire. After the defeat of the Vijayanagar king at the battle of Talkotta by the united armies of the Deccan Sultanate in 1565, generals of Vijayanagar established their own small kingdoms, the most important of which were the Wodaiyar dynasty in S. Karnataka and the Nayakas of Ikkeri/Bednur in N. Karnataka. Thus, Karwar, located in the mountainous forests of the small State of Sunda, a split-away chieftaincy of Vijayanagar, was possessed in the 17th century by Adil Shah and later by the Marathas. Onore port was initially under Portuguese control, but had lost it to Ikkeri kings in 1653; similar was the case with Barcelore port. Cannanore was initially under Arab/Persian Muslims, subsequently under Portuguese up to 1663, later under Dutch and finally under the English. Calicut was the seat of the powerful Zamorin, while Ponnani was dominated by Muslims who defied Portuguese and later the Dutch. Trichur was under Cochin king, while Cranganore was under Portuguese initially and subsequently under the Dutch after the *VOC* conquered the Cochin castle in 1663 and the port around 1674.

In the 17th century India, war was a major industry and the involved armies created demand for horses, bullocks, arms, cloth, grains and food as well as for the funds involved for the above. For example Shivaji's army required more than 4,000 to 5,000 pack oxen and 7,000 to 8,000 porters. Moving armies were often followed by hosts of marauders, *beggaries* or forced laborers, *banjara* caravans grazing on the village fields and "coolie and Bhil robbers" who severed as porters and supplemented their low pay with plunder and looting. All these people/cattle needed food/feed. Fighting in Deccan and the Konkan in 1680's and 1690's resulted in the plundering of Vengurla (at least 5 times), Candal and Punda. This resulted in dramatic reduction in trade in these regions, although Canara was prospering (Barendse, 2002). The kingdom of Ikkeri in Canara was prospering from the booming demand for Canara rice and other food stuff (Swaminathan, 1957)

British colonial rule "marked an important watershed in the ecological history of India" (Guha and Gadgil, 1989). Sophisticated arms and advanced communication facilities made it possible for the British to penetrate the dense forests, the abode of many tribals, in order to exploit the various forest resources. The technologically advanced and the dynamic and perseverant British culture produced profound dislocations at various levels in different ethnic societies of western peninsular India. The influence of the British was high on social relations around land, on conflicts over the distribution of agricultural produce, on fishing, on forests, on grazing land and on irrigation thus seriously affecting the local ethnic communities involved in all the above, since social relations and changes of local ethnic societies were dependent on the utilization of natural resources. The political dominance of the British made it possible to resort to novel modes of resource extraction, and this seriously affected resource utilization and extraction by local communities who were enjoying it all along. There were also equally dramatic changes in the forms of management and control of these various natural resources, particularly from the W. Ghats. State control on forests and forest resources was introduced leading to a State monopoly and to a curtailing of the legitimate rights of forest dwelling native ethnic communities. Thus, the imperatives of British colonial forestry were essentially commercial (Guha and Gadgil, 1989) and were at the expense of ethnic social interests, environmental homeostasis and traditional forest conservancy. The first forest product to suffer from this colonial exploitation in W. Ghats was the Malabar teak, which was used by the British to build ships, make furniture and to produce railway sleepers and which were largely shipped to UK. Around 1874 the Criminal Tribes Act was promulgated by the Britishers to keep under constant surveillance the tribals under the pretext of controlling "recalcitrant" elements. To the dismay of tribals, they introduced the doctrine of *res nullius* (rights in respect of land and land-based resources which were not conferred by the sovereign were claimed to vest with the sovereign) putting an end to the traditional legal epistemology in India, which was *lex loci ri sitae* (system by which the local people define their relation with land is the source of law) (Roy Burman, 2003). The Indian Forest Act of 1878 was essentially "designed to maintain strict control over forest utilization from the perspective of strategic imperial needs" (Guha and Gadgil, 1989). This Act imposed sharp restrictions on traditional ethnic use of forest products, denied claims of "rights-holders" on forest produce, and undermined the ecological basis of subsistence cultivation and sustainable hunting and gathering (of *kadars*, for example) from the wild. Commercial forestry also introduced monoculture of plantation species in W. Ghats to the detriment of biodiversity of mixed species. This again reduced the area of shifting cultivation so dear to some of the ethnic communities of the forest; this also affected the structured social life of shifting cultivators and their tribal cosmology.

British Forest Regulations affected the coastal ethnic communities also. For example, in the extreme southern Malabar coast, Forestry Act of 1878 restricted the local ethnic people of villages. Here the commercialization of forests and the sale of forest lands at extremely low prices to European planters, as well as the laying of roads for transportation of forest products to the coastal ports caused acute distress to local agriculturalists as they lost green manure and other forest produce. They were also denied access to pasture for their cattle. Similarly in coastal Maharashtra, British forest policies affected some tribals. For instance, an important source of income for tribals of Thane district was the sale of firewoods to *Koli* fishermen. This was greatly affected in late 19th century leading to strong protests by the tribal people. In addition to the above, British Forest Laws greatly affected the lives of traditional artisans who were dependent on specific plants, such as bamboo, rattans and woody trees, such as teak, sandal, ebony, etc. which were used to produce craft articles.

Thus, the British State's monopoly on forests of W. Ghats and mangroves of the W. Coast and its commercial exploitation ran contrary to the subsistence ethics of local ethnic communities, as in other regions of India. If we borrow the usage of Thompson (1971; see also Guha and Gadgil, 1989) we can aptly summarize this situation as follows: If the traditional use of the forest by the local ethnic communities rested on a *moral economy* of provision, the so-called "scientific forestry" claimed by British Forest Act of 1878 rested squarely on political economy of profit.

3.5   CONCLUSIONS

It is evident from the above discussion that the western peninsular India was in a great flux ever since 3rd millennium BCE, and particularly after 5th century CE, as a result of the deep impacts of trade (both maritime and land), religion and polity. It is one region in India from where maritime trade was very active. Trade brought into this region several human communities from diverse places, cultures and customs of the world and belonging to different religions and speaking different languages. There were also frequent changes in political control over the various trading ports and market towns of W. Coast. All the above changes had a very great impact on the historically older ethnic communities of this region. Their customs, cultures, worldviews and rituals changed significantly, as also their traditional professions; their original life-styles and mode of subsistence were affected. They were also subjected to religious conversions, intentionally or forcefully, especially into various brands of Christianity and into Islam with or without a change in their mother tongue. These changes caused a substantial shift in the ethnic diversity of this region.

These changes also brought a number of changes on the traditional ethnobotanical resources. Hunting and gathering of useful plants and animals changed into a sedentary agricultural mode of food production. Only a few communities that live deep in the W. Ghats forests still remain as hunter-gathers. Similar is the case with pastoral nomads. A number of plants/ plant products, which were once collected from wild were forced to be brought under cultivation. Pepper, banana varieties, cardamom, *Garcinia, Cinnamomum* species, etc. are some of the good examples for this. Ethnic communities were also made to cultivate a number of exotic crops, such as coffee, tea, rubber, eucalyptus, cashew nut, clove, nutmeg, etc.

KEYWORDS

- **East India Company**
- **Ethnic Diversity**
- **Ethnobotanical**
- **Maritime Time**
- **Periplus**
- **Ports of West Coast**
- **Portuguese Trade**
- **VOC**

REFERENCES

Barendse, R.J. (2002). The Arabian Seas. Vision Books, New Delhi.

Baxter, A.N. (1996). Portuguese and Creole Portuguese in the Pacific and the Western Pacific Rim. pp. 299–338. In: Wurm, S.A., Mühlhäusler, P. & Tryon, D.T. (Eds.). Atlas of Languages of Intercultural Communication in the Pacific, Asia and the Americas. Berlin: Mouton de Gruyter.

Boxer, C.R. (1963). Two pioneers of tropical medicine: Garcia d'Orta and Nicolas Monardes. *Diamante XIV*: 1–33.

Buchanan, F.D. (1870). A Journey from Madras through countries of Mysore, Canara and Malabar. 2 Vols. Higginbothams, Madras.

Caratini, M., Fontugne, M., Pascal, J.P., Tissot, C. & Bentelab, I. (1991). A major change at ca. 3500 years BP in thee vegetation of the Western Ghats in North Kanara, Karnataka. *Curr. Sci. 61*, 669–672.

Cardoso, H. (2015). The Indo-Portuguese creoles of the Malabar: historical cues and questions. http://www.academia.edu/3188426, downloaded on 20/7/2015.

Carney, J.A. & Rosomoff, R.N. (2009). In the Shadow of Slavery. University California Press, Berkeley, USA.

Casson, L. (1989). The Periplus Maris Erythrali. Princeton University Press, Princeton, USA.

Chaudhuri, K.A.N. (1985). Trade and Civilization in the Indian Ocean. Cambridge University Press, Cambridge, UK.

Chaudhuri, K.A.N. (1990). Asia before Europe. Cambridge University Press, Cambridge, UK.

Crosby, A.W. (1972). The Columbian Exchange: Biological and Cultural Consequences of 1492. Greenworld Press, Westpost, Ct, USA.

Dale, S. (1994). Indian Merchants and Eurasian Trade. 1600–1750. Cambridge University Press, Cambridge.

Gadgil, M. & Mehr-Homji, V.M. (1986). Localities of great significance to conservation of India's biological diversity. *Proc. Indian Acad. Sci. Suppl.* 1986, 165–180.

Gadgil, M. & Thapar, R. (1990). Human Ecology in India. Some Historical perspectives. *Interdisciplinary Sci. Rev. 15*, 209–223.

Guha, R. & Gadgil, M. (1989). State Forestry and Social Conflict in British India. Past & Present. *J. Hist. Studies 123*, 141–177.

Krishnamurthy, K.V. (2006). The Tamils and Plants (in Tamil). Bharathidasan University, Tiruchirappalli, India.

Lejju, B.J., Robershaw, R. & Taylor, D. (2006). Africa's earliest bananas. *J. Arachaeol. Sci. 33*, 102–113.

McCrindle, J, W. (1897). The Christian Topography of Cosmos. London.

McNeill, J.R. (2000). Biological Exchange and Biological Invasion in World History. Paper presented at the 19th International congress of the historical Sciences. Oslo, Aug. 6–13.

Miller, E.J. (1976). Mappila Muslims of Kerala: A Study of Islamic Trends. Bombay.

Misra, V.N. (1989). Stone Age India: an ecological perspective. *Man and Environment 14*, 17–64.

Prance, G. (2005). The seeds of Time. In: Prance, G., Nesbitt, M. (Eds.). The Cultural History of Plants. Routledge, New York and London, pp. 1–11.

Pushpangadan, P. & Atal, C.K. (1986). Ethnomedical and ethnobotanical investigations among some scheduled caste communities of Travancore, Kerala, India. *J. Ethnopharmacol. 16*, 175–190.

Ray, H.P. (2003). The archaeology of seafaring in ancient South Asia. Cambridge University Press, Cambridge, UK.

Roy Burman, B.K. (2003). Indigenous and Tribal Peoples in World System perspective. *Stud. Tribes Tribals 1*, 7–27.

Sali, S.A. (1989). The Upper Palaeolithic and Mesolithic Cultures of Maharashtra. Pune.

Schroff, S.D. (1966). Letters and documents on the India trade. *Islamic Culture 37*, 188–205.

Subash Chandran, M.D. (1997). On the ecological history of the Western Ghats. *Curr. Sci. 73*, 146–155.

Sukumar, R., Ramesh, R., Pant, R.K. & Rajagopalan, G. (1993). A δ 13C record of late Quaternary climate change from tropical peats in southern India. *Nature 364*, 703–706.

Sundara, A. (1991). In: Perspectives in Dakshina Kannada and Kodager. Mangalore University Decennial Volume. Mangalore: Mangalore University, pp. 4–63.

Swaminathan, K.D. (1957). The Nayakas of Ikkeri. Madras.

Thapar, R. (2000). Cultural Pasts: Essays in Early Indian History. Oxford Univ. Press, New Delhi.

Thompson, E.P. (1971). The morol Economy of the English Crowd in the Eighteenth Century. Past & Present. *J. Hist. Studies. 50*, 76–136.

Wink, A. (1990). Al-Hind, the Making of the indo-Islamic World. Vol. I. Oxford University Press, New Delhi.

CHAPTER 4

EUROPEAN CONTRIBUTIONS TO THE ETHNOBOTANY OF WESTERN PENINSULAR INDIA DURING 16th TO 18th CENTURIES

K. V. KRISHNAMURTHY[1] and T. PULLAIAH[2]

[1]Consultant, R&D, Sami Labs, Peenya Industrial Area, Bangalore–560058, Karnataka, India

[2]Department of Botany, Sri Krishnadevaraya University, Anantapur–515003, India

CONTENTS

ABSTRACT

This chapter summarizes the information available on the European contributions to the ethnobotany of Western Peninsular India after the 16th century. Works of Garcia da Orta, Acosta, Gerard, EcCluse and Van Rheede are summarized. These works essentially emphasized the traditional way of classification of plants and drugs obtained from them as well as of using them for therapeutic purposes instead of emphasizing the Hippocratic approach. They also indicate how valuable are the drug sources recognized by the ethnic communities, particularly by the Ezhava community, of the Western peninsular India for human utilization.

4.1   INTRODUCTION

Starting from the 15th century the Europeans began to show much interest on the extra-European plants and ventured actively to collect useful taxa, including medicinal, compile information relating to them and to exploit them. They started to systematize and rationalize assessments of the plants of non-western world to streamline this process of exploitation during their voyages to different parts of the world and there was scarcely a part of the world unrepresented in their maritime travels. Considering the plant collection made by them as a whole, it appears that India, and South East Asia were the most active areas followed by West Indies, Central and South America, Africa and South Pacific and Australasia. Among these, India was more easily accessible to European exploitation (Mackay, 1996). This interest of Europeans on knowledge of non-Western plants, particularly of Indian plants, was driven dominantly by the Hippocratic agendas (Grove, 1996). The scholarly interests shown by these European authors who wrote about Indian ethno-medico-botanical plants were also tended to emphasize the Hippocratic approach. These works, barring two contributions that contain information drawn from ancient and classical ethno-botanical traditions of India are not widely known (Spudich, 2008). These works seriously attempted to present traditional Indian botanical knowledge classified as per western knowledge systems and are, therefore, complementary to that available from contemporary Indian sources. As an example for the latter, we may cite the work of Krishnamurthy (2006), who has brought out the traditional ethno-botanical knowledge of ancient Tamils of South India. These European writings recorded regional folk botanical knowledge often unavailable in their original localities (Spudich, 2012). This chapter deals with works of those European authors that concern with the ethnobotany of western peninsular India during 16th to 18th centuries.

4.2 GARCIA DA ORTA

Long before Dutch became dominant in western Peninsular India and shown interest in its ethnobotany, the Portuguese were involved in a complex, but not highly organized, act of collecting plants of economic value and transferred them to Lisbon (Kapil and Bhatnagar, 1976); they also introduced some economically important plants into India (Mehra, 1965). In this process, the Portuguese extended the much older patterns of distribution and trade in ethnobotanicals that had long existed in the Arabian Sea region (King, 1899; see also Krishnamurthy's article in this volume). Portuguese travelers to West Coast of India (and other parts of the world) were soon advised to observe indigenous practices and collect and supplement data on ethnobotanicals to extend European *Materia Medica* and this type of advice elicited the preparation of the first major European book on Indian ethnobotany (Grove, 1996). The preparation of a book of this type was undertaken by Garcia da Orta. Garcia, born in 1490, arrived at Goa, on the west coast of peninsular India, from Portugal on September 1534 after leaving on 12th March 1534. He came to Goa with "a great desire to know about the medical drugs… as well as all the fruits and pepper… their names in all the different languages, as also the countries where they grow, and the trees or plants which bear them and likewise how the Indian physicians use them" (Gaitonde, 1993). His intention was also to compile a description of plants from which medicines sold in Europe and the Portuguese colonial possessions were extracted (Grove, 1996). Garcia wrote his famous book in 1563 in Goa in Portuguese, which was quickly translated into a Latin edition in 1567 and into English with an introduction and index in 1913 by Markham (Garcia, 1913). The first Portuguese edition was the third book printed by the Portuguese in India.

Garcia compiled his book, whose structure is in the form a dialog between him and an imaginary interrogator called Ruano, not only from information collected from local physicians and folk healers, but also after critical experimentation and substantiation for his own prescription as well as for physicians in Europe. His book served as the first source of information on the description of medicinal plants and the drugs obtained from them. The therapeutic effects of these drugs were carefully recorded by him. A number of these plants were also illustrated through line drawings. His book also contained very valuable information on the source locations from where the plant drugs were obtained, and included those available on the West coast of India and those that came from abroad before being grown in India. Garcia's book can be "said to lie at the core of the relationship between European colonial expansion and the diffusion of botanical knowledge," particularly the ethnobotanical/ethnomedical knowledge. In this book, the contemporary Hippocratic emphasis on accuracy and efficacy tended to privilege strongly the folk ethnomedical knowledge of India and, thus, led to "effective discrimination" against the older Arabic, Aryan and classical European texts and cognition systems (Grove, 1996). Orta's work is profoundly an indigenous text reflecting traditional knowledge of local ethnic people, although written by an outsider. It is organized essentially on non-European precepts. These aspects, although first recognized by Clusius (see Section 4.5), have been later emphasized by Boxer (1963) who draws attention to the book's wider historical value.

According to Garcia the privileging of European traditions and preconceptions would lead to medical failure and hence Indian ethnic concepts on medicine should be given importance while practicing Indian drugs and procedures of taking them. However, as Grove (1996) has rightly pointed out, his text had also been affected by the "delicate balance in power relations between the European physician, (his) Moslem patron and the local Arab or Persian doctor" (parenthesis by the authors of this article), although his own medical knowledge was better as it was more pluralistic. He was an European physician (although he shifted his basic allegiance away from Portuguese government), his patron was the Moslem king of Deccan (Burham Nizam Shah) and his friends were Arab or Persian doctors practicing under this king. Garcia's personal intimacy and love with local ethnic physicians of West Coast of India is abundantly evident in the approbation he grants specifically to Malayali and Canarese doctors and their medicines, although it was often difficult to get information from them (Grove, 1996). The most important aspects of Garcia's role as a doctor using contemporary traditional ethnobotanical and medical knowledge of West Coast of India "prefigured the pioneering role played by other lone European doctors employed by Indian potentates in promoting and utilizing indigenous technical knowledge, Honigsberger and Johann Konig being the two outstanding examples of this in the first decades of East India Company rule" (Grove, 1996; see also Grove, 1995).

4.3   CHRISTOVAL ACOSTA

Christoval Acosta (some refer to him as Cristobel Acosta), another Portuguese physician came to Goa of West Coast of Peninsular India in 1568 with "a desire to see the diversity of plants God has created for the human health" (Gaitonde, 1993). His Spanish work followed that of Garcia da Orta (Acosta, 1578). This book describes 69 plants and other sources of drugs and medicines. The text runs to 448 pages along with illustrations of 46 plants. Most of the descriptions are believed to have been copied from Garcia's book mentioned above. Acosta also collected information on plants and drugs from local physicians and folk healers of West Coast of peninsular India, experimented with them, and used them in writing his book.

4.4   JOHN GERARD

Johan Gerard, an apothecary and horticulturalist working in London, reported on medicinal uses of plants of India, including those from Western Peninsular India, used in Europe at the end of the 16th century (Gerard, 1597). The very accurate wood-cut illustrations of Gerard's book suggest that knowledge on medicinal plants and other plants of ethnobotanical interest (particularly of India) was highly sought after and collected by Europeans.

4.5   EcCLUSE

EcCluse (also called Clusius), professor and director of the Leiden Botanic Garden at Netherlands, combined and annotated earlier texts (Clusii, 1605). He extended particularly the works of Garcia da Orta (1563) and of Gerard (1597) through additions of commentaries and illustrations and created a two-volume book. These were widely used in the 17th century Europe as definitive texts on Indian Ethnobotanical medicines. Like all the earlier works, EcCluse's work also botanical empiricism, helped by commercial pressures. Orta's book work was immediately translated of the medicinal history of the New World with knowledge into a Latin edition in 1567 by EcCluse. Clusius also included with Orta's text a translation gained from Orta's book, Clusius went on to establish both the Hortus Medicus of Emperor Maxmilian in Vienna and, in 1593, the Leiden Botanic Garden (Grove, 1996). The close association between Clusius and Garcia da Orta ensured the diffusion of botanical knowledge between S.W. India and Leiden botanic garden. Clusius' early adoption of Orta's work "reinforced the primacy of the Leiden garden in tropical botany, particularly Indian botany."

4.6 VAN RHEEDE

Hendrik Adriaan Van Rheede tot Draakenstean (hereafter Van Rheede), the then Dutch governor of Malabar at Cochin and Commander of the Dutch East India Company (VOC) compiled *Hortus Malabaricus* in 12 volumes between 1678 and 1693 and published it in Netherlands (Van Rheede, 1678–1693). This book, written largely in response to the medical needs of VOC, covered 1595 pages of double folio size and detailed 742 useful plants. Van Rheede was collaborated in this work by *Collatt Vaidyan* Itty Achuden of Carrapuram near Cherthala in Kerala, a famous traditional physician of that time, as well as by three Konkan Brahmin-Priest physicians Ranga Bhat, Vinayaka Pandit and Appu Bhat, who were residing at Cochin then. This book of thirty years of work also involved a team of experts that included European physicians, Professors of medicine and botany, Indian scholars and vaidyas, technicians, illustrators, engravers and VOC officials. The material of the book drafted in Malayalam was translated into Portuguese after thorough verification, then rendered into Dutch and finally into Latin. The English translation, with annotations, was made by professor Manilal (Manilal, 2003). There are many research publications on *Hortus Malabaricus* detailing on many aspects of this stupendous work (Grove, 1998; Heniger, 1986; Manilal et al., 1977; Manilal, 1989, 1996, 2005, 2012; Mohan Ram, 2005, 2012; Nicolson et al., 1998) and the interested readers are advised to refer to these works for getting more details. In this article, the importance of *Hortus Malabaricus* in reference to aspects related to ethnobotany of South Western Peninsular India and their relevance to European botany alone will be discussed.

The writing of Hortus Malabaricus was first facilitated by the establishment of Dutch power in Cochin on the decline of Portuguese power in Malabar coast and subsequently by the connections between Van Rheede and the Dutch botanical establishment (Grove, 1996). This text also seemed to lie at the core of the relationship between European colonial expansion and the diffusion of botanical knowledge as stated earlier for Garcia's book. In Van Rheede's text also the contemporary Hippocratic emphasis on accuracy and efficacy tended to privilege strongly traditional ethnobotanical knowledge," particularly *Ezhava* community knowledge, and "to lead to effective discrimination against older Arabic, Brahminical and European

Classical texts and systems of cognition in natural history" (Grove, 1996). Hortus Malabaricus is profoundly indigenous in its content and mode of construction, is far from being inherently European and is organized on "essentially non-European percepts." The works of Heniger (1986) and Manilal (1989), according to Grove (1996), have not really been concerned to identify the wider historical significance of the power of the *Ezhava* community's affinities within the text of the Hortus Malabaricus "with all that it implies for the assertion of the *Ezhava* classificatory superiority" (Grove, 1996). Grove also emphasized that Van Rheede was largely responsible for elevating *Ezhava* ethnobotanical knowledge, with the main aim of acquiring the highest quality of traditional knowledge and expertise. Van Rheede followed a very rigorous adherence to Ezhava systems of plant /drug classification (Heniger, 1986). Van Rheede, in fact, rejected the methodologies of plant description followed by Father Mathew of St. Joseph, an European Botanist. For useful and critical field identification of Plants Van Rheede and his team heavily depended on the knowledge of *Ezhavas*, the best of whom was Itty Achuden. According to Grove (1996) the contents of *Hortus Malabaricus* was far more influenced by the Ezhava collaborators of Van Rheede than his own accounts suggest. Grove also stated that, for the subsequent history of tropical botany, "the insight of the *Ezhavas* into the affinities between a large number of plants in the Hortus Malabaricus is revealed by the names they gave to those species which have the same stem and to which one or more prefixes are added." The names of plants also provide us a considerable amount of cultural and social material about the *Ezhavas* as well as of other ethnic communities of Malabar coast.

The Ezhava traditional nomenclatural and classificatory knowledge, as contained in Hortus Malabaricus, had direct influence on many botanically important texts that followed it: Linnaeus, Adanson, Jussieu, Dennstedt, Haskarl, Roxburgh, Hooker, Gamble and many more texts. Hence, Hortus Malabaricus apparently remains as the only record of the accumulated traditional/ethnobotanical knowledge of the Ezhava community. This book mentions medicinal plants which form the sources of 2789 prescriptions for more than 210 diseases which was rampant in Malabar in the 15th to 17th century (Manilal, 2012). If Van Rheede had not compiled this information and published it as Hortus Malabaricus, the hereditary ethnobotanical knowledge of Ezhava community would have been totally lost. Since Van Rheede, Manilal and others have stated that the *Ezhava* ethnomedical knowledge (as exemplified by Itty Achuden) is pre-Ayurvedic knowledge (this is evident from a perusal of the various uses of plants listed in this book indicate that they are quite different from those listed in Ayurveda) and since *Ezhavas* were traditionally not native to Kerala and migrated there during historic times from Tamil Nadu (see Chapter 2 of this book), their ethnic knowledge may belong to the Siddha system of Medicine so dear and traditional to the ancient Dravidians or was the outcome of local folk empirical knowledge.

Hortus Malabaricus is also important in another respect. Information on the medical uses of plants described in this book is of immense importance and current relevance in the context of the growing global demand for natural drugs as well as the intellectual Property Rights regime and Biological Patent Laws. The publication of medicinal uses of Malabar plants will help in the fight for protection of traditional botanical knowledge of India, particularly of Western Peninsular India. This book can be used as an effective legal weapon in our efforts to prevent patenting of the over 650 important medicinal plants of Malabar by anybody with vested interest, and preserve and protect our traditional knowledge on plants for use by our people (Manilal, 2012).

4.7 CONCLUSIONS

The basic theme of European colonialism which started by the late 14th century has been the subject of debate for a very long time. This theme has been continuously debated from the points of view of discovery, exploration, curiosity, commerce, scientific exchanges, indigenous cultures, imperial expansion, brutal exploitation of resources of the colonized countries, and study of biodiversity. Many believed that imperial and commercial purposes predominated, while others believed that scientific curiosity was more important. India is one country which was frequently visited by different colonial powers of Europe, particularly after the 16th century under the guise of trade. These included the Portuguese, Dutch, Danish, French and English. They all tried to assess the natural resources of India and the traditional knowledge of various ethnic communities on the useful plants (and animal). They also made all efforts to systematically study them and to document them so that they can be used by them. This chapter summarizes the European contributions to the ethnobotany of Western peninsular India during the 16th and 17th centuries. The works of Garcia da Orta, Christoval Agusta, John Gerard, EcCluse (Clusius) and Van Rheede are discussed. Almost all these workers have been impressed by the traditional way of handling the ethnobotanical resources, particularly the systematic manner in which they were named, classified, protected and sustainably used, in contrast to the way in which these resources were earlier used by European and Arabic scholars.

KEYWORDS

- Acosta
- Christoval
- Ezhava Traditional Knowledge
- John Gerard
- Gerard EcCluse
- Hortus Malabaricus
- Van Rheede
- L'Esduse
- Clusius

REFERENCES

Acosta, C. (1578). Trato de las drogas, y Medicinas de las India's Orientalis. Burgos, Spain.

Boxer, C.R. (1963). Two pioneers of tropical medicine: Garcia d'Orta and Nicolas Monardes. *Diamente 14*, 1–33.

Clusii, C. [L'Esduse]. (1605). Exoticorum libre decem: quibus animalium, Plantarum, aromatum, aliorumque peregrinorum fructum historiae describuntur: Item Petri Bellonis observationibus. Plantin. Antwerp.

Gaitonde, P.D. (1993). The Portuguese Pioneers in India. Bombay.

Garcia, Da Orta. (1913). Colloquies on the simples, drugs of India and some of the fruits found, and wherein matters are dealt with concerning practical medicine and other goody things to know. Goa 1563. New Edition translated with an introduction and index by Markham, C. Hokluyt Society, London. Indian Edition published by Periodical Expert Book Agency, Delhi.

Gerard, J. (1597). The Great Herbal or General History of Plants. John Norton, London.

Grove, R.M. (1995). Green Imperialism. Cambridge Univ. Press, Cambridge.

Grove, R.M. (1996). Indigenous knowledge and the significance of South-West India for Portuguese and Dutch constructions of Tropical Nature. *Modern Asian Studies 30,* 121–143.

Grove, R.M. (1998). Indigenous knowledge and the significance of South–West India for Portuguese and Dutch constructions of tropical nature. In: Grove, R.H., Damodaran, V., & Sangwan, S. (Eds.) Nature and the Orient: the Environmental History of South and South East Asia. Oxford Univ. Press. Oxford.

Heniger, J. (1986). Henrik adriaan Van Reede tot Drakenstein and Hortus Malabaricus; a contribution to the study of Dutch colonial botany. A.A. Balkema, Rotterdan/Boston.

Kapil, R.N. & Bhatnagar, A.K. (1976). Portuguese contributions to Indian botany. *Isis 67,* 449–452.

King, G. (1899). The early history of Indian botany. Report of the British Association for the Advancement of Science. 904–919.

Krishnamurthy, K.V. (2006). Tamilarum Thavaramum (=The Tamils and the Plants). Bharathidasan Univ. Publication, Tiruchirappalli, India.

Mackey, D. (1996). Agents of Empire: the Banksian collectors and evaluation of new lands. pp. 38–57. In: Miller, D.P. & Reill, P.H. (Eds.). Visions of Empire. Cambridge Univ. Press, Cambridge.

Manilal, K.S. (Ed.) (1989). Botany and history of *Hortus Malabaricus*. Oxford. New Delhi.

Manilal, K.S. (1996). *Hortus Malabaricus* and Itty Achuden: A study on the Role of Itty Achuden in the compilation of *Hortus Malabaricus* (in Malayalam). Mentor Books, Calicut.

Manilal, K.S. (2003). Van Rheede's Hortus Malabaricum (Malabar Garden). English Edition with annotations and modern botanical nomenclature. Vols. 1–12. Univ. Kerala, Thiruvananthapuram.

Manilal, K.S. (2005). *Hortus Malabaricus*, a book on the Plants of Malabar, and its impact on the religions of Christianity and Hinduism in 17th Century Kerala. *Indian. J. Bot. Res. 1*, 13–28.

Manilal, K.S. (2012). Medicinal Plants described in *Hortus Malabaricus*, the first Indian regional flora published in 1678 and its relevance to the people of India today. pp. 558–565. In: Maiti, G. & Mukherjee, S.K. (Eds.) Multidisciplinary Approaches in Angiosperm Systematics. Vol. 2. Publication Cell, Univ. Kalyani, Kalyani, India.

Manilal, K.S., Suresh, C.R. & Sivarajan, V.V. (1977). A reinvestigation of the Plants described in Rheede's *Hortus Malabaricus*; an introductory report. *Taxon 26*, 549–550.

Mehra, K.L. (1965). Portuguese introductions of fruit plants into India. *Indian Horticulture 10(1)*, 8–12; *10 (3)*, 9–12; *10 (4)*, 23–25.

Mohan Ram, H.Y. (2005). On the English edition of Van Rheede's *Hortus Malabarricus* by K.S. Manilal. *Curr. Sci. 89*, 1672–1680.

Mohan Ram, H.Y. (2012). The Story of *Hortus Malabaricus* (1678–1693). In: Indian Traditional Knowledge (Abstracts). NCBS/Piramal Life Sci. Bangalore. p. 07.

Nicolson, D.H., Suresh, C.R. & Manilal, K.S. (1998). An Interpretation of Van Rheede's *Hortus Malabaricus*, International Association of Plant Taxonomists. Scientific Books, Konigstein, Germany.

Spudich, A. (2008). Such Treasure and Rich Merchantize: Indian Botanical Knowledge in 16th and 17th century European Books. Exhibition Catalogue. NCBS/TIFR, Bangalore, India.

Spudich, A. (2012). 16th and 17th Century European writings on Indian Botanical Medicines. Abstract of Talk. In: Indian Traditional Knowledge: history, influences and new directions for Natural science: NCBS/TIFR and Piramal Life Sciences, Bangalore, p. 8.

Van Rheede, H.A. (1678–1693). Hortus indicus Malabaricus, Continens Regioni Malabarici opud Indos Celeberrimi omnis generis Plantas rariores. 12 Vols. Amsterdam.

CHAPTER 5

LISTENING TO A FAIRY TALE ON A MOONLIT NIGHT...SOME REFLECTIONS ON THE HUMAN AFFINITIES WITH PLANTS IN THE WORLDVIEWS OF INDIGENOUS COMMUNITIES ALONG THE WESTERN GHATS OF KARNATAKA

B. S. SOMASHEKHAR

Fulbright Fellow, Associated with: 1) School of Conservation of Natural Resources, Foundation for Revitalisation of Local Health Traditions, 74/2, Jarakabande Kaval, Attur P.O., via Yelahanka, Bengaluru 560 106, India; and 2) IINDICUS (Institute for Indigenous Cultures and Studies), Bengaluru.

E-mail: bssomashekhar@hotmail.com, bs.somashekhar@frlht.org, cultureindicus@yahoo.in

CONTENTS

ABSTRACT

Natural landscapes are known to be the 'first' homes of human groups, while the diverse elements present in it, become the sole source of meeting the needs of human beings for their living. Human groups in their attempt to habituate the landscapes explore, examine and evaluate the natural elements; discover significance and use value in them and in the course of applying them to meet their needs assign specific meanings. Sustained interactions with nature and plants over generations have helped human groups to build many unique affinities and affiliations with them, which have eventually become a part of their worldviews and belief systems. These belief systems often direct the indigenous communities to consider natural elements, as an extended part of their *community*; assign human attributes and 'non-materialistic' values; consider them divine and treat with high degree of esteem. Such an intense level of significance assigned to nature and plants has secured them culturally indispensible positions, which is often 'irreplaceable.' Such human understanding ardently followed by the entire indigenous community over generations, becomes ingrained in their cultures. These worldviews and belief systems have served human communities as the learning templates for understanding the world, the learning from which gets crystallized as indigenous human knowledge.

During the course of interactions with nature and plants in their landscapes over generations for living, many indigenous communities along the Western Ghats are found to have forged a strong bonding with them. Such unique affinities while helping the communities to evolve different vocations and occupations based on the natural resources also bring to light an astounding spectrum of bio cultural manifestations of plant usage in many life contexts, in the form of nature veneration, sacred groves, beliefs and myths, rituals, taboos, cultural memories, stories, songs, and so on. Such specific forms of affinities also denote the communities' worldviews, cosmovision and indigenous knowledge which are integral to their cultures. A fascinating range of human affinities with plants and nature evolved by the indigenous communities across the Western Ghats is visible, especially in the form of ancestor worship, plant veneration, celebration of nature's bounty, human values attributed to plants, ritual specific and festival specific usage of plant produce, prudent use of resources, indigenous ecological knowledge in the form of story, poetry, beliefs, maxims, and many more. In a way, these human affinities indicate the enormous amount of passion, indigenous communities have developed towards the plant wealth in their surroundings. It also points at an awe-inspiring cultural lattice that amalgamates natural elements, and indigenous communities' performative knowledge, on which unique profiles of human understanding of plants and nature are constructed, away from the conventional 'use value' and western perspectives. However, such worldviews which are not only unique to indigenous communities in India, but are also a part of 'Indian way of knowing the nature,' are poorly studied, less understood, and remain smothered in the glare of western science, and therefore warrant urgent attention to study and understand them, lest their sheer brilliance and enormous merit go unused.

5.1   INTRODUCTION

Storytelling has been an integral element of human cultures; although a form of monolog, it requires two actors – a narrator and a listener- for an active rendition, that engages both with equal intensity of interest and involvement. Vigorous bouts of fun, surprise, excitement, disbelief, and many other human emotions flow quite profusely throughout a storytelling session, while making the listener identify with the context and characters in the story, and finally to leave him with a piece of lesson to reflect on. Storytelling has been an age old ingenious means of communication and human learning, augmented by the indigenous communities, while it also facilitates reiteration of oblique instructions, reaffirming a human group's worldly knowledge and encapsulation of human experience of the world, and so on. We all have been a part of different storytelling episodes during one or the other stages of our lives. Many kinds of stories – fantasy filled, adventurous, heroic, moral, supernatural, mythological, mystic, mysterious and such – have greatly enlivened our lives. Although specific contexts and time periods would facilitate a storytelling, at least two prominent seasons in a year, in the context of indigenous communities of southern India, appear to have been traditionally earmarked for the purpose: one, immediately after harvesting of crops, during *Kaartika* and *Maargashira* months, and two, during summer in *Vaishakha* and *Jyestha* months. While for the first season, it is the threshing yards in the crop fields, and for the second season, it is the open fields or village shrines that usually become the storytelling venues. Family members in small groups or the entire village community in large gatherings would assemble in these venues, based on the context. Simple oral renditions become spectacular when the elements of dance, drama or music unique to the communities are dabbed appropriately, to unveil a powerful performance of a story.

To be a part of a simple storytelling session or to witness a spectacular performance, especially on a moonlit night, would leave one spell bound, as the rendition would unfurl a delightful experience that elevates one to a different state of mind. We all can recount very distinct and everlasting memories associated with such storytelling episodes which we were a part of, and which we fondly cherish.

Having inspired from such storytelling episodes, if it was possible to use the template here for this chapter to reflect on the indigenous communities' affinities with plants which have been assimilated into their worldviews and belief systems, then, altogether different story would unfold here.... In a much similar fashion of a fairy tale, that was narrated many years ago, on a moonlit night, to a group of young people, who sat in the threshing yard, leaning against the haystack and bundles of freshly harvested farm produce, who had transfixed their eyes and were absorbed. While the divine sheets of milky white moonlight from the sky had covered the surroundings, with stars twinkle in the sky, cool breeze wafting from across the fields and a mellowing silence slowly embraced, then, the magical moments descend on the group to make the storytelling a mesmerizing meet...

...the Story unfolds thus...

5.1.1 IN THE NORTHERN WESTERN GHATS OF KARNATAKA...

On the first Monday of *Bhadrapada* month, *Kunbis*, a forest dwelling community in Yellapur and Joida taluks of Uttara Kannada district in the Western Ghats of Karnataka, observe a community ritual which is termed 'Gaaphri,' in *Kunbi* parlance. By the noon of that day, members from different *Kunbi* hamlets in the region, start assembling near the community shrine of their ancestor spirits, to perform the ritual. Different wild flowers, wild fruits and inflorescence of Arecanut are specifically brought for the purpose, and hung in front of the ancestor idols (only amorphous stones, though). Special dishes are offered to the spirits and a puja is performed, followed by a feast. This ritual is exclusively performed to propitiate the ancestor spirits and seek their approval to commence the harvesting of different roots and tubers in the forest, the community elders assert. After the day's proceedings, when the assembly is ready to disperse, the headman presents a bundle of small tufts of fresh green paddy straw, to every member of the community as 'prasada,' which the members receive with utmost reverence. This bundle which might appear abstract and unintelligible to an outsider, is actually a festoon made with tufts of paddy straw. Every tuft is made with a small sheaf

Figure 5.1a. Kunbi headman holding a bundle of tassel-tufts made of paddy straw

Figure 5.1b. Close-up view of a bundle of tassel tufts of paddy straw

Figure 5.1c. Festoon of tassel-tufts hung before the figurines of ancestral spirits

Figure 5.1e. Festoon of tassel-tufts of paddy straw hung across the doorway of a Kunbi house

Figure 5.1d. Kunbi headman offering a tassel-tuft to ancestral spirits inside the shrine_ note the tassel-tufts and other forest produce hung

Figure 5.2a. Way side open shrine of 'Hulidevaru'(Tiger God) and the huge mango tree

Figure 5.2b. Idol of 'Hulidevaru' installed on an open platform

Figure 5.2c. A way side clump of sacred trees

Figure 5.2d. Temporary shelter and platform built around a sacred tree

Figure 5.3a. A vendor in a local market sells bundles of fresh floral stocks of wild plants for Balindra puja

Figure 5.3b. A vendor in a local market sells fresh floral stocks of many wild plants as ritual ingredients for Balindra puja

Figure 5.3c. Fresh stocks of *Diplocyclos palmatus* and other wild plants for sale as ritual ingredients in a local market

Figure 5.3d. Close-up view of a bundle of fresh floral stocks of wild plants used for Balindra puja

Figure 5.3e. Twining shoot of Diplocyclos palmatus wound around the neck of a water pot, for Balindra puja

Figure 5.4. Fruit bearing twigs of Gooseberry on sale in a local market for Tulsi puja

Figure 5.5. Bhoomi Hunnime celebration -
Puja to the earth goddess

Figure 5.7. Long poles of Bilva tree
spread before the temple car

Figure 5.6. Offerings of arecanut inflorescence to a
local spirit in its hideout, as a part of tuluva rituals

of fresh paddy straw folded into a scroll tied at the base and wrapped with a mango leaf. The swaying long blades of straw are trimmed so as to make it look like a 6-inch long tassel-tuft. About 8–10 such tassel-tufts are stringed together to make one festoon, and several such tufts would be prepared for the day. Back at their hamlets, the community members would tie this festoon across the main doorway of their homes, without fail. *Kunbis* consider this ancestor puja ritual and tying the festoon are inseparable elements in their lives. The whole event might look quite unusual to an outsider.

From Joida, travel south to reach Sirsi, a bustling town in Uttara Kannada, and proceed in the direction of Nirnalli, a small village in the midst of dense forest. As one approaches the village, what welcomes a visitor is an open platform with a small idol of tiger, next to a huge mango tree. One would remain awe struck looking at the tiger's idol in the midst of the forest, while the villagers who pass by are seen offering their obeisance to the idol. As if to add to the bewilderment, they would inform that it is the idol of 'Hulidevaru' (the tiger God), and the huge mango tree is its hideout. Feeling quite surprised and intrigued about such unique human perception of nature, subsequent inquiries along the highway in the direction of Siddhapura, a small town in the southern part of the district, would reveal many such local deity-tree pairs. Isolated patches of wild vegetation or tree clumps are commonly encountered, in which one would find small idols and terracotta figurines installed around the trees; one would even find small shelters built around some tree clumps. These trees are locally referred to as "Devaramara", meaning "tree of the local deity". Villagers consider such trees to be the seats of village spirits and guardian deities.

Continue the journey further south, to reach Soraba, another small town on the fringes of Malnad. One would come across the local people often talking about *Idiga* community and *Bhoomi Hunnime*. As is well known, *Idigas*, are traditionally the Toddy tappers and agriculturists with their unique living styles. On the full moon day of *Ashvayuja* month, members of this community observe with much grandeur a farming ritual called *Bhoomi Hunnime* (meaning, full moon day of the earth goddess). *Idigas* believe that, it is a requirement for this ritual, to prepare a special dish of 108 different potherbs (the number 108 is not a figure of speech, though) to appease the deities. Accordingly, on the preceding day they would scout the croplands and fallow fields in the neighborhood and gather tender shoots, growing tips and nibbles from 108 different plant species; herbage thus procured is used as potherb and greens in the homesteads to prepare a special stew which is offered to goddess earth on the ritual day. While the celebration of *Bhoomi Hunnime* forms a major socio cultural event ardently practiced in the region, it is to be noted that the 108 potherbs are not chosen in an *ad-hoc* manner, neither all the wild plants available in the vicinity are gathered at random, but only those plants that constitute 108 herbs are carefully identified and collected while discarding those which are not in the list. It is thus interesting to note that the practice of gathering 108 herbs and the preparation of a special stew for the ritual offering actually connote the *Idiga* community's acquisition of rich indigenous knowledge about wild seasonal foods (Somashekhar, 2008b).

5.1.2 AND IN THE CENTRAL WESTERN GHATS OF KARNATAKA...

From Soraba, turn towards Sagara-Shivamogga-Tirthahalli, the towns that form the hub of Malnad of Karnataka. With many forest and garden based livelihoods and professions in place, one would find that the local communities here have adopted culturally vibrant lifestyles, which is evident during many occasions. For instance, on the day of Diwali festival, the market place would be literally flooded with head loads of different floral stalks and wild produce brought for the festival sale. One would find the local people throng the market and exuberantly engage in buying a small bundle of floral stalks and wild produce on sale. To one's surprise, these fresh floral stalks are not the usual flowers and garlands found in the market. A seasoned botanist would readily identify the constituents of this bundle as the floral stalks of wild plants, such as *Achyranthes aspera, Alternanthera sessilis, Cassia auriculata, Celosia argentea, Cymbopogon* spp., fruit bearing slender branches of *Diplocyclos palmatus* (*Mahalinganaballi*), flowering twigs of Black Niger (*Guizotia abyssinica),* and a few panicles of Ragi and Paddy. This bundle of floral stalks is exclusively purchased by the local people for performing 'Balindra puja' ritual as a part of Diwali celebration. As is known in Indian mythology, *Balindra* who was an ambitious king, had performed 99 *yajnas* and was about to complete the 100th, which was a prelude for him to secure the position of lord *Devendra*. Having sensed this threat, and unwilling to relinquish, *Devendra* pleaded with the 'trimurtis' to save him. At his behest, lord Vishnu, came to Bali in the guise of 'Vamana' (dwarf Brahmin), and asked him for a piece of land, though as a present, which could be covered by his three strides. Agreeing to this, Bali showed Heaven and Earth, on which *Vamana* placed his first and second strides. And when *Vamana* was looking for a place for the third, Bali requested him to place his foot on his head. In the attempt, lord Vishnu, pushed Bali to 'Patala,' and thus saved *Devendra*. However, pleased by the generosity and sheer obedience of Bali, lord Vishnu appeared before him in his true stature and granted a boon, according to which, Bali could come back to earth on one day during Diwali, to dispel the darkness of ignorance. And, 'Balindra puja' is thus a ritual to welcome back king Bali.

Accordingly, on the day of Diwali, a small lump of cow dung is kneaded into an irregular pyramid as the figurine of *Balindra*, and kept in the courtyard of the houses. Floral stalks brought from market are inserted as a special offering and a puja is performed. While the floral stalks are meant to be offered to *Balindra*, the slender shoot of *Mahalinganaballi* is coiled around the neck of a pot-filled with water and is offered puja. This bundle of floral stalks and slender twig of *Mahalinganaballi* become the indispensible ritual ingredients ardently brought by the local people for performing *Balindra puja*. This has been a customary practice observed by the local people over many generations, the elders assert.

Subsequent to Diwali, many villages in Tirthahalli region celebrate *Tulsi puja* with much fervor on the 12th day of the new moon cycle of *Karthika* month. On that occasion, a *Tulsi* plant (Sacred Basil, *Ocimum sanctum*) retained in the courtyard of the house becomes the focus of attention; family members erect a small pandal around it; coconut fronds are spread over it, while it is decorated with different garden and forest produce; fruit bearing twigs of *Amla* (*Emblica officinalis*) and Tamarind (*Tamarindus indica*) are exclusively brought and planted next to *Tulsi* plant. In the evening, earthen lamps are kept around the plant; family members and neighbors gather, special dishes are offered to Tulsi and a puja is performed. People believe that, *Tulsi* is a consort of lord Krishna and this ritual is observed as a mark of paying respect to *Tulsi* and lord Krishna. *Tulsi puja* as a ritual is predominantly performed by the *Malnad Vokkaliga* community in this region.

From Tirthahalli, cross over to Udupi, a coastal town on the west coast of Karnataka. One would encounter a widespread practice of drinking a special decoction of a tree bark, amongst the local populace. This bitter decoction is made with the bark of *Maddaale* tree (*Alstonia scholaris*) and every household adheres to this practice on the new moon day of *Aati* month (the day and the ritual are locally called 'Aatiamavasye'). On this day, before sunrise, a member from the household would go to a *Maddale* tree in the neighborhood, extract the bark with the help of a clean sharp stone, and bring the pieces home. Inner fleshy portion of the bark is separated, pounded with pepper, cumin and garlic, and boiled with milk to prepare the decoction. Subsequently, all the members of the family take bath and drink a glassful of this decoction on empty stomach. It is believed that the decoction drinking is a time tested preventive healthcare practice against malaria, and is in vogue amongst the local communities for many generations, in the guise of a ritual. Additionally, it is a customary practice among the households to use certain wild plant produce for preparing many season-specific dishes on the day, and all through *Aati* month. *Pathrode*, a native variant of steam cooked rice cake, is one such exclusive dish prepared by using select wild plants. As a botanist would confirm, the wild plant produce is composed of tender shoots of *Cassia tora*, leaf stalks of *Colocasia esculenta* and *Amorphophallus peoniifolius*, and unripe fruits of *Cayratia mollissima*. Additionally, fresh leaves of Turmeric (*Curcuma longa*), Jack *(Artocarpus heterophyllus)*, Coral tree *(Erythrina indica)*, and Screw pine (*Pandanus unipapillatus*) are mandatorily used for making small cups, into which the batter is filled to prepare *Pathrode* (Bhagya et al., 2013).

5.1.3. FINALLY, IN THE SOUTHERN WESTERN GHATS OF KARNATAKA...

From Dakshina Kannada enter Kodagu, the district bordering Kerala in the Western Ghats. One would find that, on the 18th day of *Aatimaasa*, a wide spread celebration called 'Kakkada Padinetta' is observed by Kodavas, the native inhabitants of Kodagu. On this day, it is a customary practice in all the households of Kodavas to use a wild herb called 'Aatisoppu' or 'Maddusoppu' to prepare a special sweet dish. Consumption of this herb is an age old custom and is invariably followed on the 18th day of this month, the Kodavas assert. This practice is so popular and ardently followed all over the region that, the plant gets its local name 'Aatisoppu' from 'Aatimaasa' and corresponds to the species, *Justicia wynaadensis*. Local people believe that, consumption of this herb is to be done invariably on this day, as the bio-availability of 18 medicinal principles found in the plant would be maximum. The herb is used to prepare special delicacies, such as 'Maddu Puttu' (Sweet cake) or 'Payasam' (indigenous variant of sweet Porridge). This ritual celebration in the recent years has caught the attention of the local mass media too (Rajendra, 2007).

A form of ancestor puja, a wild tree being the seat of a guardian deity, veneration of wild trees, unique ritual ingredients, *Tulsi* puja, festival specific offerings, healthcare herbal decoction- range of illustrations as above, which to an outsider might appear quite peculiar, abstract, unintelligible and unusual, only denote the unique usage and indispensable cultural significance assigned to different plants and plant produce, by the indigenous communities in their culture, which are often expressed during a ritual, celebration, cultural process or a system. All of these accounts drawn from different socio-cultural contexts in the Western Ghats of Karnataka depict on one hand, an astounding display of bio-cultural manifestations of plant usage assimilated by the indigenous communities, while on the other, point in the direction of a rich repertoire of human affinities, affiliations and associations with the plant world. It is also to be noted that an exclusive set of plants or plant produce is employed in all these community processes, which is quite unique and endemic to the community or the region and stands distinct. These illustrations thus bring to light the 'irreplaceable' cultural position enjoyed by 'specific' plants. India abounds such unique pieces of human aspects of plants, each one standing out from the other, in respect of their affinities and associations with plants. These interesting accounts do not merely add up to interesting elements of community cultures, but point at an awe-inspiring cultural lattice that amalgamates elements of plant world, indigenous human groups, and their understanding of plant world.

Such human aspects of plants have remained integral to indigenous cultures, which often appear in the form of beliefs, myths, legends, customary practices and rituals, memories, narratives, stories, songs, taboos and guidelines, proverbs and riddles, trying to reflect the communities' worldviews, cosmovision and indigenous knowledge. However, despite being the integral elements of indigenous cultures, such worldviews have largely remained in the dark- either less studied and therefore less understood, or labeled irrational, or smothered in the glare of the western science. The present chapter is thus an attempt to look at such interesting multiplicity of human affinities and associations with plants, as found embedded in the cultures of many indigenous communities along the Western Ghats. Hitherto less studied though, these affinities do help construct interesting profiles of indigenous human understanding of plants in India, away from the materialistic 'usufruct value' and western perspectives. For the sake of the discussion in the chapter, examples of generic nature are drawn from different socio-cultural contexts, while specific illustrations correspond to the indigenous communities and plants inhabiting the Western Ghats and Transition zone of Karnataka.

5.2 COMMUNITIES' PERCEPTIONS OF NATURE: FACTORS IN THE FOREGROUND

It is known that, natural environment is the basic setting for the life of human groups around the world. With different natural formations in the form of hillocks, mountains, water bodies, streams and rivers, open fields and variety of vegetation, plants, animals, birds, wildlife, people, coupled with natural phenomena, such as wind storms, rain, hail, lightning and thunderstorms, the human groups generally perceive the environment as a whole and not into separate and distinct entities (Pretty et al., 2009). However, environment is the sole resource base for the human groups to secure all the needs and requirements for their lives, and in turn becomes their 'home' and 'world.'

In the process of familiarizing with their surroundings, indigenous communities around the world have literally explored, examined and evaluated different elements of the environment. Accordingly in this endeavor, different kinds of human actions, interactions and processes with the natural elements ensue in an iterative fashion; human groups would continue to initiate, attempt, modify, strengthen and realign their interactions with the nature; their experience from such interactions would get transformed into many learning lessons, while such knowledge gets accrued into perception, understanding, visualization, and insights; the learning and experience of the community members in a generation would guide the actions of the members from successive generations; common understanding and shared knowledge amongst the group members would slowly blossom into collective wisdom and community culture. It is quite obvious to expect that, the indigenous communities have passed through these sequential stages in their attempt to make a comprehension of their surroundings or nature. Scholars studying comparative sciences of cultures thus assert that, when a person is born, he is bound to be under the influence of two settings: Natural and Cultural (Balagangadhara, 1994, 2005).

It is said that such an understanding becomes a way of these communities' going about in the world (Balagangadhara, 1994), and human ecologists have recognized this strong association with the natural world (Geertz, 1973; Gadgil et al., 1993; Posey, 1999; Pretty et al., 2009). It is generally recognized that, different human cultures interact with nature in different ways and forge different relationships with the local environments (Gadgil et al., 1993; Milton, 1998; Berkes, 2008). Thus, nature and culture converge on many planes to span values, beliefs, practices, norms, livelihoods, knowledge and languages amongst the indigenous communities (Milton, 1998; Posey, 1999; Turner and Berkes, 2006; Berkes, 2008). The interesting aspect of human dimensions of plant world is the unique nature of this community knowledge, which has remained endemic to small geographic regions or to ethnic groups.

The sustained association of indigenous communities with their environment over many generations enables them to accumulate a variety of perceptions about the natural elements around them; accumulation of such perceptions in turn would enable them to construct an understanding of their surroundings- this understanding becomes the basis for the unique 'Viewpoints' about nature and 'Worldviews' to emerge.

It is said that, meanings and interpretations of the natural world by indigenous communities are most diverse, who tend to perceive their surroundings with innumerable memories, encounters, and associations (Jain, 2000). Indian folklore is replete with such illustrations of meanings and interpretations, in the form of tales and oral narratives, faiths, values and guidelines, beliefs and taboos, myths, legends, proverbs, lullabies, songs, hymns, games, dance, all of which reiterate such inseparable human associations with nature. There have been a few methodical inquiries in recent decades to understand such intricate relations (Frazer, 1960; Ramanujan, 1989, 1990, 1997).

5.3   FORGING A BOND: NATURE IS DIVINE

One of the most passionate articulations about the significance of Nature, in recent times was by Aldo Leopold in his *A Sand County Almanac* (1949), in which he wrote thus:

> "We abuse land because we regard it as a *commodity* belonging to us. When we see land as a *community* to which we belong, we may begin to use it with love and respect.... That land is a community is the basic concept of ecology, but that land is to be loved and respected is an extension of ethics. That land yields a cultural harvest is a fact long known but latterly often forgotten" (italics mine).

While it is understandable that, this concern assumes prominence in the light of global environmental crisis and alarming depletion of natural resources in the modern world, it is however true that, indigenous communities in India for long, are known to have imbibed many conservation friendly attitudes and perspectives into their traditions. It is to be noted that nature and its elements here are traditionally regarded by them more as *community* and not as *commodity* as cautioned by Leopold.

Indigenous communities are known to base their views of nature on spiritual beliefs, while the modern day industrialized communities tend to base their beliefs on Science (humans as a biological species), although many modern people in industrialized countries still acknowledge a spiritual or affective relationship with nature.

Indian philosophy relates nature and its fundamental elements fire, water, earth, space and air, as *panchmahabhutas* and considers them to be the basic constituents of all forms of life. Human affiliations with nature here, are more of ritualistic and religious nature than materialistic, and are deep-rooted in cultures. It is not uncommon to see mountains and rivers held in high esteem by the indigenous communities who consider them as abodes of deities, spirits and ancestors. It is said that, in most Hindu traditions, Earth is to be revered, for she is the mother. Mother Earth, known by several names (*Bhu, Bhumi, Prithvi, Vasudha, Vasundhara, Avni*) is considered to be a *devi*, or a goddess. She is seen in many temples together with Lord Vishnu ("all-pervasive") in South India and is worshiped as his consort. She is to be honored and respected. The earliest classical texts, the Vedas, have inspiring hymns addressed to Earth (Narayanan, 2001).

Ritual and devotional resources that privilege the natural environment abound in the Indian traditions. Many Indian *Puranas* and epics mention specific places in India as sacred and charged with power. The philosophical visions of the various Hindu traditions portray the earth, the universe, and nature in many exalted ways. Nature is sacred; for some schools, this *Prakriti* ("nature", sometimes translated as "cosmic matter") is divine immanence and has potential power (Narayanan, 2001).

It is interesting to note that, this spiritual way of understanding the nature is quite innate and vibrant at many places along the Western Ghats. For instance, the entire region that spans Konkan and Uttara Kannada district is traditionally believed to be the land of lord Parasurama of Indian mythology. It is because of such legacy, several wilderness areas, hill ranges and rivers along the Western Ghats are viewed by the local people as the abodes of different deities of 'Hindu' traditions. One can recount the names of rivers and mountains associated with different deities. *Yana*, for instance, a unique geological formation of gigantic columns of igneous rocks, in the dense forests of Uttara Kannada district, is considered by the local residents as the abode of lord Bhairava, an accomplice of Lord Siva. Likewise, the entire hillock, *Karikanammanagudda* in Honnavara taluk of Uttara Kannada, in the midst of evergreen forests, is venerated by the local communities as the abode of the goddess *Karikanamma*. Similarly, Devimane Ghat here is considered as the home of the Forest goddess. Many other hillocks and wilderness areas along the Western Ghats are venerated and such striking examples may be found at Saundatti (Goddess *Yellamma*), Ulavi (Lord *Ulaveesha*), Daareshwara (Lord *Daareshwara*), Someshwara

(Lord *Somanatha*), Koteeshwara (Lord *Koteeshwara*), Narasimha Parvatha, Sringeri (Goddess *Shaarade*), Talakaveri (Goddess *Kaveri*), Brahmagiri, Koodlu Tirtha, Kodachaadri (Goddess *Mookambika*), Kumara parvata (Lord *Subrahmanya*), which are held in high esteem as the abodes of different deities, while annual fairs and festivals are held on a particular day. Incidentally, many of these sites, over the years have been transformed into pilgrim centers and have become cultural landscapes of the region. Interestingly, a pilgrimage route stretching from Hanagal to Ualvi in Uttara Kannada passes through forested area and at select places enroute, where sporadic wild populations of *Bilva* (*Aegle mormelos*) are seen, devotees consider these to be the extended portions of the sacred *Bilva* trees of mount Kailas, and offer puja during their pilgrimage.

Similarly, many rivers in the Western Ghats are associated with Indian epics and myths. For instance, river *Sharavathi*, one of the major west flowing rivers of Karnataka, is said to have sprung when lord Rama (of the epic Ramayana) shot an arrow to the ground in his attempt to get water to quench the thirst of Sita, during their exile into forests; *Ambutirtha* (literally a sacred pond created by an arrow) is the place in the Western Ghats where the river *Sharavati* originates. *Sitanadi* is a small river in Udupi district, named after Sita. *Aghanaashini*, is another west flowing river in the Western Ghats, and is believed to alleviate the sins of those who take a dip in it, and this belief is the factor behind naming the river so (*agha* meaning sin, *naashini* meaning one who alleviates it). *Mukti hole* is a tributary of this river which literally means a stream that bequeaths salvation. *Kaalinadi* is the name of another river in Uttara Kannada named after goddess *Kaali*. Rivers in the Western Ghats are not considered as mere water courses, thus assert the indigenous communities, through such beliefs.

5.4   ONENESS WITH NATURE

It is said that, landscapes are a product of the connection between people and a place; these are spaces which people feel they have a relationship with, and of which they hold memories and build history (Pretty et al., 2009).

Examples of indigenous communities venerating nature in India are many. It is not uncommon to see amongst indigenous communities, who reckon mother earth as a goddess and name it *Bhoodevi*, while all the human values and virtues are attributed to it. Particular mention may be made about the beliefs of *Male Kudiya* community in Udupi, Dakshina Kannada and Kodagu districts, in this regard. So as a girl attains puberty, *Bhoodevi* too attains puberty, assert the *Male Kudiyas*. The community believes that, the earth goddess attains puberty during monsoons and to mark this development, a regional ritualistic festival called "Keddasa" is celebrated after the monsoons. On the day, the community members go on a ritual hunting, prepare special dishes with select wild tubers and offer puja to earth goddess (Rai and Gowda, 1996). Likewise, the members of *Idiga* community in Shivamogga district celebrate *Bhoomi Hunnime* ritual, when the paddy crop starts flowering and panicles are sufficiently developed, who consider this ritual as the celebration of the goddess earth's pregnancy, akin to a pregnant woman in their community (Nagesh, 2011).

The *Halakki Vokkaliga* community of Uttara Kannada goes one step further in extolling the goddess earth. For them, the well grown standing crop of paddy in the fields is akin to a *Halakki* bride, who is just married, with the nuptial ceremony due. And thus, when the paddy crop is sufficiently grown in the field, they celebrate its nuptial, in the form of "Gadde Gandana Habba" (literally, the festival of Paddy field's Husband). For this purpose, well grown branches of Screw pine (*Pandanus unipapillatus)* or *Bakula* tree *(Mimusops elengi)* are brought and planted at select places amidst the standing paddy crop. *Halakkis* believe that bringing these twigs is analogous to a Halakki son-in-law coming to his in-laws place to meet his wife, while the planting of twigs signifies the union of "Gadde Ganda" with his wife (branches of Screw pine and *Bakula* being the "Gadde Ganda" and Paddy field being the female). The standing paddy crop would yield a good harvest, only if this nuptial ceremony was completed, assert the *Halakkis*. Likewise, it is not uncommon to see many villages in Shivamogga district performing the marriage of Peepal and Neem trees (Ramesh, 2010).

Cultural anthropologists have pointed out that, such oneness with nature is likely to have evolved from a continuous and direct dependence upon nature (Milton, 1998; Berkes, 2004). Such dependent relationships and the resultant understanding of the environment are more often depicted through actions rather than words and, thus often taken for granted in everyday lives. However, Ellen (1996) proposed three definitions of nature that accommodates this oneness with nature as: nature as a category of 'things'; nature as space that is not human; and nature as inner essence. Some believe that modernist views have gone beyond viewing nature and culture as two separate entities, instead view them as opposing entities whose interaction generally leads to one or the other being damaged in some way (Pretty et al., 2009).

5.5   MANY MEANS OF MAKING A MEANING

Away from such spiritual and higher forms of perception, indigenous communities' understanding of nature and plants generally vary. Three broad premises responsible for such variations, may however be recognized: (a) there are different landscapes and habitats in nature with different bio-resources; (b) different human communities occupy these landscapes which have evolved different ways of going about in the world and accordingly adhere to different worldviews; (c) needs and requirements of these human groups vary with their understanding of the bio-physical environment and the underlying worldviews.

These premises appear to largely govern the frequency and intensity of indigenous communities' learning in the world, and subsequent development of a bonding with the plant world. Human understanding of plants appears to be largely *context driven* and even within a community, the nature of learning differs amongst its members. However certain broad stages that precede such a learning may be recognized as: Recognition of a plant and such natural elements in the living environment; describing and making a comprehension of it in order to familiarize oneself with; exploring its inherent features for its possible usefulness; examining and applying it's features to meet the diverse needs of one's life; replication of such attempts in subsequent contexts, to secure the required usufructs as prompted by earlier success or failures; assigning definite importance and value to it based on the experience; attempting to explore and identify the same element across the environment; and finally assigning a specific place to it in the communities' life- these would be the key stages that enable different human communities' learning about the natural resources in the environment (Somashekhar, 2008b).

Indigenous communities are known to attribute meaning to natural systems and processes through different ways, including livelihoods, cosmologies, worldviews and spiritual beliefs (Berkes 2008). Such cultural understandings fundamentally govern both individual and collective actions, which in turn shape the nature and composition of human landscapes. As a result of such interconnections, a feedback system exists, whereby a shift in one system often leads to a change in the other (Berkes and Folke, 2002; Maffi and Woodley, 2007).

It is said that, knowledge about nature continues to get accumulated within a human community and gets transferred to next generation through cultural modes of knowledge transmission, especially through rituals, performative learning, stories, narratives and close observations (Balagangadhara, 1994; Pilgrim et al., 2007, 2008; Pretty, 2009). It therefore, comprises of a non-static compilation of observations and understandings contained within the social memory that constantly evolves to try and make sense of the way the world behaves. Balagangadhara (1994) asserts that these processes serve as 'configurations of learning' in the Indian traditions, and terms it as 'performative knowledge' as against the textual knowledge of the western world. It is however argued that a definite meaning to a seemingly abstract nature of this human action is difficult to assign (Staal, 1979). Berkes (2008) recognizes this as "knowledge-belief-practice" complex which is central to linking nature with culture.

5.6   FOREMOST OF THE MEANINGS: VENERATION OF NATURE AND PLANTS

Indigenous communities are known to perceive nature and plants as the extended manifestations of the divine power. The community perspective of venerating plants elevates human association of plants to an altogether different plane. Two distinct forms of communities' veneration of nature are seen across the Western Ghats: one, Venerated trees and two, Sacred groves.

It is not uncommon to come across along the Western Ghats, several trees being held in high degree of esteem by many indigenous communities, who venerate them as the seats of different deities. Enthoven (1915) about a century ago documented more than 30 different trees and plants being venerated across the west coast districts of Maharashtra.

Many striking examples are available to illustrate this perception. Halakki Vokkaliga community of Uttara Kannada for instance, traditionally holds in high esteem many plants including cluster fig, coconut, areca nut, jackfruit, mango, screw pine, *Bakula* (*Mimusops elengi*), and *Tulsi*, since several cultural processes of this community invariably make use of different produce obtained from these plants (Nayaka, 1993). Likewise, many communities in Dakshina Kannada district hold *Alstonia scholaris* tree in high esteem. As a part of the ritual *Balindrapuja*, green branches of this tree are planted next to a *Tulsi* plant in the courtyard, and offered Puja in the households here.

Many venerated plants are also found to be associated with different socio-cultural contexts of the region/ community along the Western Ghats. In a recent field study of bio-culturally significant plants in 50 select localities of Karnataka, Somashekhar and Gowda (2014) identified 35 tree species being venerated by the local communities on 15 different socio-cultural premises. In addition to serving as the seat of village deities, these trees were found to be venerated on account of: being the seat of a clan deity, guardian deity, healing deity and of accomplices/consorts of village deities; being the hideouts of ancestral spirits, sages and mythological figures; being associated with community ancestors and cultural heroes; being a memorial tree; being cultural and heritage landmarks; being a boundary mark to ascertain the cultural spaces and demarcate village boundaries; being the socio-cultural venue, and being the focal point of select rituals and customs amongst the local communities. These trees are the true "bio-cultural assets" of the region, and are thus not to be equated with 'Sthalavrikshas' of Tamil Nadu in southern India, it was asserted. One such tree species was *Ficus mysorensis,* as brought out by the study, which was found to be considered as an embodiment of a local deity, 'Antaraghattamma' (literally meaning the goddess of the transition zone), by the local communities at Antaraghatte of Chikkamagaluru district in the transition zone of Karnataka. The local communities believe that the deity resides in this tree and therefore venerate it for long. A small shrine is built around the base of this tree; a huge community fair and a procession of about 1500 bullock carts are held annually during the *Chaitra* month. The shrine dates back to the days of Tarikere chieftains, of post Vijayanagara empire period in southern India (1600 CE).

The second conspicuous form of veneration of nature is seen with sacred groves; these are small patches of vegetation preserved in the name of a village deity or a spirit, and are held in high esteem by the local people. Traditionally many village communities in the Western Ghats consider village forests as the property of village gods, in which they are situated. Such groves are referred to by different names as: *Devarabana, Devarakadu, Hulidevarakadu, Nagabana, Bhutappanabana, Jatakappanabana, Chowdibana.* Studies in the recent years have examined the size, extent, distribution, biodiversity, ecosystem services, cultural significance, and other related aspects of sacred groves at many places along the Western Ghats (for instance, see Gadgil and Vartak, 1981; Chandran et al., 1998; Malhotra et al., 2001). These studies have put the total number of sacred groves in the Western Ghats (Uttara Kannada, Shimoga, Udupi, Dakshina Kannada and Kodagu) at 1424. Sacred groves are known to serve many socio-cultural functions, many scientists assert, such as: Providing cultural space to local communities; a common property resource; assertion of group identity; assertion of group solidarity; propitiation of deity or ancestral spirits; propitiation of totems.

There is yet another form of culturally important groves in Dakshina Kannada district. *Marathi Naika, Male Kudiya* and *Malnad Vokkaliga* Communities here, follow a unique practice of earmarking certain portion of their crop land as 'Bhairavavanas,' in the name of lord Bhairava, in which the trees of *Strychnos nux-vomica* and *Alstonia scholaris* are invariably maintained (Ramesh, 2010).

There are many places along the Western Ghats, where the wild populations of culturally important trees are found, are treated as cultural groves. Such groves are especially seen with the trees, Ashoka (*Saraca asoca*), Kadamba (*Anthocephalus cadamba*), Surgi (*Mammea suriga*), and Bilva (*Aegle mormelos*). Many wild populations of these trees are protected and referred to respectively as 'Ashokavana' (at Kasage, Mugilukoppakaan and Bisugoda villages of Uttara Kannada), 'Kadambavana' (at Shaalmala in Uttara Kannada), 'Suragibana' (at Katagala, Hiregutti and Masur villages of Uttara Kannada), and 'Bilvatopu' (at Kalmaradimatha and Hunasaghatta villages of Chikkamagaluru).

5.7 MULTIPLE MEANINGS: BIOCULTURAL MANIFESTATIONS IN THE TRUE SENSE!

Many indigenous communities inhabiting the Western Ghats are found to have assigned high degree of importance to select plants and plant produce, in their lifestyles and traditions. As a part of the communities' belief base, this understanding is quite ingrained in their cultures and often remains specific to a community. Jain (2000) asserts that, mythological association of plants enjoys such high degree of importance, and this ancient practice forms a major part of human affinities with plants in India.

While from a common man's perspective, assigning a name to a natural element becomes a prelude for all kinds of interactions with that element, since such a name serves as the point of reference and an identification term. Naming of plants in a local context would be usually influenced by the striking features of a plant associated with its occurrence, habitat, life form, size, color, and such morphological features, or properties and functions, such as taste, aroma, and usefulness, or similarity with other elements of nature. In the Indian context, it is said that, plant naming is either 'Swarupabodhaka'— based on morphological features, or 'Gunabodhaka'—based on properties, uses, and functions (Jain, 2000). It is quite interesting to note that, the local communities have found out the innovative scheme of plant nomenclature as a means of registering their understanding of the plant world. However, very little attention has been paid in India to study the cultural significance of vernacular plant names.

5.7.1 CULTURAL UNDERSTANDING OF PLANT NAMES

An interesting interpretation of the Kannada names of native plants by Somashekhar (2008a, 2012), points at a unique scheme of plant nomenclature evolved by the indigenous communities, who have associated certain plants with many deities and mythological figures in the Indian traditions, and consequently a higher degree of esteem reposed on such plants. Some such delightful Kannada names of plants with the respective mythological figures are: 'Ramapathre' (lord Rama, Rama's mace, *Myristica dactyloides*), 'Lakshmanaphala' (Lakshmana, Lakshmana's fruit, *Annona muricata*), 'Sitadande' (Sita, Sita's garland, *Dendrobium* spp.), 'Bhimanakaddi' (Bhima, Bhima's stick, *Sida acuta*), 'Pandavara Adike' (Pandavas, Betelnut of Pandavas, *Pinanga dicksonii*), 'Krishnanakireeta' (Lord Krishna, Krishna's Crown, *Clerodendrum paniculatum*), 'Shivana Jate' (Lord Shiva, Shiva's knotted hair, *Caryota urens*), 'Gaurimudi' (Gauri's Braid, *Lepidagathis cristata*), 'Gangethaali' (Gange's pendant, *Andrographis serpyllifolia*), 'Brahmadanda' (Lord Brahma, Brahma's Baton, *Celosia argentea*), 'Vishnukanthi' (Lord Vishnu, Aura of Vishnu, *Evolvulus alsinoides*), and 'Nagadanthi' (Naga, Serpent's tooth, *Baliospermum montanum*).

5.7.2 CULTURAL ORIGIN OF PLANTS

While associating a plant with a deity or a mythological figure through its local names makes one way of assigning cultural significance to plants, there are different stories, beliefs and myths that substantiate the high degree of cultural value assigned to some plants in the indigenous cultures. Usually in the form of origin stories of plants, these anecdotes bring out many interesting dimensions of cultural understanding about the origin and occurrence of plants on earth, as held by indigenous communities. It is a common belief that the plant Lotus originated from the navel of Lord Vishnu. *Brahma's Hair* (Gandhi and Singh, 1989) presents a compilation of interesting accounts of mythological association of many plants including Coconut, *Parijatha*, Silk Cotton, *Kadamba*, *Ashoka* and *Tulsi*, as prevalent in different parts of India. Many indigenous communities in the Western Ghats are found to narrate such origin myths of certain plants associated with their culture. Interesting anecdotes are available in respect of: *Tulsi* (Sacred Basil, *Ocimum sanctum*) with *Halakki Vokkaliga* community (Hegde, 1977); Bottle gourd and 'Pavattila tree' with *Male Kudiya* community (Rai and Gowda, 1996); Indian Spinach and *Amaranthus spinosus* with Tulu speaking communities (Shetty, 2000);

and *Champak* (*Michelia champaca*) with *Soliga* community (Madegowda, 2009). The local people in Hunasaghatta village of Chikkamagluru district narrate an interesting myth about *Bilva* tree (Wood Apple, *Aegle mormelos*). According to the legend, in ancient times there used to live a sage in the nearby forests doing 'tapas'; one day when he was deeply engrossed, a traveler passing through the forest who had lost his way, stopped and repeatedly asked the sage in his attempt to seek directions to reach the next village; in the process, the sage was forced to break his silence. Having realized that his tapas was disturbed, the sage, in a fit of anger pulled out a rosary from his neck, tore it and threw its beads in all directions. Wherever the beads were strewn, a *Bilva* plant sprang up from the beads. That is how the natural population of *Bilva* is so scanty, they assert, pointing at its sporadic occurrence in the forests nearby.

5.7.3. MIGRATION MYTHS AND ORIGIN STORIES OF INDIGENOUS COMMUNITIES

While such legends indicate the origin of those plants important to the communities, a much stronger version of human association with plants is seen in the form of cultural memories related to their origin or migration, which is passionately carried on by them over generations. For instance, *Male Kudiya* tribe of Dakshina Kannada, narrates a story of their origin, in which it is said that, the founder of the community was sent to earth by Lord Shiva and goddess Parvati long ago, who named him as 'Kudiya,' and sent with him, different tubers of Dioscoreas.

Similarly, *Kunbis* of Uttara Kannada narrate an anecdote associated with their migration. According to the story, Kunbis who were the original inhabitants of Goa and Konkan region, had to flee the land owing to the oppression by the Portuguese rule in the early 16th century AD. While migrating from their homeland, they carried with them Sweet potato, Cashew nut, Indian spinach, and Chrysanthemum and planted them wherever they settled in Uttara Kannada and Dakshina Kannada (Thurston and Rangachari, 1909).

Likewise, the *Gangadikara Vokkaligas*, a farming community inhabiting the transition zone of Karnataka passionately relate their migration to a *Jamun* tree (*Syzigium cuminii*), though of mystic nature. According to this cultural anecdote, several generations ago, their ancestors were living under the rule of a local chieftain who was not so people friendly. He was bent on marrying a beautiful girl from this community, to which the community elders opposed. Not willing to heed to the advice, the chieftain decided to wage a war in his attempt to secure the girl, and soon a fight ensued between his men and this community. Owing to the oppression and lack of necessary tactics to survive the overpowering chieftain, the elders of the community decided to escape from the village. And on a day, when the chieftain's men surrounded them, the community attempted to flee the village in the night but in vain, as they were unable to cross the swollen river flowing at the village border. Then the community members frantically prayed to the celestial bodies to save them from the chieftain's army. As if to answer their plea, a *Jamun* tree which was on the distant bank of the swollen river, grew its branches to gigantic stature and in a miraculous manner sent out a branch as a bridge to this side of the river where the community members were desperately waiting; the distressed members quickly walked on the gigantic branch and safely crossed the swollen river. As soon as all the members reached the other bank of the river, the *Jamun* tree regained its original size, and withdrew its branch, while the chieftain's men were stranded tasting the defeat. Having learnt that the *Jamun* tree had literally saved them from the pangs of the chieftain, the fleeing community members offered their obeisance to the *Jamun* tree God. And ever since *Jamun* tree became a savior in the eyes of the community members, who started venerating it during the family rituals while respectfully recounting this miracle. Accordingly, it became a customary practice for the community members to ardently bring a branch from a *Jamun* tree during the marriages and plant it along with other poles in the marriage pandal erected in front of the bride's house, and the newlywed couple offers puja to this branch. Not bringing the branch and offering a puja, would be like inviting the curse of the *Jamun* God, and the community would never dare do it, the elders admit, while recounting this mystic legend during the marriage rituals (Somashekhar, 2007).

Likewise, the origin story of the *Kadamba* dynasty (435–550 CE) which ruled the greater parts of Uttara Kannada, Belgaum, Dharawad, Shivamogga, and Goa in the Western Ghats, is associated with *Kadamba* tree (*Anthocephalus indicus*). It is said that, one Mayuravarma, a courageous Brahmin youth was humiliated and exiled by the Pallava rulers in Tamil Nadu over a trivial feud. Unwilling to give up, Mayuravarma was frantically looking for help to fight back, but was struggling. And on one such day of distress, while he was sitting under a *Kadamba* tree near his home and was half asleep, he saw in his dream his family goddess, who advised and motivated him to establish an independent kingdom. Subsequently, when Mayura gathered necessary help from his well wishers, built his army, successfully fought back the Pallavas and founded his kingdom, it was named *Kadamba*, as a mark of respect to the tree under which he received the blessings from his goddess in the dream, which was instrumental in motivating him for founding the dynasty (Mehamood, 1986). Similar anecdotes are available with other indigenous communities, which associate their origin with different plants.

5.7.4 PLANT NAMES IN PLACE NAMES

Another fascinating form of human association with plants and nature that can be recognized in the Western Ghats is reflected through place names. It is interesting to note that many wilderness areas, habitats and human habitations are appropriately named after a predominant plant species found in the area. Interesting reviews of toponymy and place names denoting the practice of identifying a place with a predominant landscape element or a plant species are available for Uttara Kannada (Somashekhar, 2014a) and other regions of Western Ghats (Bhat, 1979; Murthy, 1981). In Uttara Kannada, it is not uncommon to find villages named after one or the other plants. This practice is so profuse here that one can frequently come across villages named after 'Baale,' 'Maavu' and 'Kanchi' the Kannada names for Banana, Mango and Citron, respectively. It is interesting to note that these three names appear in 12, 8 and 4 village names respectively, and seem to be the popular choices for such naming. No wonder in this choice, as these plants are encountered more than frequently in the landscapes here. Similarly, Kannada names of common plants, such as 'Halasu' (Jack, *Artocarpus heterophyllus*), 'Hunase' (Tamarind, *Tamarindus indica*), 'Mundige' (Screw pine, *Pandanus unipapillatus*), 'Bidiru' (*Bambusa arundinacea*) and 'Honne' (*Pterocarpus marsupium*) appear in 3 village names each. Likewise, names of wild plants, such as 'Matthi' (*Terminalia tomentosa*), 'Anale' (*Terminalia chebula*), 'Thega' (*Tectona grandis*), 'Bagine' (*Caryota urens*), 'Suragi' (*Mammea suriga*), 'Nurakalu' (*Buchanania lanzan*), 'Kyadgi' (*Pandanus tectorius*), are frequently seen embedded in the many village names of this district, registering the local communities' ingenious means of establishing passionate affiliations with the local vegetation. It is noteworthy to mention that of the 1300 revenue villages in this district, more than 80 villages are named after either a tree or a plant, predominantly found in the region. This high degree of inclination for naming a place after a plant in the neighborhood, is perhaps innate to the communities of this region (Somashekhar 2014a).

5.7.5 PLANT NAMES FOR HUMAN GROUPS

Now from place naming to human group naming. It is interesting to note that, many communities across the Western Ghats consider certain plants as their totems. For instance, *Komatis* who have 16 septs, have named them after different plants. Likewise, *Bants* of Dakshina Kannada district, name their endogamous septs *Kayerthannayya* and *Kochattabannayya*, respectively after the plants, *Strychnos nux-vomica* and Jack fruit (Thurston and Rangachari, 1909). Madegowda (2009) reports 5 clans of the *Soliga* community which take different species of Jasmine (*Jasminum* spp.) as their respective clan symbols: a) *Sanna Mallige* for *Selikiru Kula*, b) *Boddaganna Mallige* for *Teneyaru Kula*, c) *Suthu Mallige* for *Suriru Kula,* d) *Halu Mallige* for *Halaru Kula,* and e) *Bili Mallige* for *Belliru Kula.* Likewise, a sub division of *Tigala*, the gardening community in the transition zone of Karnataka is named after *Ulli*, the onion. Similarly, *Kombara*, meaning a cap made of the spathe of Arecanut (*Areca catechu*) is the name of an exogamous sept of *Kelasi*, the barber community of Dakshina Kannada district.

5.8 FROM MEANING MAKING TO VALUE ASSIGNING!

The striking importance of plants in a community culture would be conspicuous especially during rituals and customary observances amongst the indigenous communities, in which certain plants and plant produce are found to have been ardently secured for the performance; significance of such plants on that particular day is so high that, the rituals would remain incomplete in the wake of any non-availability of such plants or plant produce. This high degree of importance has allowed certain plants to enjoy an irreplaceable position in many community cultures (Somashekhar, 2008a).

Many striking examples of such irreplaceable value assigned to plants are found with indigenous communities along the Western Ghats. Thurston and Rangachari (1909) for instance, reported many culturally significant usages of plants amongst different communities here: it is an ardent practice amongst the *Kurubas*, to set up a post made of the wood of silk-cotton tree (*Bombax malabaricum*) beneath the marriage pandal on the second day of marriage; amongst the *Jogis*, when a girl reaches puberty, she is put in a temporary hut which is thatched invariably with the branches of *Jamun*, Neem, Mango and *Vitex negundo;* amongst the *Billavas* of Dakshina Kannada district, the wood of *Strychnos nux-vomica* is forbidden from using for the funeral pyre. Likewise, Gomme (1889) reported that, amongst the *Kodavas*, when a boy is born, a little bow made of castor plant stick, along with an arrow made of its leaf-stalk is thrust into his little hands, and a gun fired at the same time in the courtyard.

Nayaka (1993) reports many interesting accounts of the usage of different plants during rituals amongst the *Halakki Vokkaliga* community. For instance, as a part of performing post death rituals on the 11th day of death, an elderly person of this community, chops off the stems of *Tinospora cordifolia, Achyranthes aspera* and Banana, which are brought for the purpose; during a *Halakki* marriage, a festoon of cluster fig twigs is tied to the marriage pandal; the batons used by *Halakki* men for performing the "Suggikolaata" dance during the harvest ritual, are invariably made from the wood of 'Bagine' (*Caryota urens*), 'Maddaale'(*Alstonia scholaris*), and 'Kaare' (*Canthium parviflorum*). Shetty (2000), reports innumerable cultural usages of coconut and arecanut inflorescence amongst the *Tuluva* communities of Dakshina Kannada and the beliefs associated with such usage. Likewise, during the marriage rituals amongst the *Mogaviras*, the fisherman community of Dakshina Kannada, a branch of 'Maddaale'(*Alstonia scholaris*), is planted in the marriage pandal as the '*Kanyakamba*' (Bride's post); the groom ties *thaali* to this pole before tying it to his bride (Shetty, 2000). It is customary for the *Gowlis* of Uttara Kannada, to offer special dishes to the community deities only on the leaves of *Dillenia indica* tree, during a community festival called 'Shillengaana' (Ramesh, 2010).

Somashekhar (2008a), reports an array of plants preferred for the performance of rituals related to birth, puberty, marriage and death, amongst select communities in the Western Ghats. It was found that, these communities despite sharing a common landscape identify themselves with a unique set of culturally significant plants, each one remaining distinct from the plants of the other. The study also discussed how these plants through certain rituals play significant role in enabling a community secure distinct socio-cultural identity and a specific name akin to the present day caste.

Perhaps, the culturally significant role assigned to plants in the context of community rituals is best displayed during the 'Sirisandhi' ritual performed with much grandeur amongst the *Tuluvas* of Dakshina Kannada. The performance makes predominant use of fresh Arecanut inflorescence brought in several head loads. It is believed that, the person performing the ritual will remain possessed and passes on the powers of 'Siri,' the spirit, to these Arecanut inflorescences; during the ritual, he performs several fierceful rounds of circumambulation on the ritual platform. At every circular movement, he grabs one inflorescence offered to him, and beats his face with it in a state of trance. Inflorescences are offered to him one after the other, as he moves around until the thirst of the spirit is quenched. It is believed that, the power of the spirit will not be dissipated unless Arecanut inflorescences were offered, which thus becomes an inseparable ingredient for this ritual (Shetty, 2000).

Another prominent ritual of such grandeur which warrants the usage of specific plants is seen with the harvest ritual of *Kodavas*, called 'Huttari.' The ritual assumes the shape of a feast, celebrated by the community members in honor of the annual rice-harvest and takes place on the full moon day. Early in the morning on that day, Leaves of *Ashvatha* (*Ficus religiosa*), 'Kumbali' and 'Keku' trees are collected and tied into small scrolls, with a piece of a creeper called 'Inyoli,' and a piece of fibrous bark called 'Achchi.' Later in the night, younger member from the family, who will serve as the sheaf cutter during the ritual reaches the paddy field along with his family members; he binds one of these leaf scrolls to a clump of paddy, pours milk to it and offers puja. He then cuts armful of paddy from select clumps in the neighborhood and distributes it amongst his members. The assembly subsequently returns, the scrolls and freshly harvested paddy straw is fastened at select places inside the house; subsequently the members savor the *Huttari* feast (Richter, 1870).

Likewise, at a place called Bangaramakki in Sagar taluk of Shivamogga district, during the annual car festival of lord *Hanumantha*, it is customary to bring the long poles of Bilva tree (*Aegle mormelos*) and spread them on the ground before the 'Ratha' (temple car, in which the procession idol of lord *Hanumantha* is seated), as a necessary prelude to initiate the procession and other rituals on the occasion. Subsequently, the 'Ratha' treads over these branches and makes circumambulation around the shrine (Somashekhar and Gowda, 2014).

Away from such socio-cultural contexts, however from an emotional perspective, trees are often considered as Kinship symbols in many parts of the world, as reported by Russel (1979). Equating tree with kinship lineage is a relatively late social symbolism, which acquired its significance in the Neolithic, at the swidden farming stage of cultural evolution, it was argued. From then on, however, the relationship between trees and kinship has been expressed in a great variety of beliefs and ritual practices all over the world.

5.9 INDIGENOUS CULTURAL KNOWLEDGE BLOSSOMS!

All the diverse perceptions and worldviews held by indigenous communities over many centuries have allowed the acquisition of precious indigenous knowledge base about nature and its elements which is quite unique. These elements may be clustered into the following forms.

5.9.1 FOLK LITERATURE

Nayaka (1993) reports many striking facets of such a rich ecological knowledge base sequestered amongst the *Halakki Vokkaliga* community. One of the noteworthy elements is the unparalleled kind of riddles called 'Kanni' that showcase the community's insights about the dynamics of plant world, and offer an intelligent articulation of their understanding of local ecology. Likewise, 'Gumate Pada,' a special form of folk poetry, sung by the men folk during many community gatherings is endemic to the community; several couplets of such songs extol the virtues of plants used by the community. The 'Gumate Pada' on *Tulsi* and other wild potherbs is noteworthy in this context. Chandru (1993) reports an equally rich ecological knowledge found amongst the *Soliga* community. "Goruke" is the special form of *Soliga* songs that encapsulate their unique understanding of forest resources in a 'Soligaseeme' (forest province of *Soliga* community). A few of such Goruke songs in fact describe and enlist many trees and wild potherbs found in the forests. *Soliga* community's ecological knowledge is also encapsulated in many riddles about plants. Such riddles may be considered as the community derived indigenous means of testing, augmenting and reaffirming one's knowledge of plants and other forest dynamics. Lalitha (1993) reports the unique knowledge of seasonal wild foods and potherbs embedded in the folk poetry of *Jenu Kuruba* and *Kadu Kuruba* tribes in Mysore and Kodagu districts of Karnataka.

5.9.2 ECO-DYNAMICS

The *Idigas*, who are traditionally toddy tappers, possess an excellent knowledge of the ecological dynamics of forest palms, especially of *Caryota urens*. Some of the dense wild populations of this palm are found in Honnavara taluk of Uttara Kannada district and two insights of the community denote such ecological excellence. One, fruits of this palm form a predominant and obligatory food resource for Lion Tailed Macaques in this region and those trees frequented by these animals are generally spared from toddy tapping and kept untouched by this community, as a means of sustaining the food resource for the animal. Second, once the tapping life is over and the palms are considered not yielding, they are felled and the wood is allowed to season.

The community has found out that the palm wood is quite hard, and this quality is popular amongst the locals as a metaphor and an indigenous yardstick, to gauge the hardness of other timbers. Thus, seasoned wood from tapped trees is harder than the wood from untapped trees, assert the elders of this community (Somashekhar, 2014b).

5.9.3   LESSONS FROM NATURE FOR COMMUNITY LIFE

Understanding about nature and its dynamics as deduced by the indigenous communities has even prompted them to transform and extend their learning to real life situations as a kind of philosophical understanding. *Niti Slokas* (traditional hymns about virtues and good conduct) may be considered as an intelligent articulation of many valuable lessons from Nature. Dubois (1906) in his *"Hindu manners and Customs,"* documented several *Niti Slokas*, which readily point at such perceptions of nature, extended to real life situations, as prevalent amongst the people of Karnataka. Some such hymns are:

> *"Birds do not perch on trees where there is no fruit; wild beasts leave the forests when the leaves of the trees have fallen and there is no more shade for them; insects leave plants where there are no longer flowers... Thus it is that self-interest is the motive of everything in this world."*

> *"Great rivers, shady trees, medicinal plants, and virtuous people are not born for themselves, but for the good of mankind in general."*

> *"A virtuous man ought to be like the Sandal-tree, which perfumes the axe that destroys it."*

> *"When one sees blades of Darbha grass on white –ant heaps one can tell at once that snakes are there."*

5.10   INDIGENOUSNESS OR INDIANNESS?: FEATURES AND DEFINITION

Being the knowledge base sequestered in community cultures and considered a part of their traditions, indigenous knowledge about nature is variously termed as traditional ecological knowledge (TEK), local ecological knowledge (LEK), indigenous knowledge (IK), ecoliteracy, or more generally ecological knowledge. It remains as an important part of people's capacity to manage and conserve both wild and agricultural systems over extended periods (Pilgrim et al., 2008).

It represents human experience acquired over thousands of years of direct contact with the environment. Although the term TEK came into widespread use in the 1980s, the practice is as old as ancient hunter gatherer cultures (Berkes, 1993). This knowledge tends to be socially embedded, often contributing to cultural traditions, identities, beliefs, and worldviews. It differs from modern knowledge by being dynamic, adaptive, and locally derived, thus coevolving with the ecosystem upon which it is based. This ecological knowledge has substantial environmental, human, and economic value (Pilgrim et al., 2007).

It is necessary to note here that, the indigenous knowledge is not merely a system of knowledge and practice, but is an integrated system of knowledge, practice and beliefs. Houde (2007) accordingly recognized six "faces" of TEK, e.g., factual observations, management systems, past and current land uses, ethics and values, culture and identity, and cosmology.

However, scientists assert that, the term *ecological knowledge* poses definitional problems of its own. If ecology is defined narrowly as a branch of biology in the domain of western science, then strictly speaking there can be no TEK, they argue; most traditional peoples are not scientists. But on the contrary, if ecological knowledge is defined broadly to refer to the knowledge, acquired by indigenous communities, of relationships with one another and with their environment, then the term TEK becomes tenable. It is what Levi-Strauss called the "science du concert", native knowledge of the natural milieu. In this context, *ecological knowledge* is not the term of preference for referring to indigenous knowledge (Berkes, 1993).

Indigenous environmental knowledge can be defined as a body of knowledge built up by a group of people through generations of living in close contact with nature. It includes a system of classification, a set of empirical observations about the local environment, and a system of self-management that governs resource use. The quantity and quality of this indigenous knowledge varies amongst the community members. With its roots firmly in the past, traditional environmental knowledge is both cumulative and dynamic, building upon the experience of earlier generations and adapting to the new technological and socio-economic changes of the present.

Martha (1992) summarizes the salient features of indigenous ecological knowledge and makes a quick comparison with the features of western science for better comprehension as below.

- TEK is recorded and transmitted through oral tradition. Western science employs the written word.
- In TEK, mind and matter are considered together (as opposed to a separation of mind and matter in western science).
- TEK is based on the understanding that the elements of nature (earth, air, fire, and water), which are considered inanimate by western science, also have a life force. All parts of the natural world—plant, animal, and inanimate elements—are therefore infused with spirit.
- TEK is rooted in a social context that sees the world in terms of social and spiritual relations between all life-forms. Western science is hierarchically organized and vertically compartmentalized.
- TEK is intuitive in its mode of thinking; Western science is analytical.
- TEK is mainly qualitative; western science is mainly quantitative.
- TEK is holistic; western science is reductionist. It deliberately breaks down data into smaller elements to understand whole and complex phenomena.
- TEK is diachronic (based on long time series of information on one locality); western science is largely synchronic (short time series over a large area).
- TEK is learned through observation and hands-on experience; western science is taught and learned in a situation usually abstracted from the applied context.
- TEK does not view human life as superior to other animate and inanimate elements: all life-forms have kinship and are interdependent. Unlike western science, humans are not given the inherent right to control and exploit nature for their interests at the expense of other life-forms.
- TEK is moral (as opposed to supposedly value-free).
- TEK's explanations of environmental phenomena are often spiritual and based on cumulative, collective experience. It is checked, validated, and revised daily and seasonally through the annual cycle of activities. In direct contrast, Western science employs methods of generating, testing, and verifying hypotheses and establishes theories and general laws as its explanatory basis.

Berkes and Turner (2006) recognized that, indigenous ecological knowledge develops through a combination of long-term ecological understanding and learning from crises and mistakes. Two broadly conceptualized pathways of knowledge development were identified: the depletion crisis model (depends on the learning that resources are depletable), and the ecological understanding model (based on the incremental elaboration of environmental knowledge). Turner and Berkes (2006) extended this argument and identified other pathways leading to ecological understanding. These include: lessons from the past and from other places, perpetuated and strengthened through oral history and discourse; lessons from animals, learned through observation of migration and population cycles, predator effects, and social dynamics; monitoring resources and human effects on resources (positive and negative), building on experiences and expectations; observing changes in ecosystem cycles and natural disturbance events; trial and error experimentation and incremental modification of habitats and populations.

However, despite such lauded significance, scientists are often reluctant to accept indigenous knowledge as valid because of its spiritual base, which they may tend to regard as superstitious. Berkes (1993) reiterates that "the attitude of many biological scientists and natural resource managers to indigenous knowledge has frequently been dismissive". There is a great deal of evidence that traditional people do possess scientific curiosity, which has been however questioned by those who regard the knowledge of other cultures as *pre-logical* or *irrational,* thus playing down the validity of indigenous knowledge. It is thus quite logical to consider that, while these traits on one hand are able to characterise the different dimensions of indigenous knowledge, they characterise simultaneously, though implicitly, the knowledge systems originated in India. Thus, indigenousness in this context also connotes the Indianness of cultural knowledge.

5.11   INTRICATE CORRELATIONS BIND NATURE AND CULTURE

Mathez-Stiefelet (2007) proposed that the relationships between a human group and its natural environment are mediated by various interrelated facets of culture, including the *ontology* (conception of being, what and how of the natural, spiritual and human world), *epistemology* (conception of knowing, what is knowledge, what and how we know about nature, society and spirits), *normative orientations* (principles that guide our actions, the conception of what is good or bad, what is the value of nature, humans and the spiritual world), and *practices* (activities of everyday life, what we do, how we use nature, relate to people and to the spiritual entities). All these cultural aspects together form a framework of interpretation and interaction with the natural world that we call a 'worldview.'

In the Indian context, this precious knowledge base innately reiterates the correlation between nature and indigenous cultures, as the natural elements are traditionally perceived with high degree of esteem. A passage from the *Matsya Puranam* which is instructive, would illustrate the high level of importance assigned to plants and nature in the Indian culture. It goes as:

"...(once) Goddess Parvati planted a sapling of Asoka tree and took good care of it. She watered it, and it grew well. The divine beings and sages came and told her: "O [Goddess]... almost everyone wants children. When people see their children and grand children, they feel they have been successful. What do you achieve by creating and rearing trees like sons...? Parvati replied: "One who digs a well where there is little water lives in heaven for as many years as there are drops of water in it. One large reservoir of water is worth ten wells. *One son is like ten reservoirs and one tree is equal to ten sons (dasaputrasamodruma).* This is my standard and I will protect the universe to safeguard it..." (Narayanan, 2001; italics mine).

Scientists have noted that, Ethnobotanical studies over the past century have focused on documenting the role of particular plants in traditional human cultures. However, none of these studies enlists all plant species to be found in the study region. But only those species of relevance to the native people, or which are considered salient for some reason, were considered culturally important. Cultural relevance, ranges from extremely highly relevant to minimally relevant. The problem is this: *How to gauge the cultural importance of a particular plant taxon? Is there a mechanism whereby cultural significance can be evaluated and compared in a meaningful way?*—are the kind of questions that have emerged in the recent years as an offshoot of human ecology studies (Pretty, 2009), that require convincing responses. Alcorn (1981) as if to voice these concerns was inquisitive as: "*What kinds of resources are necessary and what makes "plant X" a specific kind of resource?*"

Different theories and propositions highlighting the significance and concerns in this direction have emerged. Some of them are: (a) Those plants that are "used" within a culture, whether as food, material, medicine, religious object, subject in mythology, must be considered to have importance in that culture. Furthermore, the more widely or intensively a plant is used, the greater it's cultural significance (Turner, 1988); (b) Berkes and Folke (1993) proposed that the indigenous knowledge is the "Cultural capital" which refers to factors that provide human societies with the means and adaptations to deal with the natural environment and to actively modify it: how people view the world and the universe, or cosmology; environmental philosophy and ethics, including religion; traditional ecological knowledge; and social/political institutions. Cultural capital includes the wide variety of ways in which societies interact with their environment; (c) Skutnabb-Kangaset (2003) proposed that it can therefore be viewed as a form of cultural insurance; and (d) The value of accumulated ecological knowledge, termed ecoliteracy, is vital to both human and ecosystem health and hence its maintenance is essential (Pilgrim et al., 2007).

In the Indian context, the indigenous ecological knowledge throws open many different possibilities in addition to being a means of understanding the eco-cultural dynamics. Its importance to resource management has been well described and has been a key theme of cultural ecology and holds much promise. Gadgil et al. (1993) recognized that indigenous communities with a historical continuity of resource-use practices often possess a broad knowledge base of the behavior of complex ecological systems in their own localities. This knowledge gets accumulated through a long series of observations transmitted from generation to generation. Such "diachronic" observations can be of great value and complement the "synchronic" observations on which western science is based.

Jain (2000) asserts that, a very useful tradition relating to plant diversity has been the indigenous tradition of maintaining kitchen gardens or planting trees in the courtyards. Perhaps this is the smallest and very primordial form of copying the natural diversity of plants onto an accessible piece of land. Very vibrant illustrations depicting this association are encountered all across the Western Ghats. The kitchen gardens maintained by *Halakki Vokkaligas* (for native vegetables), *Havyaks* (for spices, wild fruits and flowers), *Kunbis* (Tubers and root vegetables), and *Idigas* (potherbs and leafy vegetables) along the Western Ghats are astounding, as each one amounts to a distinct endeavor, in respect of the biodiversity it houses and the cultural practices the communities adopt. They also signify the communities' understanding in respect of seed exchange, seed storage, nursery techniques, harvesting calendar and traditional recipes, in addition to being the assured year round supply of nutrient supplements for the family.

Likewise, the seemingly abstract and cryptic nature of many rituals amongst the communities in the Western Ghats, appear to carry an innate philosophy of prudent use of natural resources, and formation of human groups, while assigning a definite cultural identity to them in a landscape (Somashekhar, 2008a). Heinen and Ruddle (1974) proposed that the rituals of the *Warao* Indians of the Orinoco delta of Venezuela, that involve the distribution of *Moriche* (*Mauritia flexuosa*) starch, constitute an important survival mechanism by channeling surplus food between subgroups, and by institutionalizing the storage. Reichel-Dolmatoff (1976), proposed that many adaptive behaviors of native Amazon Indians have taken the shape of rituals which actually contribute to maintaining equilibrium and avoid frequent relocation of settlements. Such implicit elements point at many interesting dimensions of plants playing a crucial role in sustaining the identity of human groups, and a possibility of expanding the dimensions of human perceptions of plants.

Ramanujan (1990) recognized the innate intricacies of such 'Indian' knowledge, which is largely oral with specific genres (such as proverb, riddle, lullaby, tale, ballad, prose narrative, verse, or a mixture of both, and so on), nonverbal materials (such as dances, games, floor or wall designs, objects of all sorts), and composite performing arts. Recognizing its inherent uniqueness in the light of classical forms of human knowledge, he cautions that both need to be viewed complimentary to each other. He asserts thus:

"… one often identifies the "classical" with the written and the "folk" with the oral. But, for India, we should distinguish between three sets of independent oppositions. The three are classical vs. folk, written vs. spoken, fixed vs. free or fluid. The classical, the written, and the fixed do not necessarily belong together…."

5.12   CONCERNS, CONSEQUENCES AND AFTERTHOUGHTS

Plants are not perceived as mere natural forms of vegetation available in the neighborhood for meeting different needs of human life, many indigenous communities along the Western Ghats have copiously demonstrated thus. To them, plants are more of noble creatures with spirit and life, hence are a part of their *community*, value-belief systems and cultures. Higher degrees of significance and esteem are assigned to them. Perhaps it is a community evolved way of ensuring harmony with other forms of life around them, and this has become a salient feature of many indigenous communities in India.

However, this situation is changing in the modern world. As these indigenous communities become less reliant on local resources and begin to adopt modern lifestyles, their indigenous knowledge suffers, as it is either supplanted by modern knowledge or no longer gets transmitted and remains unattended. With a departure from its cultural traditions and accepting market-based lifestyles, combined with a growing disconnection from the land and intrinsic concerns, the unique indigenous knowledge of resources held by the communities is becoming diluted and often becomes isolated. For the want of appropriate users, it is not growing, but becoming obsolete.

As scientists have noted, many of these drivers originate from capitalist economies that favor non-stop economic growth, resulting in shifts in consumption patterns, globalization of markets, and commoditization of natural resources. Consequently, culturally inappropriate modernization and abrupt changes in the living styles have crept in causing much damage.

Indigenous communities in the Western Ghats are not insulated from such consequences of rampant modernization as is happening elsewhere in India. Large scale influence of, and easy access to modern and commodity based lifestyles in the recent decades, combined with widespread disliking amongst the communities to continue their traditional and indigenous ways of living, have resulted in large scale departures from their indigenous roots. Access to western style of school education, has further alienated many present generation members of these communities, from their traditional knowledge and skills which are largely rooted in the natural resources and plants in their homelands. Such disconnections with their homelands and cultures have led to sudden disappearance of many cultural contexts which were active till recently. Shift in local knowledge bases, and less time spent experiencing the nature, along with community elders and family members, result in less transmission of cultural knowledge. Consequently the communities are deprived of opportunities to practice their traditions, including the cultural-context specific language. This in turn would lead to erosion of language which is quite unique and endemic to the region and the community. Thus, rituals, customary observances and other traditions of many indigenous communities are becoming discrete and stand–alone elements, being practiced out of 'obligation' rather than the 'need.' The 'whole' of indigenous cultures is thus getting fractionated or mutilated, leading to its disintegration. Pretty et al. (2009) have warned that, such abrupt changes might lead to cultural collapse as the stories, narratives, ceremonies and rituals start losing their meaning when placed out of context.

Disconnection from the traditions would mean disconnection from the plants too, and such disconnections would in turn lead to neglect of those culturally significant plants in the wild, leading to slow disappearance of such plant populations too. Scientists have pointed out that, the combined loss of biodiversity and indigenous ecological knowledge has long-term implications as we lose the uses and future potentials of species, for instance in curing diseases or feeding populations. At the same time, we are losing adaptive management systems embedded in pre- and non-industrialized cultures that may offer insights. Plant diversity along the Western Ghats too is reeling under such threats already.

5.13   TO END THE STORY…

In the Indian context, it is to be remembered that, plants have been considered not as mere natural forms of vegetation, but have been the vital cultural resources which have influenced, inspired, challenged, nurtured and sustained the indigenous communities over the millennia, while enabling the communities forge many different kinds of affinities and affiliations with nature, simultaneously stimulating the proliferation of human knowledge, and sustaining the indigenous cultures. This unique bonding is reflected in all the indigenous worldviews, cosmovision, livelihoods, knowledge and wisdom. It is precious and irreplaceable, and for the same reason becomes vulnerable. One needs to be aware of this precarious situation and needs to take measured steps to augment, expand, sustain and conserve these traditions, for the future of indigenous communities, for, the existence of these communities in their true stature depends on these very cultures, rooted in this land. It is quite a challenge given the pandemic nature of the influence of urban and modern lifestyles, but it is not difficult.

KEYWORDS

- **Ancestor Puja**
- **Assigning Human Attributes to Nature**
- **Belief**
- **Biocultural Manifestation**
- **Community Origin Story**
- **Cultural Knowledge**
- **Departure from Cultures**
- **Festival Specific Offerings**
- **Human Affinity with Plants**
- **Indigenous Communities**
- **Indigenous Cultures**
- **Indigenous Ecological Knowledge**
- **Irreplaceable Cultural Value**

- **Irreplaceable Ritual Ingredients**
- **Meaning Making**
- **Migration Myth**
- **Nature**
- **Plants in Place Names**
- **Ritual**
- **Sacred Groves**
- **Spiritual Understanding of Nature**
- **Venerated Trees**
- **Veneration of Plants and Nature**
- **Vernacular Plant Names**
- **Village Deities**
- **Western Ghats**
- **Worldview**

REFERENCES

Alcorn, J.B. (1981). Factors influencing botanical resource perception among the Huastec: Suggestions for future ethnobotanical inquiry *J. Ethnobiol, 1(2),* 221·230.

Balagangadhara, S.N. (1994). The Heathen in His blindness....Asia, the West and the Dynamic of Religion. E.J. Brill, Leiden. Netherlands.

Balagangadhara, S.N. (2005). How to Speak for the Indian Traditions: An Agenda for the Future. *J. American Academy of Religion, 73(4),* 987–1013.

Berkes, F. (1993). Traditional Ecological Knowledge in Perspective. In: Inglis, J.T. (Ed.). Traditional Ecological Knowledge Concepts and Cases. International Program on Traditional Ecological Knowledge, International Development Research Centre, Ottawa, Canada.

Berkes, F. (2004). Rethinking community-based conservation. *Conservation Biology, 18(3),* 621–630.

Berkes, F. (2008). Sacred Ecology, 2nd Edition. Routledge, New York.

Berkes, F. & Folke, C. (1993). A systems perspective on the interrelations between natural, human-made and cultural capital. *Ecological Economics, 5(1),* 1–8.

Berkes, F. & Folke, C. (2002). Back to the future: Ecosystem dynamics and local knowledge. In: Gunderson, L.H., Holling, C.S. (Eds.). Panarchy: Understanding transformations in human and natural systems. Island Press, Washington, DC. pp. 121–146.

Berkes, F. & Turner, N.J. (2006). Knowledge, Learning and the Evolution of Conservation Practice for Social-Ecological System Resilience. *Human Ecology, 34(4),* 479–494.

Bhagya, B., Ramakrishna, A., Sridhar, K.R. (2013). Traditional seasonal health food practices in southwest India: nutritional and medicinal perspectives. *Nitte University J. Health Sci. 3(1),* 30–35.

Bhat, R.K. (1979). *TulunaadinaSthalanaamagalu* (Kannada). Kannada SahityaParishath, Bangalore, India.

Chandran, M.D.S., Gadgil, M. & Hughes J.D. (1998). Sacred groves of the Western Ghats. In: Ramakrishnan, P.S., Saxena, K.G. & Chandrasekara U.M. (Eds.) Conserving the Sacred for Biodiversity Management. Oxford and IBH, New Delhi, India. pp. 211–32.

Chandru, K. (1993). *Soligara Sanskrithi:* Upasanskrithi Adhyayana (Kannada). Kannada Sahithya Academy, Bangalore, India. pp. 272–48.

Dubois, J.A.A. (1906). Hindu Manners, Customs and Ceremonies. Clarendon Press, Oxford: England.

Ellen, R.F. (1996). The cognitive geometry of nature: A contextual approach. In: Palsson, G. & Descola, P. (Eds.) Nature and society: Anthropological perspectives. Routledge, London and New York.

Enthoven, R.E. (1915). Folklore Notes, Vol. II-Konkan. British India Press, Bombay, India.

Frazer, J.G. (1960). The Golden Bough. Abridged edition, London.

Gadgil, M., Berkes, F. & Folke, C. (1993). Indigenous knowledge for biodiversity conservation. *Ambio, 22,* 151–156.

Gadgil, M. & Vartak V.D. (1981). Sacred groves in Maharashtra – An inventory, In: Jain, S.K. (Ed.) Glimpses of Indian Ethnobotany. Oxford & IBH Publishers, New Delhi, India. pp. 279–294.

Gandhi, M. & Singh, Y. (1989). Brahma's Hair – On the Mythology of Indian Plants. Rupa and Co. New Delhi, India. pp. 175.

Geertz, C. (1973). The Interpretation of Cultures. Basic Books, New York.

Gomme, G.L. (1889). Coorg Folklore. *The Folk-Lore Journal, 7(4),* 295–306.

Hegde, L.R. (1977). *Janapada Jivanamatthu Kale* (Kannada). Samaja Pusthakalaya, Dharawad, India.

Heinen, H.D. & Ruddle, K. (1974). Ecology, Ritual, and Economic organization in the distribution of Palm starch among the Warao of the Orinoco delta. *J. Anthropological Res., 30(2),* 116–138.

Houde, N. (2007). The six faces of traditional ecological knowledge: challenges and opportunities for Canadian co-management arrangements. *Ecology and Society, 12(2),* 34–50.

Jain, S.K. (2000). Human aspects of Plant diversity. *Economic Botany, 54(4),* 459–470.

Lalitha, A.C. (1993). *Kaadukurubaru matthu Jenukurubara Sanskrithi*-Upasanskrithi Adhyayana (Kannada). Kannada Sahithya Academy, Bangalore, India. pp. 124–48.

Leopold, Aldo. (1949). *A Sand County Almanac.* Oxford University Press, New York.

Madegowda, C. (2009). Traditional Knowledge and Conservation. *Economic & Political Weekly, 44(21),* 65–69.

Maffi, L. & Woodley, E. (2007). Biodiversity and culture. UNEP's 4th Global Environment Outlook Report. UNEP, Nairobi.

Malhotra, K.C., Gokhale, Y., Chatterjee, Y. & Srivastava, S. (2001). Cultural and Ecological Dimensions of Sacred groves in India. Indian National Science Academy, New Delhi, India and Indira Gandhi Rashtriya Manav Sangrahalaya, Bhopal, India.

Martha, J. (1992). Research on Traditional Environmental Knowledge: Its Development and Its Role. In: Martha, J. (Ed.) Lore-Capturing Traditional Environmental Knowledge. Dene Cultural Institute and International Development Research Centre, Canada.

Mathez-Stiefel, S., Boillat, S. & Rist, S. (2007). Promoting the diversity of worldviews: An ontological approach to bio-cultural diversity. In: Endogenous Development and Bio-cultural Diversity. pp. 67–81.

Mehamood, T.K. (Ed.). (1986). Uttara Kannada Darshana (Kannada). Kannada Sahitya Parishath, Bangalore, India.

Milton, K. (1998). Nature and the environment in indigenous and traditional cultures. In: Cooper, D.E. & J.A. Palmer (Eds.). Spirit of the environment: Religion, value and environmental concern. pp. 86–99. Routledge, London and New York.

Murthy, C. (1981). Vaagartha (Kannada). Baapco Publishers, Bangalore, India.

Nagesh, H.K. (2011). Krishi-Suggi. In: Ramesh S.C. (Ed.). Krishi Paaramparika Jnaana (Kannada). Kannada University, Hampi, India. pp. 85–95.

Narayanan, V. (2001). Water, Wood, and Wisdom: Ecological Perspectives from the Hindu Traditions. *Daedalus, 130(4),* Religion and Ecology: Can the Climate Change? (Fall, 2001), pp. 179–206.

Nayaka, N.R. (1993). *Haalakki Vokkaligara Sanskrithi* – Upasanskrithi Adhyayana (Kannada). Kannada Sahithya Academy, Bangalore, India. pp. 370+48.

Pilgrim, S., Cullen, L., Smith, D.J. & Pretty, J. (2008). Ecological knowledge is lost in wealthier communities and countries. *Environmental Science and Technology, 42(4),* 1004–1009.

Pilgrim, S., Smith, D.J. & Pretty, J. (2007). A cross-regional assessment of the factors affecting ecoliteracy: Implications for policy and practice. *Ecological Applications, 17(6),* 1742–1751.

Posey, D.A. (Ed.). (1999). Cultural and spiritual values of biodiversity. UNEP and Intermediate Technology Publications, Nairobi, Kenya.

Pretty, J., Adams, B., Berkes, F., Ferreira de Athayde, S., Dudley, N., Hunn, E., Maffi, L., Milton, K., Rapport, D., Robbins, P., Sterling, E., Stolton, S., Tsing, A., Vintinner, E. & Pilgrim, S. (2009). The Intersections of Biological Diversity and Cultural Diversity: Towards Integration. *Conservation and Society, 7(2),* 100–112.

Rai, B.A.V. & Gowda, D.Y. (1996). Male Kudiyaru. Mangalore University, Mangalore, India.

Rajendra, G. (2007). Yummy Coorg Food. *Shakthi* Daily, Madikeri, India. Dated 28 November 2007 (http://www.geocities.ws/shakthidaily/articles/kodavaDelicacies.htmlaccessed 20th June 2015).

Ramanujan, A.K. (1989). Telling Tales. *Daedalus, 118(4)* Another India (Fall, 1989), 238–261.

Ramanujan, A.K. (1990). Who needs Folklore? The Relevance of Oral Traditions to South Asian Studies. South Asia Occasional Papers Series No.1. Centre for South Asian Studies. University of Hawaii. Manoa. pp. vii, 32.

Ramanujan, A.K. (1997). A Flowering Tree and Other Oral Tales from India. University of California Press, Berkeley. USA, pp. xv + 270.

Ramesh, S.C. (Ed.). (2010). Karnatakada Janapada Aacharanegalu (Kannada). Kannada University, Hampi, India.

Reichel-Dolmatoff, G. (1976). Cosmology as Ecological Analysis: A View from the Rain Forest. *Man, New Series, 11(3),* 307–318.

Richter, G. (1870). A Manual of Coorg. Basel Mission Book Depository, Mangalore.

Russel, C. (1979). The Tree as a Kinship Symbol. *Folklore, 90(2),* 217–233.

Skutnabb-Kangas, T., Maffi, L. & Harmon, D. (2003). *Sharing a World of Difference: The Earth's Linguistic, Cultural, and Biological Diversity.* UNESCO, Paris, France.

Shetty, B.S. (2000). *Sasya Jaanapada* (Kannada). Kannada University, Hampi, India. pp. 36.

Somashekhar, B.S. (2007). Five Poles of Banyan and a Twig of Euphorbia: Role of community specific cultural knowledge as a means of habituating the landscapes-case study from southern India. Paper submitted to "The XV International Conference of the Society for Human Ecology-Local Populations and Diversity in a Changing World," October 4–7, 2007, Rio de Janeiro, Brazil.

Somashekhar, B.S. (2008a). Plants and Cultural Configurations. Paper presented at the "National Seminar on Plants People and Culture," 14–15th March 2008, Sringeri, India. JCBM College, Sringeri and Foundation for Revitalization of Local Health Traditions, Bangalore, India.

Somashekhar, B.S. (2008b). *Taavareyele Melitta Tutthu, Desi Krishi Maale: Ahara-Arogya* (Kannada). Kannada Pustaka Pradhikara, Govt. of Karnataka, Bangalore, India. pp. 146.

Somashekhar, B.S. (2012). *Aranya-Janateyanaduvina Saanskritika Phalashruti* (Kannada). Paper presented at the "State level Seminar on Forests, People and Culture." 29 May 2012, Sirsi, India. College of Forestry, Sirsi and Western Ghats Task Force, Govt. of Karnataka, Bangalore, India.

Somashekhar, B.S. (2014a). *Aghanaashini Nelada Akshathanelegalu* (Kannada). Green India-IINDICUS, Dandeli. India. pp. 74.

Somashekhar, B.S. (2014b). Traditional Ecological Knowledge prevalent among select indigenous communities of Uttara Kannada district, Western Ghats, Karnataka. Technical Report No. 3. IINDICUS, Bangalore, India.

Somashekhar, B.S. & Gowda, K.V. (2014). Baseline Profiling of Cultural-Heritage Biodiversity Resources in select Eco zones of Karnataka. Project completion Report, submitted to Karnataka Biodiversity Board, Govt. of Karnataka, Bangalore. IINDICUS, Bangalore, India.

Staal, F. (1979). The Meaninglessness of Ritual. *Numen, 26 (Fasc. 1)*, 2–22.

Thurston, E. & Rangachari, K. (1909). Castes and Tribes of Southern India. Vols. 1–7. Government Press, Madras, India.

Turner, N.J. (1988). "The Importance of a Rose": Evaluating the Cultural Significance of Plants. *American Anthropologist, New Series, 90(2)*, 272–290.

Turner, N.J. & Berkes, F. (2006). Coming to understanding: Developing conservation through incremental learning in the Pacific Northwest. *Human Ecology, 34*, 495–513.

CHAPTER 6

ETHNOBOTANICALS OF THE WESTERN GHATS

S. NOORUNNISA BEGUM, K. RAVIKUMAR, and D. K. VED

Centre of Repository of Medicinal Resources, School of Conservation of Natural Resources, Foundation for Revitalization of Local Health and Traditions, 74/2, Jarakabande Kaval, Attur PO, Via Yelahanka, Bangalore 560–064, India. E-mail: noorunnisa.begum@frlht.org, k.ravikumar@frlht.org, dk.ved@frlht.org

CONTENTS

ABSTRACT

Western Ghats comprises rich bio-cultural diversity and is home to varied ethnomedicinal practices. There are several eminent scholars and scientists who have documented ethnobotany of the Western Ghats. The present chapter is based on the survey of the existing literature studies and highlights the therapeutic uses of medicinal plants used by various tribal communities living in various pockets of the Western Ghats. A brief account is given on threat- ened medicinal plants and efforts on in situ conservation of medicinal plants in the Western Ghats. The enumeration includes listing of 1116 botanicals in alphabetical order, part used and name of the disease/medicinal uses along with 6 plates.

6.1   INTRODUCTION

In India, though the records of plant utilization for medicine, food and other purposes date back to ancient times but research work under the title 'ethnobotany' were initiated by 1954 and till date more than 9000 plant species used by tribals or aboriginals for different purposes from different regions of India have been documented. The studies on ethnobotany India were initiated by E.K. Janaki Ammal, as an official program of the Botanical Survey of India, studied subsistence food plants of certain tribals of South India. This was followed by S.K. Jain, who took the lead and streamlined the subject with his several books and numerous publications on the subject.

Ethnobotanical knowledge is rich among about 50 million tribals of 160 linguistic groups in 227 ethnic groups belonging to 550 tribal communities of about 5,000 tribal hamlets in India (Anonymous, 1994). According to a status report of the All India Coordinated Research Project on Ethnobiology (AICRPE) covering 24 research center in India enumerates over 9,500 wild species used by the tribals. About 7,500 of them are used for medicinal purposes (Anonymous, 1994). Roughly, more than 3,000 plants are used in different purposes of food, fodder, fiber, for house building, musical instruments, fuel, oil seeds, narcotics and beverages, in material culture and also in magico-religious beliefs.

Although the concept and definition of the science of ethnobotany was clearly defined respectively by Powers (1874) and Harshberger (1896), the elements of this science appeared in India even before it. Written records of the use of plants for curing several human and animal diseases in India can be traced back to the earliest (4500–1600 BC) scripture of Hindus, the

Rigveda. The juice of the legendary plant 'Soma' is widely believed to be a mushroom, *Amanita muscaria*, mentioned as '*Oshadhi*,' e.g., heat producer. A good number of plants were described in '*Atharvaveda*.' '*Vrikshayurveda*' of Rishi Parashara (ca. 1 AD) comprises huge information on various aspects of plants. Notable ancient scriptures on medicinal plants include, Charak Samhita (1000–800 BC), Sushrut Samhita (800–700 BC) and Vagbhatta's '*Astanga Hridya*.' But the Indian System of Medicine has declined with the invasion of Greeks, Scythians, Huns, Moghuls and Europeans. *Garcia da Orta* (1563) published a book '*Coloquios dos simples' e drogas medicinas da India*' which mentioned about 50 common taxa of medicinal significance and other utilities as gathered around Goa and in Malabar. Acosta (1578) also published a book '*Tractado de las drogas y medicinas de las Indias Orientalis*' on Indian Medicinal Plants from Malabar. Van Rheede (1678–1693), the compiler of '*Hortus Malabaricus*' gave an excellent and accurate introduction of medicinal plants. The science of ethnobotany began taking shape during the British India regime. They surveyed wild and cultivated plants as part of their floristic studies. William Roxburgh (1832) recorded medicinal uses of herbs during his floristic investigations from south of India. Sir George Watt published 'Dictionary of the Economic Products of India' (1889–1896), in which he provided nearly 3000 local names of the plant products and their uses as obtained from various regions of India. His work is not only a monumental one, but also reflects true ethnobotany and indigenous knowledge of Indian societies. Subsequently Bodding (1925, 1927, and 1940) published medicines used by Santal tribe.

6.2   ETHNOBOTANY IN INDIA

E.K. Janaki Ammal studied subsistence food plants of certain tribes specially of South India (Janaki Ammal, 1956). S.K. Jain compiled ethnobotanical works in India under the title "Dictionary of Indian Folk Medicine and Ethnobotany" (Jain, 1991), Notable plants of Ethnobotany of India (Jain et al., 1991), A Handbook of Ethnobotany[2] (Jain and Mudgal, 1999) and Tribal Medicine (Pal and Jain, 1998). He published several books and more than 300 research papers on different aspects of ethnobotany and folklore medicine in India. Several reviews have also been published on ethnobotany of plants of India including the Western Ghats. Dey and De (2011) reviewed ethnobotanical aspects of *Rauvolfia serpentina* in India, Nepal and Bangladesh.

6.3   ETHNOBOTANICAL WORK ON THE WESTERN GHATS

The region comprises the states of Gujarat, Goa, Maharashtra, Karnataka, Kerala and Tamil Nadu. Some of the tribals inhabiting these states are *Goa*: Dhodia, Dubla (Halpati), Naikda (Talavia), Siddi (Nayaka), Varli, Kunbi, Gawda, Velip, *Kerala*: Adiyan, Arandan, Aranadan, Eravallan, Hill Pulaya, Mala Pulayan, Kurumba Pulayan, Karavazhi Pulayan, Pamba Pulayan, Irular, Irulan, Kadar, Wayanad Kadar, Kanikaran, Kanikkar, Kattunayakan (Kochuvelan), Koraga, Kudiya, Melakudi, Kurichchan, Kurumans, Kurumbas, Maha Malasar, Malai Arayan, Malai Pandaram, Malai Vedan, Malakkuravan, Malasar, Malayan, Nattu Malayan, Konga Malayan, Malayarayar, Mannan, Muthuvan, Palleyan, Paniyan, Ulladan, Uraly, Mala Vettuvan, Ten Kurumban, Jenu Kurumban, Thachanadan, Thachanadan Moopan, Cholanaickan, Mavilan, Karimpalan, Vetta Kuruman, Mala Panickar. *Karnataka*: Jenukuruba, Bettakuruba, Soliga, Yerava, Malekudia, Panjariyerava, Thammadi, Medha. *Dadra, Nagar, Haveli & Daman*: Warli, Konkana, Dhodia, Dubala (Halpati), Kathudi (Katkari), Naika and Koli. In Thane district of *Maharashtra*, The Warli tribe are mainly spread out in the villages of Dahanu, Talasari, Mokhada, Vada, Palghara and extends up to the *Gujarat* border. Warli Tribe has become famous because of their traditional folk painting art.

The Tamil Nadu state: Irular, Kadar, Kanikaran, Kattunayakan, Kurumbar, Malayali, Mudhuvan, Palliyan, Paniyan and Soligas.

There are many reports on ethnomedicinal plants of the Western Ghats. Indian council of medical research has prepared a database on ethno=medicinal plants of the Western Ghats (Kholkunte, 2008). Centre for Ecological Sciences, Indian Institute of Science, Bangalore, reported the "Medicinal Plants of Western Ghats", their importance and the medicinal values (Sahyadri e-news CES-ENVIS's quarterly newsletter). Medicinal plant species of Western Ghats include Algae, Lichens, Pteridophytes, climbers, herbs, shrub and trees (Kumar et al., 2014; Warrier and Ganapathy, 2001; Sharma, 2014). Ethnobotanical research has been intensified at different research stations, many researchers have tried to give details on medicinal plants and notable contributions are described as under.

6.3.1   KERALA

An ethnobotanical exploration in the Trivandrum forest division in Kerala was carried out by John (1984) and reported one hundred useful raw drugs of the Kani tribe. Ethnobotany of rice of Malabar was given by Manilal (1981). Ethnobotany of the Nayadis of Kerala was given by Prasad and Abraham (1984). In a study on ethnobotany of Cannanore district, Ramachandran (1987) listed over 30 species of ethnomedicinal plants. Nagendra Prasad et al. (1987) published an account of 56 medicinal plants. The unripe fruits of Ginseng, *Trichopus zeylanicus* Gaertn. subsp. *travancoricus* (Bedd.) Burk. of Dioscoreaceae, of Agasthiyar hills of Kerala state are eaten fresh to remain healthy and agile (Pushpangadan et al., 1988).

Details of 32 plant species used in ethnomedicine by the tribals in Naduvil panchayat in Kannur district of Kerala were given by Thomas and De Britto (1999). Ethnobotanical uses of *Momordica charantia* var. *muricata, M. charantia* var. *charantia, M.dioica* and *M. sahyadrica* in the southern Western Ghats comprising food, medicinal, cosmetic and culinary preparations were given by Joseph and Antony (2007). Ajesh et al. (2012) reported that 23 species of Fabaceae are being used for treating gynecological problems by Mannan tribes of Kerala. 50 plant species used by Kani tribe of Kottor reserve forest, Agasthyavanam in Thiruvananthapuram district in Kerala to treat 39 ailments have been enumerated by Vijayan et al. (2006). Ethnogynecological uses of 15 plant species prevalent among the tribal groups namely Mannan, Paliyan, Urali, Malayarayan and Malampandaram of Periyar Tiger Reserve in Western Ghats, Kerala were given by Augustine et al. (2010). Thomas and Rajendran (2013) provided an overview of 44 ethnomedicinal plants used by Kurichar tribe of Wayanad district, southern Western Ghats in Kerala. Information on 34 plant species used by Kuruma tribe of Wayanad district of Kerala for the treatment of cuts and wounds was given by Thomas et al. (2014).

6.3.2 KARNATAKA

Keshava Murthy and Yoganarasimhan (1990) while writing the Flora of Coorg provided data on 747 medicinal plants and phytochemical data on 471 taxa and furnished detailed information on 315 Ayurvedic drugs. Yoganarashimhan (1996) wrote a book on Medicinal Plants of India of which first volume is dedicated to Karnataka. Studies on ethnobotany of different tribes in Karnataka include Siddis of Uttar Kannada by Bhandary et al. (1995), Gawlis of Uttar Kannada district by Bhandary et al. (1996) and Kunabi tribe by Harsha et al. (2002), Khar Vokkaliga community by Achar et al. (2010), Jenu Karuba ethnic group by Kshirsagar and Singh (2000a) and Nanjunda (2010). Ethnomedicobotany of different districts in Karnataka include Dharwad by Hebber et al. (2004), Mysore and Coorg districts by Kshirsagar and Singh (2000b, 2001). Herbal therapy for Herpes in the ethnomedicine of coastal Karnataka was given by Bhandary and Chandrasekhar (2011) and skin diseases by Hegde et al. (2003) and Bhat et al. (2014). Ethnomedical survey of herbs for management of malaria in Karnataka was given by Prakash and Unnikrishnan (2013). An account of 102 medicinal plants to cure skin diseases in the coastal parts of Central Western Ghats, Karnataka was given by Bhat et al. (2014).

6.3.3   TAMIL NADU

All India Coordinated Research Project on Ethnobiology was commissioned by Man and Biosphere reserve project was sanctioned to Dr. Pushgandan as Co-ordinator. In this Tamil Nadu was given to Dr. A.N. Henry (1989-1992). This period a total of 950 medicinal plants were collected. The enumeration included 142 wild edible plant species and 553 species used for medicine (Anonymous, AICRPE-1989 to 1992 report).

Ethnobotany of the Todas, the Kotas and the Irulas of Nilgiris was given by Abraham (1981). Medicinal plants among the Irulas of Attapady and Bolunvampatti forests in the Nilgiri Biosphere Reserve was given by Balasubramanian and Prasad (1996). Folk medicine of the Irulas of Coimbatore forests was discussed by Balasubramanian et al. (1997). Ethnobotany of Kadars, Malasars and Muthuvans of the Anamalais in Coimbatore district and Irular, Kurumban and Paniyan of Nilgiris was given by Hosagoudar and Henry (1996a, b). Plants used by the tribe Kadar in Anamalai hills in Coimbatore district were discussed by Rajendran and Henry (1994). Other publications include; ethnobotany of Irulas, the Koravas and the Puliyan of Coimbatore district (Ramachandran and Manian, 1989; Ramachandran and Nair, 1981); ethnobotany of the Kanikars of Kalakkad Mundanthurai Tiger Reserve in India (Viswanathan et al., 2001), ethnomedicinal plants used by the Kanikkars of Tirunelveli district, Tamilnadu were given by Anitha et al. (2008). 52 plant species used by the Paniya tribe of Mundakunnu village of the Nilgiri hills to treat various ailments of analgesic, antidiarrheal, piles, antidiabetic, gynecological problems, vermifuge, antidandruff, venereal diseases and bone fracture were enumerated   by Manikandan (2005). Ignacimuthu et al. (2006) gave information on 60 ethnomedicinal plant species used by Paliyar tribes from southern Western Ghats in Madurai district, Tamilnadu. An ethnobotanical study of medicinal plants in Agasthiayamalai region of Tirunelveli district was done by collecting information from the experienced medicinal practitioners of Kani tribes (Sripathi and Sankari, 2010). Ten plants were collected, authenticated and information on their medicinal uses along with the parts used and mode   of administration is enumerated. Ethnomedicinal values of 52 species from Uthapuram village of Madurai district were given by Sivasankari et al. (2013). An ethnobotanical survey was conducted by Shalini et al. (2014a) to document ethnomedicinal plants which are used by a hill tribe Kanikkars of southern Western Ghats. 45 species of ethnomedicinal plants used for the treatment of skin-related ailments by the Kanikkars, an indigenous tribe inhabiting southern Western Ghats were enumerated by Shalini et al. (2014b). Viswanathan et al. (2001) provided an account on 124 ethnomedicines of Kanis in Kalakkad Mundanthurai Tiger Reserve, Tamil Nadu. Rani et al. (2011) gave an account of 174 ethnomedicinal plants used by Kanikkars   of Agasthiarmalai Biosphere reserve in Western Ghats. An ethnobotanical study of medicinal plants used by the Paliyars aborgonal community in Virudhunagar district of Tamilnadu yielded 48 medicinal plants.

6.3.4 MAHARASHTRA

Ethnomedicobotanical studies in Maharashtra include: Ethno-medicinal plants used by the tribals in Dhule district (Borse et al., 1990; Bhamare, 1998); Ethnomedicinal practices of Nasik district (Patil and Patil, 2005); Ethno-medicinal plants of Baramati region of Pune district (Deokule, 2006); Ethnomedicobotanical Survey of Ratnagiri District (Deokule and Mokat, 2004); Ethnobotany of Thakar tribe (Kamble et al., 2009); Studies on plants used in traditional medicine by Bhilla tribe (Kamble et al., 2010); Traditional medicines used by the tribes of Pune and Thane districts for the treatment of upper respiratory tract disorders (Kamble et al., 2009); Medico-ethnological studies and conservation of medicinal plants of North Sahyadri (Khairnar, 2006); Ethnobotany in Human welfare of Raigad district (Kothari and Moorthy, 1994); Folk therapies of Katkaris (Kulkarni and Deshpande, 2011); Ethno-medical traditions of Thakur tribals of Karjat, Maharashtra (Palekar, 1993); Medico-botanical and phytochemical studies on medicinal plants of Dhule and Nandurbar districts (Rajput and Yadav, 2000). Soman (2011) recorded ethnomedicinal uses of 20 plants used by tribals of Karjat Taluka in Maharashtra. 38 ethnomedicinal plants used to treat bones. Details of ethnomedicinal uses of 19 plants used by the tribes in Pannala taluka of Western Ghats were given by Soman (2014). Medicinal uses of 54 edible plant used by Warli, Katkari and Konkani tribal women of Thane district were given by Oak et al. (2015). 38 ethnomedicinal plants used to treat bone fracture in North-Central Western Ghats were discussed by Upadhya et al. (2012). Khyade et al. (2011) described 40 Angiospermic species used by the tribes Mahadev Koli, Ramoshi, Thakar and Bhills in Akola Taluka of Ahmednagar district as an antidote against snake bite.

6.3.5   DADRA-NAGAR HAVELI AND DAMAN

Sabnis and Bedi (1983) and Sharma and Singh (2001) carried out ethnobotanical studies of Dadra-Nagar Haveli and Daman.

6.3.6   GUJARAT

Kumar et al. (2007) studied plant species used by tribal communities of Saputara and Purna forests, Dangs district, extreme northern part of Western Ghats, South Gujarat. 50 plant species used for curing various diseases have been enumerated.

6.3.7   GENERAL

21 plant extracts of different species used by traditional healers from high altitude of the Western Ghats for the treatment of ulcers, cancers, tumors, warts and other disease were tested for potential anticancer activity by Garg et al. (2007).

6.4   METHODOLOGY

In writing this chapter we focused on compiling the ethnobotanical data from various published ethnobotanical and ethnographical works by various scholars. The chapter has attempted to compile information on the medicinal plants used in the Western Ghats, such as ethnobotanical aspects of medicinal plants, specific usage of plant resource by tribal communities, cultural belief, plants and their therapeutic usage, specific plant groups like orchids, chaemophytes, pteridophytes, mushrooms, practices within culture based on geographical origin. It is based on these plants compiled the following chapter is written. The Database on Medicinal Plants used in Indian Medical System developed by Foundation for Revitalization of Local Health and Traditions (FRLHT), Bangalore, Karnataka has been referred to intensively.

Based on these studies there are ca 2100 medicinal plants in the Western Ghats. Of the vast literature available for the medicinal plants in Western Ghats, we are listing few of the selected 1116 medicinal plants in the table 6.3 given below.

Some of the threatened medicinal plants and few interesting medicinal plants of the Western Ghats is provided in the plates 1 to 6.

6.5 THREATENED MEDICINAL PLANTS OF THE WESTERN GHATS

The threat of extinction of species is looming large due to their un-sustainable removal and rapid reduction/fragmentation of habitats. For such species/taxa, more intensive management becomes necessary for their survival and recovery. Increasingly, this intensive management will have to include – habitat management and restoration, intensified information gathering and possible conservation breeding. Concerns have grown, and so have efforts, to reverse the process of extinction. The IUCN, the WWF, the WCMS, RBG, KEW are some of the organization actively working for species survival at global level. In India, Zoo Outreach Organization (ZOO) and the Foundation for Revitalization of Local Health and Traditions (FRLHT) have, in collabo ration with wildlife/forest managers, pioneered conservation action in re spect of animals and medicinal plants respectively.

With so many species needing conservation action, it is imperative that a priority list of such species is drawn and the status of their wild population is documented. Attempts to address this issue led to Red Listing of species. But the process of Red listing being long and time consuming due to the need for in-depth and intensive field studies of priority taxon, it does not provide the luxury of commissioning such long drawn species specific stud- ies to initiate conservation action. What was needed was a rapid method of information gathering within available resources to reach fair level of under- standing for initiating intensive management as efficiently and effectively as possible. The development of Threat Assessment and Management plan (TAMP) process, now called Conservation Assessment and Management Prioritization (CAMP), is the result of this search.

The first such workshop for plant species was held for flora of the island of St. Helena, in the pacific, in May 1993. Organized under the auspices of Royal Botanic Gardens (RBG), Kew, the workshop assessed and prioritized plant species of St; Helena for conservation action and information gathering. The first Workshop in India for plants was held at Bangalore for medicinal plants of Southern India in February 1995. It was co-ordinated by FRLHT in collaboration with Conservation Breeding Specialist Group (CBSG) of IUCN. Since then 15 more such workshops covering wild medicinal plants of different states / regions in the country have been held.

A global analysis of a representative sample of the world's plants, conducted in 2010 by the Royal Botanic Garden, Kew together with the National Histroy Museum, London and IUCN has revealed that one in five of the worlds plants species are threatened with extinction. Around 5000 Indian medicinal plants are wild in India (as per FRLHT data base), 1/5th of these i.e. Approximately 1000 species may be Threatened with extinction. Assessment of Threat status of Indian medicinal plants, undertaken so far, is far from completion methodology and there is a need for adopting appropriate rapid for such assessments.

In this context it is relevant to review the consolidated efforts for rapid threat assessment of wild medicinal plants, through 15 CAMP workshops, which add up to 335 species assessed as Near Threatened or Threatened.

There is a need to systematically assess and identify such threatened medicinal plant species and develop appropriate conservation action plans. The species thus prioritised need to be assessed and assigned a threat status. Systematic assessment of the red list status of the prioritised medicinal plant species of the region have been undertaken through a rapid assessment process called Conservation Assessment and Management Prioritisation (CAMP) using guidelines and definitions developed by IUCN.

Out of the ten biogeographical zones of India, five CAMP workshops were exclusively conducted to assess threat status of the medicinal plant species of the Western Ghats. List of some of the endemic medicinal plants assessed in the CAMP is provided in the table 6.1. Focused CAMP workshops should be reconducted in the above said biogeographical zones in order to re look into the identity of the threatened medicinal plant species for Red Listing and their current Threat category.

TABLE 6.1   List of Endemic Threatened Medicinal Plants Assessed in the CAMP During 1995, 1996, 1997, 1999

Sl. No.	Botanical Name	Family Name
1	Adhatoda beddomei	Acanthaceae
2	Amorphophallus commutatus	Araceae
3	Ampelocissus araneosa	Vitaceae
4	Artocarpus hirsutus	Moraceae
5	Calophyllum apetalum	Clusiaceae
6	Cayratia pedata var. glabra	Vitaceae
7	Chlorophytum borivilianum	Liliaceae

TABLE 6.1   *(Continued)*

Sl. No.	Botanical Name	Family Name
8	*Cinnamomum macrocarpum*	Lauraceae
9	*Cinnamomum sulphuratum*	Lauraceae
10	*Cinnamomum wightii*	Lauraceae
11	*Curcuma pseudomontana*	Zingiberaceae
12	*Decalepis arayalpathra* [=*Janakia arayalpathra*]	Periplocaceae
13	*Decalepis hamiltonii*	Periplocaceae
14	*Diospyros candolleana*	Ebenaceae
15	*Diospyros paniculata*	Ebenaceae
16	*Dipterocarpus indicus*	Dipterocarpaceae
17	*Dysoxylum malabaricum*	Meliaceae
18	*Eulophia flava* [=*Eulophia cullenii*]	Orchidaceae
19	*Eulophia ramentacea*	Orchidaceae
20	*Garcinia indica*	Clusiaceae
21	*Garcinia travancorica*	Clusiaceae
22	*Gardenia gummifera*	Rubiaceae
23	*Glycosmis macrocarpa*	Rutaceae
24	*Gymnema khandalense* (=*Bidaria khandalense*)	Asclepiadaceae
25	*Gymnema montanum*	Asclepiadaceae
26	*Heliotropium keralense*	Boraginaceae
27	*Heracleum candolleanum*	Apiaceae
28	*Humboldtia vahliana*	Caesalpiniaceae
29	*Hydnocarpus pentandrus*	Flacourtiaceae
30	*Hydnocarpus alpina*	Flacourtiaceae
31	*Hydnocarpus macrocarpa*	Flacourtiaceae
32	*Iphigenia stellata*	Liliaceae
33	*Isodon nilgherricus* [=*Plectranthus nilgherricus*]	Lamiaceae
34	*Kingiodendron pinnatum*	Caesalpiniaceae
35	*Magnolia nilagirica* [=*Michelia nilagirica*]	Magnoliaceae
36	*Myristica malabarica*	Myristicaceae
37	*Nilgirianthus ciliatus*	Acanthaceae
38	*Ochreinauclea missionis*	Rubiaceae
39	*Paphiopedilum druryi*	Orchidaceae
40	*Piper barberi*	Piperaceae
41	*Semecarpus travancorica*	Anacardiaceae
42	*Swertia corymbosa*	Gentianaceae
43	*Swertia lawii*	Gentianaceae

TABLE 6.1   *(Continued)*

Sl. No.	Botanical Name	Family Name
44	*Syzygium travancoricum*	Myrtaceae
45	*Tragia bicolor*	Euphorbiaceae
46	*Trichopodium zeylanicum* [=*Trichopus zeylanicus*]	Trichopodaceae
47	*Utleria salicifolia*	Periplocaceae
48	*Valeriana leschenaultii*	Valerianaceae
49	*Vateria indica*	Dipterocarpaceae
50	*Vateria macrocarpa*	Dipterocarpaceae

6.6   *In situ* CONSERVATION OF MEDICINAL PLANTS IN THE WESTERN GHATS–FRLHT'S EXPERIENCE

It is widely recognized that every species has an inherent range of genetic traits that contribute to its genetic richness. It is also to be appreciated that it is impossible to gather all the genetic diversity of any one species for setting up *ex situ* conservation parts for these. The only possible way where the range of genetic diversity of any species can be conserved and allowed to follow its natural evolutionary course is to afford its protection/conservation in the wild – in the *in situ* situations. These natural sites, established especially for its *in situ* conservation of medicinal plants, were named Medicinal Plants Conservation Areas (MPCAs). Of the 55 MPCAs established in southern India, 29 fall in the Western Ghats region – Karnataka (8), Kerala (9), Tamil Nadu (7) and Maharashtra (4). All these MPCAs were established in such a way as to cover the maximum habitat diversity and viable populations of the prioritized medicinal plants species available in these states. These sites were traditionally valued as medicinal plant repositories, were easily accessible, were relatively less disturbed, formed compact micro watersheds and were likely to cause minimum interference with livelihoods of local people. These MPCA network, thus, covered 8 out of 10 major forest types (after Champion and Seth, 1968) in the states. These MPCAs also covered the available altitudinal range in the project area. The actual size of the MPCA varied from 80 ha to 400 ha with an average of 203 ha per MPCA. No MPCA could, however, be established in degraded forests for want of sizeable chunks of forest areas and in view of protection problems. Out of the detailed floristic work carried in these sites, it was noticed that viable populations of four Critically Endangered species were not getting covered under the MPCA network established in the Phase I of DANIDA funded project. Therefore, four additional MPCAs were established during the second phase of the project specifically to capture wild populations of four threatened medicinal plants species of high concern. These four MPCAs are: Kollur MPCA in Karnataka for *Saraca asoca*, Kulamavu MPCA in Kerala for *Coscinium fenestratum*, Anapady MPCA in Kerala for *Utleria salicifolia* and Nambikoil MPCA in Tamil Nadu for *Janakia arayalpathra*.

Botanical survey in different seasons were conducted to prepare comprehensive inventory of the MPCA vegetation as well as to document the ephimerals. These surveys also provided an idea of species richness in these areas at a macro level. Details of MPCAs established in states of Karnataka, Kerala, Tamil Nadu and Maharashtra along with the species recorded in respective MPCAs is provided in the Table 6.2.

TABLE 6.2 Details of MPCAs Established in States of Karnataka, Kerala, Tamil Nadu and Maharashtra (Coordinated by FRLHT)

Serial no	Name	District	Latitude (N)	Longitude (E)	Area (ha)	Recorded medicinal plant sps.
Tamil Nadu						
1.	Petchiparai	Nagercoil	8.45	77.35	210	244
2.	Mundanthurai	Tirunelveli	8.67	77.37	200	267
3.	Kutrallum	Tirunelveli	8.92	77.27	200	288
4.	Thaniparai	Tirunelveli	9.70	77.62	100	259
5.	Kodaikanal	Madurai	10.17	77.42	115	85
6.	Topslip	Coimbatore	10.42	76.83	229	189
7.	Nambikoil	Tirunelveli	8.43	77.50	400	146
Kerala						
8.	Agasthiarmalai	Thiruvananthapuram	8.67	77.18	174	217
9.	Triveni	Pathanamthitta	9.42	77.08	308	208
10.	Eravikulam	Idukki	10.17	77.08	200	83
11.	Peechi	Thrissur	10.53	76.38	156	275
12.	Athirapally	Thrissur	10.30	76.57	112	234
13.	Silent Valley	Palakkad	11.08	76.40	206	205
14.	Waynad	Wyanad	11.85	75.80	148	163
15.	Kulamavu	Idukki	9.80	76.88	215	182
16.	Anappady	Palakkad	10.45	76.81	400	271
Karnataka						
17.	Talacauvery	Madikeri	12.38	75.50	80	255
18.	Subramanya	Mangalore	12.67	75.62	200	220

TABLE 6.2   *(Continued)*

Serial no	Name	District	Latitude (N)	Longi-tude (E)	Area (ha)	Recorded medici-nal plant sps.
19.	Charmadi	Mangalore	13.08	75.47	283	310
20.	Kudermukh	Chikmagalur	13.28	75.13	110	238
21.	Kemmangundi	Chikmagalur	13.53	75.75	310	184
22.	Agumbe	Shimoga	13.48	75.11	210	270
23.	Devimane	Karwar	14.53	74.57	210	259
24.	Kollur	Udupi	13.88	74.82	275	231
Maharashtra						
25.	Amboli	Sindhudurg	15.88	74.00	267.68	146
26.	SGNP Borivali	Thane	19.24	72.92	244.96	180
27.	Honya Koli	Pune	19.00	73.58	592	183
28.	Amba	Raigad	18.30	73.72	150	118

The *in situ* (field) gene bank sites can also be used as study sites to understand the reproductive biology of the species, and guide steps towards their recovery and long-term conservation. The detailed studies undertaken at these sites, on demography of priority species can also feed into the working (management) plan prescriptions of the forest divisions.

6.7   CONCLUSION

6.7.1   OPPORTUNITIES–MEDICINAL PLANTS AND BIOSPROSPECTING

Many modern drugs have origin in ethnopharmacology and traditional medicine. Traditions are dynamic and not static entities of unchanging knowledge. Discovering reliable 'living tradition' remains a major challenge in traditional medicine. In many parts 'little traditions' of indigenous systems of medicine are disappearing, yet their role in bioprospecting medicines or poisons remains of pivotal importance (Patwardhan et al., 2004). Bioprospecting demands a number of requirement which should be coordinated, such as team of scientific expert (from all the relevant interdisciplinary fields) along with expertise in a wide range of human endeavors, including international laws and legal understanding, social sciences, politics and anthropology. In Indian context, Ayurveda and other traditional systems of medicine, rich genetic resources and associated ethnomedical knowledge are key components for sustainable bioprospecting and value-addition processes. For drug-targeted bioprospecting an industrial partner is needed, which will be instrumental in converting the discovery into a commercial product. Importance in any bioprospecting venture is the drafting and signing of an agreement or Memorandum of Understanding that should cover issues on access to the genetic resources (biodiversity), on intellectual property related to discovery, on the sharing of benefits as part of the process (short term), and in the event of discovery and commercialization of a product (long term), as well as on the conservation of the biological resources for the future generations. When ethnobotanical or ethnopharmacological approach is utilize, additional specific requirement that relate to prior informed consent, recognition of Indigenous Intellectual Property Rights as well as short- and long-term benefit sharing need to be taken into account (Parwardhan, 2005; Soejarto et al., 2005).

6.7.2 REVERSE PHARMACOLOGY

The traditional knowledge-inspired reverse pharmacology described here relates to reversing the routine 'laboratory-to-clinic' progress to 'clinics-to-laboratories' (Vaidya, 2006). It's a trans-disciplinary endeavor offering major paradigm shift in drug discovery. Instead of serendipitous findings pursued randomly an organized path from clinical observations and successes is established. Reverse Pharmacology (RP), designed as an academic discipline to reduce three major bottlenecks of costs, time and toxicity. The science has to integrate documented clinical and experiential hits into leads by interdisciplinary exploratory studies on defined targets *in vitro* and *in vivo* and conducting the gamut of developmental activities. Recently, India has amended the Drug Act to include a category of phyto-pharmaceuticals to be developed from medicinal plants by Reverse Pharmacology, with evidence of quality, safety and efficacy. These drugs will be distinct from traditional medicines like Ayurvedic, Unani or Siddha. India with its pluralistic health care system offers immense opportunities for natural product drug discovery and development based on traditional knowledge and clinical observations (Patwardhan et al., 2004). Drug discovery and development is an extremely complex, technology and capital-intensive process that is facing major challenges with the current target rich–lead poor situation. A major cause of attrition in drug discovery is due to toxicity in human trials and it is known that drugs with novel mechanisms have higher attrition rates.

Better validated preclinical targets with proof-of-concept of better efficacy and safety of drugs can, how-ever, mitigate such attrition risks. We propose that the reverse Drug Discovery Today Pharmacology approach can be useful in this process and help in reducing failure rates (Kola and Landis, 2004).

   Tabulation (Table 6.3) below provides the illustration of selected 1116 Medicinal plants reported to be used in the Western Ghats based on the compiled data from Ajesh et al., 2012; Ajesh and Kumuthakalavalli, 2013; Anitha et al., 2008; Ayyanar and Ignacimuthu, 2005 a, b; 2009, 2010, 2011; Ayyanar et al., 2008, 2010; Balasubramanian et al., 2010; Benjamin and Manickam, 2007; Bhat et al., 2012, 2014; Bose et al., 2014; De Britto and Mahesh, 2007; Deepthy and Remashree, 2014; Desale et al., 2013; Divya et al., 2013; Ganesan et al., 2004; Gayake et al., 2013; Gireesha and Raju, 2013; Henry et al., 1996; Hosmani et al., 2012; Jayakumar et al., 2010; Jegan et al., 2008; Jenisha and Jeeva, 2014; Jeyaprakash et al., 2011; Joseph et al., 2015; Jothi et al., 2008; Kadam et al. 2013; Kalaiselvan and Gopalan, 2014; Kamble et al., 2008, 2010; Khyade et al., 2008, 2011; Kshirsagar and Singh, 2007; Kumar and Manickam, 2008; P.S. Kumar et al., 2014; S.S. Kumar et al., 2014; Mahesh and Shivanna, 2004; Mahishi et al., 2005; Mali, 2012; Maridass and Victor, 2008; Mathew et al., 2006; Mohan et al., 2008; Muthukumarasamy et al., 2004; Narayanan et al., 2011; Natarajan, 2014; Navaneethan et al., 2011; Noorunnisa Begum et al., 2004, 2005, 2014; Oak et al. 2015; Pandiarajan et al., 2011; Parinitha et al., 2004; Patil and Bhaskar, 2006; Patil and Patil, 2012; Paul and Prajapati, 2014; Paulsamy et al., 2007; Pesek et al., 2008; Poornima et al., 2012; Pradheeps and Poyyamoli, 2013; Prakash et al., 2008; Prakasha et al., 2010; Rajakumar and Sivanna, 2010; Rajan et al., 2001, 2003; Rajakumar Rasingam, 2012; Rajendran et al., 1997, 2000, 2002; Rajith and Rajendran, 2010; Ramachandran et al., 2009, 2014; Ramana et al., 2011; Revathi and Parimelazhagan, 2010; Revathi et al., 2013; Rothe, 2003; Salave et al., 2011; Sajeev and Sasidharan, 1997; Samy et al., 2008; Sarvalingam et al., 2011; Sathyavathi and Janardhanan, 2011; Shanavaskhan et al., 2012; Sharma and Singh 2001; Sharmila et al., 2014, 2015; Shiddamallaya et al., 2010; Shivanna and Rajkumar, 2010; Simon et al., 2011; Silja et al., 2008; Singh et al., 2012; Sivakumar and Murugesan 2005; Smitha Kumar et al., 2014; Subramanian et al., 2010; Sukumaran et al., 2008; Sukumaran and Raj 2008; Suresh et al., 2012; Sutha et al., 2010; Thomas and Rajendran, 2013; Udayan et al., 2003, 2005; Umapriya et al., 2011; Upadhya et al. 2012; Vijayalakshmi et al., 2014; Vikneswaran et al., 2008; Viswanathan et al. 2006; Xavier et al., 2014; Yabesh et al., 2014; Yesodharan and Sujana 2007; Prasad et al., 2012.

TABLE 6.3 List of Medicinal Plants used in Western Ghats

Sl.No.	Botanical Name	Part Used	Medicinal uses
1.	*Abelmoschus manihot*	Root	Cuts, injuries, skin eruptions, wounds
2.	*Abrus precatorius*	Root, leaves, fruit, seeds	Cold, cough, diabetes, fever, inflammation, abortion, dandruff, jaundice, poisonous bite, eye problems, skin rashes, blood purifier, throat problems and bone fracture
3.	*Abutilon indicum*	Whole plant, leaves, fruit	Diabetes, rheumatism, tuberculosis, ulcers, bleeding disorders, skin roughness and prickles, piles, earache, fever, cough, diuretic, venereal diseases and scabies
4.	*Acacia caesia*	Bark	Used as soap for cleaning scabies, rashes, ringworm infection
5.	*Acacia catechu*	Root, stem, bark, leaves, flowers	Bone fracture, nervous pain, pruritus, wounds due to prickly heat, abortion, boils, throat problems and tooth problem
6.	*Acacia dealbata*	Wood, bark, leaves	Cuts and wounds
7.	*Acacia ferruginea*	Bark	Snake bite
8.	*Acacia leucophloea*	Leaves, stem-bark	Hair complaints, wounds, swellings, snake bite, fracture, burns, skin diseases
9.	*Acacia nilotica* subsp. *indica*	Stem-bark, tender leaves, gum	Sprain, dysentery, leucorrhoea, toothache, wounds
10.	*Acacia sinuata*	Leaves, pods	Dyspepsia, skin disease, dandruff
11.	*Acacia torta*	Stem-bark	Snake bite
12.	*Acalypha fruticosa*	Whole plant, roots, bark, leaves	Digestive troubles, gonorrhea, worms, snake bite
13.	*Acalypha indica*	Whole plant, leaves	Anthelmintic, bed sores, scabies, eczema, skin eruptions/protuberances, ear ache, tooth ache, severe cough, mental problem, dog bite, ringworm, burns, rat bite and wounds
14.	*Acalypha paniculata*	Leaves	Indigestion
15.	*Acampe praemorsa*	Leaves	Antibiotic, tonic, stomach-ache, ear-ache, fever, rheumatism

TABLE 6.3 *(Continued)*

Sl.No.	Botanical Name	Part Used	Medicinal uses
16.	*Achyranthes aspera*	Whole plant, root, stem, leaves, seed	Anthelmintic, bed sores, burns, dog bite, ear ache, eczema, mental problem, rat bite, ringworm, scabies, cough, skin diseases, tooth ache, wounds
17.	*Achyranthes bidentata*	Leaves	Asthma, antidote, contraceptive and night blindness
18.	*Acilepis divergens*	Whole plant	Body heat, asthma
19.	*Acorus calamus*	Rhizome, leaves	Cough, dental problems, diarrhea, diuretic, epilepsy, fertility, fever, giddiness, prevention of abortion, pyrexia, respiratory disorders, stomachache, worm infection
20.	*Acrostichum aureum*	Rhizome, fertile fronds	Wounds, boils, anthelmintic, ulcers, bladder complaints, fungal infection
21.	*Actephila excelsa*	Whole plant, leaves and flowers	Bone fracture, sprain
22.	*Actiniopteris radiata*	Whole plant	Styptic, anthelmintic, diarrhea, dysentery, helminthiasis, haemoptysis and fever
23.	*Adenanthera pavonina*	Wood, bark, leaves, seeds	Boils, inflammation, gout, rheumatism, hair wash
24.	*Adenia hondala*	Tubers, flower	Hernia
25.	*Adenostemma lavenia*	Whole plant, leaves	Giddiness
26.	*Adiantum* sp.	Aerial part	Cuts and wounds
27.	*Adiantum capillus-veneris*	Whole plant	Aphrodisiac, antibacterial, antifungal, antiviral, anticancerous, hypoglycaemic
28.	*Adiantum caudatum*	Fronds	Wounds
29.	*Adiantum lunulatum*	Leaves, root	Chest complaint
30.	*Adiantum poiretii*	Leaves	Cough, fever, diabetes, skin diseases
31.	*Adiantum raddianum*	Whole plant	Asthma
32.	*Aegle marmelos*	Root, stem, bark, leaves, fruit, gum	Diarrhea, headache, constipation, dyspepsia, intermittent fever, itches, gastric troubles, opthalmia, venereal diseases, mouth ulcers, asthma, diabetes

TABLE 6.3 *(Continued)*

Sl.No.	Botanical Name	Part Used	Medicinal uses
33.	*Aerva lanata*	Whole plant, root, leaves	Diuretic, demulcent, headache, stomachache, calculi, for easy delivery, asthma, chest pain, poison bites
34.	*Aganope thyrsiflora*	Leaves, flowers, tender fruits	Bone fracture and pain
35.	*Aganosma cymosa*	Leaves, flowers	Reduces body heat
36.	*Agave americana*	Roots, leaves, flowers	Rheumatism
37.	*Ageratum conyzoides*	Whole plant, leaves	Scabies, cuts, wounds, dyspepsia, anemia, psoriasis, ringworm, leprosy, skin diseases
38.	*Ageratum houstonianum*	Leaves	Wound healing
39.	*Aglaia elaeagnoidea*	Leaves, flowers, tender fruits	Blood pressure
40.	*Agrostistachys indica*	Leaves	Antidotes for poisons
41.	*Albizia amara*	Bark	Used as hair shampoo
42.	*Ailanthus excelsa*	Leaves	Diabetes, menstrual disorders, liver diseases, asthma, bone fracture
43.	*Alangium salvifolium*	Bark, fruits, leaves	Eye diseases, diabetes, fever, tuberculosis
44.	*Alangium salvifolium* var. *hexapetalum*	Bark	Snake bite, paralysis
45.	*Albizia amara*	Leaves, stem, bark	Hair growth, cuts, burns, stomachache and dizziness
46.	*Albizia lebbeck*	Bark, seed	Skin diseases, leucorrhea, jaundice, dysentery
47.	*Albizia odoratissima*	Stem bark, leaves	Dysentery, stomachache
48.	*Albizia procera*	Bark	Snake bite
49.	*Alhagi maurorum*	Leaves, flowers	Cuts, fever
50.	*Allium cepa*	Bulbs, young shoot	Asthma, baldness due to fungal infection, blood clot, body heat, cold, carminative, cuts, eye pain, fainting, foot sores, headache, nail disease, psoriasis and rheumatism
51.	*Allium sativum*	Bulbs, plant	Cuts, swellings, gastric troubles

TABLE 6.3   *(Continued)*

Sl.No.	Botanical Name	Part Used	Medicinal uses
52.	*Allophylus serratus*	Leaves	Fracture, sprain
53.	*Aloe vera*	Leaves	Edible, asthma, boils, burns and swellings, coolant, dandruff, dyspepsia, expel worms from the body, foot thorns, eye complaints, gynecological complaints, hair complaints, headache, insect bites, kidney stones, laxative, menstrual problem, paronychia, skin disorders, stomachache
54.	*Alphonsea zeylanica*	Leaves, flowers and tender fruits	Body pain, blood purify
55.	*Alpinia calcarata*	Rhizome	Stomach disorders, cold, cough, fever, inflammation and reducing pain
56.	*Alpinia galanga*	Rhizome	Hypnotism, asthma, cold, cough, fever, headache, stomach pain and skin disorders
57.	*Alseodaphne semecarpifolia*	Leaves, flowers	Body heat, blood discharge
58.	*Alsophila gigantea*	Fronds, rhizome	Anti-inflammatory, snake bite
59.	*Alstonia scholaris*	Stem bark, latex, leaves	Asthma, cuts, burns, boils, improve lactation wounds
60.	*Alternanthera sessilis*	Leaves, whole plant	Aphthae, coolant, snake bite, eye pain, indigestion
61.	*Alysicarpus bupleurifolius*	Root	Abortifacient
62.	*Amaranthus caudatus*	Leafy twigs	Purify blood and piles
63.	*Amaranthus cruentus*	Leaves	Leucorrhoea, body heat
64.	*Amaranthus graecizans*	Roots	Ear problem
65.	*Amaranthus spinosus*	Whole plant, root, leaves, inflorescence	Allergies during pregnancy, athletes foot disease, blister, boils, diuretic, eczema, hair complaints, other skin diseases, promoting digestion, stomach pain, used to remove thorns wound
66.	*Amaranthus tricolor*	Twigs, seeds, whole plant, root	Edible, dandruff, ringworm, vomiting

TABLE 6.3 (Continued)

Sl.No.	Botanical Name	Part Used	Medicinal uses
67.	*Amaranthus viridis*	Whole plant, leaves	Edible, coolant, eczema, stomach ache, scabies
68.	*Amaranthus paniculatus*	Leaves, seed	Digestion, intestinal worms
69.	*Ammannia bac-cifera* subsp. *aegyptiaca*	Whole plant	Eczema
70.	*Amomum zerumbet*	Rhizome	Digestive problems
71.	*Amorphophallus commutatus*	Tubers	Piles, bacterial infections
72.	*Amorphophallus paeoniifolius*	Rhizome, fruit, corms	Piles, used to remove thorns, bone fracture
73.	*Amorphophallus paeoniifolius* var. *campanulatus*	Tuber	Bleeding piles
74.	*Amorphophallus sylvaticus*	Corm	Asthma, menstrual disorders, liver diseases
75.	*Anacardium occidentale*	Fruit, seeds, stem-bark, flowers, seed oil	Cholera, foot sores, corn, dyspnoea, skin ailments, dysentery, foot crack, asthma, headache, wound, piles
76.	*Anagallis arvensis*	Whole plant, leaves	Fever, anorexia, fertility
77.	*Ananas comosus*	Fruit	Pregnancy, venereal diseases
78.	*Anaphalis elliptica*	Whole plant	Fever
79.	*Anaphalis neelgerryana*	Leaves	Stomach pain, fertility
80.	*Anaphyllum beddomei*	Rhizome	Eczema
81.	*Ancistrocladus heyneanus*	Leaves	Rheumatism
82.	*Andrographis alata*	Whole plant, leaves	Antidote for poisons, skin wounds
83.	*Andrographis paniculata*	Root, leaves	Constipation, contraceptive purposes, diabetes, diarrhea, fever, indigestion, jaundice, leprosy, liver disease, liver problem, insect bites, ringworm infection, scabies, scorpion-sting, snake bite, sore throat

TABLE 6.3   *(Continued)*

Sl.No.	Botanical Name	Part Used	Medicinal uses
84.	*Andrographis rothii*	Stem	Injury healing
85.	*Andrographis serpyllifolia*	Roots	Snake bite
86.	*Anethum graveolens*	Seeds	Boils
87.	*Angiopteris evecta*	Spores, leaves	Dysentery, leprosy, other skin diseases
88.	*Anisomeles indica*	Leaves	Ringworm infection, fever
89.	*Anisomeles heyneana*	Leaves	Headache, urinary complaints
90.	*Anisomeles malabarica*	Whole plant, leaves	Cough, cold, bowels complaints, intermittent fever, leucorrhoea, rheumatic pains
91.	*Annona reticulata*	Fruit, seed	Diarrhea
92.	*Annona squamosa*	Leaves, fruits, seed	Cuts, wounds, dysentery, inflammation, tuberculosis, induce abortion, lice, worms
93.	*Anogeissus latifolia*	Stem-bark	Dyspepsia, vomiting, dysentery, stomach ache, cough, paralysis
94.	*Anredera baselloides*	Leaves	Snake bite, eczema
95.	*Anredera cordifolia*	Leaves	Edible
96.	*Antiaris toxicaria*	Bark	Rheumatic problems
97.	*Antidesma acidum*	Leaves	Bone fracture
98.	*Antidesma bunius*	Leaves	Ulcers, indigestion
99.	*Antidesma montanum*	Whole plant	Purify blood, rheumatism
100.	*Ardisia solanacea*	Root	Itching
101.	*Areca catechu*	Leaves, young leaves, young fruits, fruits, young seeds	Tinea, masticatory, burns and sprains, cough, expel tape worm from the body, laxative
102.	*Argemone mexicana*	Root, leaves, flower, seed	Scabies, jaundice, ringworm infection, eye disease, wounds, skin rashes, tooth problem

TABLE 6.3 *(Continued)*

Sl.No.	Botanical Name	Part Used	Medicinal uses
103.	*Argyreia cuneata*	Leaves	Fever, rheumatism, skin wounds, anti-inflammation
104.	*Argyreia elliptica*	Root	Stomachache, toothache
105.	*Argyreia pomacea*	Leaves	Cooling effect to the eyes
106.	*Argyreia speciosa*	Leaves, root	Fever, headache, cuts, cooling effect, urinary disorders
107.	*Argyreia strigosa*	Root	Stomach pain
108.	*Arenga wightii*	Spadix	General health
109.	*Arisaema leschenaultii*	Roots, whole plant	Snake bite, paralysis
110.	*Arisaema tortuosum*	Rhizome	Ear complaints
111.	*Aristolochia bracteolata*	Whole plant, leaves	Inflammation, boils, snake bite, eczema, antifertility, fungal diseases, scabies, ring-worm infection
112.	*Aristolochia indica*	Aerial part, roots, stem, leaves	Headache, skin disease, snake bite, skin cuts & wounds, stomachache, scabies, blood clotting, cholera, diarrhea, poison, fever, insect bite, dandruff, scorpion-sting, menstruation, piles
113.	*Aristolochia krisagathra*	Leaves	Rheumatism, to reduce excessive body heat
114.	*Aristolochia tagala*	Whole plant, leaves, root	Antidote for poisons, snake bite, stomachache
115.	*Artabotrys zeylanicus*	Flower	Vomiting
116.	*Artemisia nilagirica*	Roots, flowering twig, leaves, flowers	Asthma, bronchitis, insecticide, fever, vet-erinary, leprosy, skin diseases and wounds
117.	*Artocarpus go-mezianus* subsp. *zeylanicus*	Stem bark	Vomiting
118.	*Artocarpus heterophyllus*	Leaves, young fruits, fruit and seed	Gingival wounds, ulcers, vomiting and mouth ulcers
119.	*Artocarpus hirsutus*	Bark, fruit and seed	Edible, pimples and cracks
120.	*Artocarpus lakoocha*	Bark	Astringent and skin burn

TABLE 6.3 *(Continued)*

Sl.No.	Botanical Name	Part Used	Medicinal uses
121.	*Asclepias curassavica*	Whole plant, leaves, flower	Hair complaints, wounds, good styptic
122.	*Asparagus racemosus*	Whole plant, tubers, stem	Fever, tuberculosis, galactogogue, diarrhea, dysentery, aphrodisiac, dysuria, fits, gynecological complaints, rickets, skin ailments, dyspepsia, toothache, infertility, loss of libido, bed-wetting, cracks in heels, spermatorrhoea, epilepsy, digestion, inflammation, leucorrhoea, excess body heat, tastelessness, relieves ulcer induced pain, appetizer, promotes lactation, liver tonic and stomach pain
123.	*Asparagus racemosus* var. *javanicus*	Root	Dysentery, swellings
124.	*Asplenium nidus*	Rootstock, leaves	Fever, elephantiasis, cough, chest diseases, cold
125.	*Asplenium polydon* var. *bipinnatum*	Whole plant, crosiers	Promote parturition, anticancerous,
126.	*Asystasia dalzelliana*	Leaves	Sprain
127.	*Asystasia travancorica*	Leaves	Rheumatism
128.	*Atalantia monophylla*	Fruit	Rheumatism, paralysis, cold, poison bites
129.	*Atalantia racemosa*	Leaves	Itching skin, paralysis, chronic rheumatism
130.	*Azadirachta indica*	Root, stem bark, twig, leaves, flower, fruit, seed, gum	Jaundice, rheumatism, worms, ulcers, skin ailments, cuts and scratches, boil, chicken pox, anti diabetic, anti bacterial, anti viral, contraceptive, fever, leprosy, scabies, stomach ache, poison bites, blood purifier, wounds, skin diseases and tooth problem
131.	*Azanza lampas*	Root	Bone fracture, gonorrhea, oral problems
132.	*Azima tetracantha*	Leaves	Fever

TABLE 6.3 *(Continued)*

Sl.No.	Botanical Name	Part Used	Medicinal uses
133.	*Baccaurea courtallensis*	Tender fruits, ripe fruits	Edible, fever, infertility problem, mouth and stomach ulcers
134.	*Bacopa monnieri*	Whole plant	Enhances memory power, eczema, asthma, epilepsy, knee pain, inflammation of mucous membrane and severe cough
135.	*Balanophora fungosa* ssp. *indica*		Piles, internal hemorrhage
136.	*Baliospermum montanum*	Root	Indigestion, vomiting, constipation, treating bone fracture.
137.	*Bambusa bambos*	Root leaves, young leaves, terminal bud, seed	Rheumatism, skin cuts, sprain, detachment of placenta; de worming, fractured bones, rheumatism and urinary problem
138.	*Bambusa vulgaris*	Leaves	Fertility problem
139.	*Barleria buxifolia*	Root	Stomach pain
140.	*Barleria cristata*	Root	Headache
141.	*Barleria involucrata* var. *involucrata*	Leaves, flowers	Scabies
142.	*Barleria prionitis*	Leaves	Gingival wounds, urinary irritation, wounds, eczema
143.	*Barringtonia racemosa*	Fruits	Septic due to skin allergy, dog bite
144.	*Basella alba*	Whole plant, leaves and stem	Coolant, inflammation, mouth ulcers, boils, piles, constipation in children and pregnant women
145.	*Bauhinia acuminata*	Bark, leaves	Urinary discharge, throat troubles and skin diseases
146.	*Bauhinia malabarica*	Bark	Abdominal pain
147.	*Bauhinia purpurea*	Bark, leaves	Bone fracture
148.	*Bauhinia racemosa*	Bark, leaves, flowers	Dysentery, stomach pain, to remove particles like dust from eyes, jaundice, indigestion

TABLE 6.3   *(Continued)*

Sl.No.	Botanical Name	Part Used	Medicinal uses
149.	*Bauhinia tomentosa*	Bark	Diarrhea
150.	*Bauhinia variegata*	Stem-bark, roots, bark	Boils, wounds, skin diseases, piles, and ulcer
151.	*Begonia malabarica*	Aerial part, leaves	Fever, giddiness, rheumatism, scores, pimples, nasal infection
152.	*Begonia subpeltata*	Leaves	stomachache
153.	*Benincasa hispida*	Fruit	Haemoptysis, burns
154.	*Berberis tinctoria*	Roots, Fruit	Jaundice
155.	*Bidens biternata*	Whole plant, leaves	Leg pain
156.	*Bidens pilosa*	Whole Plant	Wounds, antiseptic and cough
157.	*Biophytum reinwardtii*	Whole Plant	Scabies
158.	*Biophytum sensitivum*	Whole plant, leaves, seeds	Wound, bleeding, chest pain, excess of heat, migraine, fever, skin rashes, eczema, diuretic, dissolves calculi, insect or snake bite
159.	*Bischofia javanica*	Bark, leaves, fruit	Sores, tooth ache, eye diseases, throat troubles
160.	*Bixa orellana*	Flowers, fruit	Dysentery, boils, prevent blisters and scars
161.	*Blachia calycina*	Leaves, flowers, tender fruit	Throat pain, toothache, blood purifier, body pain, nervous weakness
162.	*Blainvillea acmella*	Leaves, seed	Gynecological complaints, skin cuts
163.	*Blechnum orientale*	Rhizome, fronds	Boils, urinary complaints, diaphoretic, anthelmintic, intestinal wounds
164.	*Blepharis integrifolia*	Whole plant	bone fracture
165.	*Blepharis maderaspatensis*	Leaves	wounds
166.	*Blumea barbata*	Tender leaves	Edible
167.	*Blumea eriantha*	Root, bark	Snake bite
168.	*Blumea lacera*	Whole plant, leaves and flowers	Blood discharge, diabetes, tooth pain, cold, weakness, giddiness, leukemia, nervous problem

TABLE 6.3 *(Continued)*

Sl.No.	Botanical Name	Part Used	Medicinal uses
169.	*Boerhavia diffusa*	Whole plant, root, leaves, seeds	Edible, cuts, wounds, snake bite, diabetes, eye, jaundice, anemia, urinary complaints, renal disorders, kidney stones, nephritis, scabies, ring worm infections, jaundice and wounds
170.	*Bombax ceiba*	Tender leaves, flower, root, stem-bark, gum	Aphrodisiac, dizziness, dysentery, dyspepsia, wound, snake bite, menorrhagia and general tonic
171.	*Bombax malabarica*	Flowers	Sore, boil, burns
172.	*Borassus flabellifer*	Rhizome, fruit	Fever, headache, rheumatism
173.	*Boswellia serrata*	Resin	To ward off germs and mosquitos, chronic laryngitis, fever, jaundice, ringworm.
174.	*Botrychium lanuginosum*	Whole plant	Antidysentric, antibacterial
175.	*Brassica juncea*	Seeds, leaves	Headache, dysentery, eye diseases (white patches in pupil), gingival wounds
176.	*Brassica nigra*	Leaves	Wormicide
177.	*Breynia retusa*	Whole plant, stem	Body heat, conjunctivitis
178.	*Breynia rhamnoides*	Root	Breast cancer
179.	*Breynia vitis-idaea*	Roots, leaves	Cuts, wound, tonsils, mouthwash for toothache
180.	*Bridelia retusa*	Stem-bark, fruit	Dysentery, hypertensive properties, ear complaints, jaundice
181.	*Bridelia scandens*	Stem bark, fruit	Jaundice
182.	*Bridelia stipularis*	Leaves	Toothache, bone fracture
183.	*Bryophyllum pinnatum*	Leaves	Boils, sprain, dysentery, cholera, wounds, bruises, boils and bites of venomous insects
184.	*Buchanania axillaris*	Stem-bark	Stomachache
185.	*Buchanania lanzan*	Bark	Diarrhea, rheumatic pain, asthma
186.	*Bulbophyllum albidum*	Bulb, leaves	Strengthening of weak uterus for conception
187.	*Bulbophyllum distans*	Pseudobulbs	Rheumatism

TABLE 6.3   *(Continued)*

Sl.No.	Botanical Name	Part Used	Medicinal uses
188.	*Bulbophyllum sterile*	Pseudobulbs	Rheumatism, inflammation
189.	*Butea monosperma*	Seed, stem-bark, gum, leaves, flower	Dysuria, dysentery, diarrhea, all types of skin diseases, bed sore, cough, cold, ulcers, cleaning uterus after delivery, induce sterility, intestinal worms in children, rheumatic swelling, bone fracture, cough, dhobis itch and ringworm
190.	*Cadaba fruiticosa*	Root	Scorpion sting
191.	*Cadaba trifoliata*	Whole plant	Rheumatism
192.	*Caesalpinia bonduc*	Leaves, tender fruits, Seeds	Gastric, stomach pain, asthma, chest pain, jaundice, headache, easiness of delivery, fever, diabetes, anthelmintic
193.	*Caesalpinia mimosoides*	Leaves	Chest pain, boils, epilepsy
194.	*Caesalpinia pulcherrima*	Bark	Induce abortion
195.	*Cajanus cajan*	Leaves, immature fruits	Dysentery, galactagogue
196.	*Calamus brandisii*	Flower, seed	Whooping cough, earache, hydration and headache
197.	*Calamus rotang*	Roots	Scorpion sting
198.	*Calanthe masuca*	Tuberous roots	Acne, sebaceous cysts
199.	*Calendula officinalis*	Flower	Stomachache
200.	*Callicarpa tomentosa*	Leaves, flowers	Stomach ache, body heat, scabies, skin diseases, boils, eczema.
201.	*Calophyllum calaba*	Seeds	Skin problems, antibacterial property
202.	*Calophyllum inophyllum*	Leaves, seed	Scabies, rheumatism, purgative, fish poisoning, ringworm infection, baldness due to fungal infection, skin irritation, earache, headache, hair growth and graying of hair
203.	*Calotropis gigantea*	Root, stem, leaves, latex, flower	Abortion, asthma, dog-bite, eczema, expelling thorn from foot, gingival wounds, headache, herpes, intestinal worms, migraine, otorrhoea, pruritus, psoriasis, reduce swelling on rheumatic joints, scabies, scorpion sting, skin wounds, snake bite, stomach pain

TABLE 6.3 *(Continued)*

Sl.No.	Botanical Name	Part Used	Medicinal uses
204.	*Calotropis procera*	Root bark	Snake bite, for bone hardening and pain reliever
205.	*Calycopteris floribunda*	Branches of leaves	Gangrene, snake bite, reduce swelling on rheumatic joints, laxative
206.	*Camellia sinensis*	Root, leaves	Dysentery, diabetes
207.	*Canarium strictum*	Resin	Rheumatism, fever and cough, ward off insects, headache, cold
208.	*Canavalia gladiata*	Leaves, fruits	Relieve pain in external piles
209.	*Canavalia virosa*	Leaves	Body pain
210.	*Canna indica*	Rhizome, root, leaves, flowers	Digestion, diuretic, paralysis
211.	*Canscora diffusa*	Whole plant	Throat problems
212.	*Canscora roxburghii*	Whole plant	Ulcer, Vermifuge
213.	*Canthium coromandelicum*	Bark, root, leaves, flowers, tender fruits	Stomach disorders, fever, diuretic
214.	*Canthium parviflorum*	Root, leaves, fruits	Boils, coolant, dysentery, reduce hemorrhage in just delivery mother, scabies and skin inflammation, snake bite, strengthen the whole body
215.	*Capparis rheedii*	Leaves, tender fruits	Hemicrania
216.	*Capparis sepiaria*	Roots, stem bark, leaves	Dandruff, dysentery, eczema, headache, hip pains, swelling
217.	*Capparis zeylanica*	Leaves	Breathing problems, immune stimulant, anti-inflammatory
218.	*Capsicum annuum*	Leaves, fruit	Anti-inflammation, cold, cough, otorrhoea, rheumatic complaints
219.	*Caralluma bicolor*	Whole plant	Weight loss
220.	*Caralluma umbellata*	Whole plant	Indigestion

TABLE 6.3 *(Continued)*

Sl.No.	Botanical Name	Part Used	Medicinal uses
221.	*Cardiospermum canescens*	Leaves	Rheumatic pain
222.	*Cardiospermum halicacabum*	Whole plant, root, stem, leaves	Asthma, scorpion sting, centipede bite, dandruff, antiseptic, delivery pain, diarrhea, fever joint pain, muscle pain, rheumatic pain, skin cuts, sprain, treating stomachache in pregnant women
223.	*Carex filicina*	Whole plant	Antidote for poisons
224.	*Careya arborea*	Stem bark, flower	Aphthae, bed sore, gynecological complaints, infertility, piles, scorpion sting, toothache, intestinal ulcers
225.	*Carica papaya*	Root, leaves, fruit, latex	Abortion, body pain, constipation, gynecological complaints, rickets, skin wounds, speedy recovery of bone fracture, swellings, toothache, ringworm
226.	*Carissa carandas*	Fruit	body heat, stomachache
227.	*Carissa congesta*	Latex, root, bark	Jaundice, fever, stomach pain, wound,
228.	*Carissa spinarum*	Fruit	Quenching thirst
229.	*Carthamus tinctorius*	Seed	Diuretic
230.	*Caryota urens*	Tender leaves, inflorescent	Hair growth, body heat, health tonic, Jaundice
231.	*Casearia graveolens*	Leave, Flowers	Giddiness, stomach troubles
232.	*Casearia tomentosa*	Leaves	Bone fracture
233.	*Casearia wynadensis*	Leaves	Used as fish poison for fishing.
234.	*Cassia fistula*	Root, bark, leaves, fruits, fruit pulp	Boils, burning sensation, constipation, eczema, skin diseases, gangrene, herpes, induce antifertility, itching due to impure blood, leprosy, ringworm, scabies, snake bite, stomachache, bone fracture, wormicide
235.	*Cassine albens*	Exudate	conjunctivitis.
236.	*Cassine glauca*	Fruits	Increase bone in nose, snake bite

TABLE 6.3 *(Continued)*

Sl.No.	Botanical Name	Part Used	Medicinal uses
237.	*Cassytha filiformis*	Whole plant, stem	Rheumatism, health restorer, hair growth, leucorrhea, general diseases, sprain fever, body fluid, wounds and intestinal ulcers
238.	*Catharanthus roseus*	Leaves, root	Anticancer, menorrhagia, diabetes, cancer, Wasp sting
239.	*Catunaregam spinosa*	Root, leaves, flowers, fruit	Nasal trouble, headache, piles, dyspepsia, stomachache, vomiting, herpes, pruritus, scabies, snake bite, leech bite, itching
240.	*Cayratia elongata*	Root	Inter trigo
241.	*Cayratia pedata*	Whole plant	Antiseptic, cancer, ulcer
242.	*Cayratia trifolia*	Root	Boils
243.	*Celastrus paniculatus*	Root, leaves, flowers, tender fruits	Cold, itching, snake bite, weakness
244.	*Celosia argentea*	Roots, leaves, seed	Coolant, dysuria, fever, diarrhea stomach ache
245.	*Celtis cinnamomea*	Leaves	Cuts and wounds
246.	*Celtis philippensis* var. *wightii*	Bark	Digestion problems
247.	*Centella asiatica*	Whole plant, stem, leaves	Swelling of the body, anemia, bodyache, boils, brain tonic, dandruff, diarrhea, excellent remedy for ulcer, fever, gynecological complaints, hair growth, headache, jaundice, menstrual pain, ringworm, skin diseases, syphilis, to enhance memory power
248.	*Centotheca lappacea*	Whole plant	Tooth pain
249.	*Centranthera indica*	Leaves	Fever
250.	*Centratherum anthelminticum*	Seeds	Diuretic, snake bite and scorpion sting
251.	*Ceratopteris thalictroides*	Fronds	Skin diseases
252.	*Ceropegia acuminata*	Leaves, roots	Wounds

TABLE 6.3   *(Continued)*

Sl.No.	Botanical Name	Part Used	Medicinal uses
253.	*Chassalia curviflora*	Aerial, leaves flowers, tender fruit	Sinusitis, body pain, syphilis, gastric complaints, nasal blockage
254.	*Cheilanthes tenuifolia*	Rhizome, roots	General tonic
255.	*Cheilocostus speciosus*	Rhizome, stem, leaves	Eye and ear complaints, mumps, otorrhoea, ringworm infection, diabetes, wounds
256.	*Chenopodium album*	Leaves, seeds	Laxative, easy digestion
257.	*Chionanthus mala-elengi* subsp. *linocieroides*	Leaves, flowers, tender fruits	Asphyxia, gastric complaints, stomach pain
258.	*Chionanthus mala-elengi* subsp. *mala-elengi*	Root	Mouth ulcers
259.	*Chlorophytum laxum*	Leaves	Indigestion
260.	*Chlorophytum tuberosum*	Leaves, root	Edible, centipede bite, other poisonous bite
261.	*Chloroxylon swietenia*	Leaves, inner bark	Wounds, cuts, prickles, tooth problems, psoriasis, wounds
262.	*Christella parasitica*	Whole plant	Gout and rheumatism
263.	*Chromolaena odorata*	Leaves	Cuts, gangrene, wounds
264.	*Chrysopogon fulvus*	Bark peel	Cough and chest pain
265.	*Cinnamomum riparium*	Leaves, flowers	Edema
266.	*Cinnamomum tamala*	Leaves, bark	Diaphoretic
267.	*Cinnamomum verum*	Bark, root, root bark, leaves	Boils, cough, stomachache, flatulence, nausea and vomitting
268.	*Cinnamomum wightii*	Bark	Expectorant, bone fracture
269.	*Cipadessa baccifera*	Stem, tender stem, leaves, flowers, tender fruits	Blood purifier, fever, snake bite, stomach disorders, throat pain, wasp sting

TABLE 6.3 *(Continued)*

Sl.No.	Botanical Name	Part Used	Medicinal uses
270.	*Cissampelos pareira*	Vine, roots, whole seed, leaves	Abdominal pain, asthma, arthritis, body pain and body heat, dropsy, headache, heart problems, kidney stones, kidney infections and pains, leucorrhoea, , loose motion, lumbago, muscle cramps, snake bite, stomachache, menopause
271.	*Cissus discolor*	Whole plant	Ringworm, fractures
272.	*Cissus javana*	Whole plant	Fracture
273.	*Cissus quadrangularis*	Whole plant, tender shoots, roots, stem and leaves	ear diseases, Bone fracture, indigestion, piles, pregnancy, to relieve tumor like swellings caused by wind exposure to the naval region, worms, regularize menstruation, digestion troubles, stimulant, rheumatism, ringworm infection, leucorrhoea, skin diseases and stomachache
274.	*Cissus trilobata*	Leaves	Itching
275.	*Citrus aurantifolia*	Fruit	Aphthae, boils, cuts, dog bite, indigestion, scorpion sting, snake bite
276.	*Citrus limon*	Fruit	Bone fracture, tinea infection, whitlow, reduce obesity, skin diseases,
277.	*Citrus medica*	Root, fruit	Pimples, nail disease, skin ailments, snake bite, sun stroke, indigestion
278.	*Clausena indica*	Leaves, flowers	Eye sight
279.	*Cleisostoma tenuifolium*	Whole plant	Anemia
280.	*Clematis gouriana*	Whole plant, root, leaves	Boils and itches, skin ailments, skin burns, rheumatism, cough
281.	*Cleome gynandra*	Whole plant, roots, leaves, flowers	Headache, diabetes, earache
282.	*Cleome viscosa*	Leaves, seed	Headache, earache, eye troubles, skin diseases, inflammation and boils, arthritis, loss of appetite, constipation, deworming, wounds, swelling, and fever
283.	*Clerodendrum inerme*	Leaves	Fever, psoriasis, scabies, ringworm infection, insect bite, malaria
284.	*Clerodendrum indicum*	Root	Snake bite

TABLE 6.3 *(Continued)*

Sl.No.	Botanical Name	Part Used	Medicinal uses
285.	*Clerodendrum infortunatum*	Aerial part, leaves	Rheumatism, skin cuts, weaning child, wounds, scabies, gangrene
286.	*Clerodendrum phlomidis*	Leaves, flowers	Cuts, wounds, paralysis, sprain, body pain, blood vomiting, arthritic pain
287.	*Clerodendrum serratum*	Leaves, flowers root	Asthma, wheezing, fertility, rheumatism, snake bite, dyspepsia
288.	*Clerodendrum viscosum*	Leaves	To prevent excessive bleeding due to injury
289.	*Clidemia hirta*	Root	Varicose pain
290.	*Clitoria ternatea*	Whole plant, roots, stem-bark, flower, leaves	Body swelling, diarrhea, aphrodisiac, fever, rheumatism, wormicide, anti-inflammation, otorrhoea, poison, clean the uterus after delivery, antifertility, scorpion sting, scabies, ringworm infection, ulcer, throat infection
291.	*Coccinia grandis*	Tubers, leaves, fruits	Antidote, throat pain, mouth ulcers, dyspepsia, boils, jaundice, leprosy, psoriasis, body pain, diabetes, diarrhea, toothache
292.	*Cocculus hirsutus*	Leaves	Sprain, cooling effect, headache, scorpion sting, skin burns, dyspepsia, bone fracture
293.	*Cocos nucifera*	Leaves, fruits	Cold, fever, head injury, indigestion, rheumatism, ringworm infection, burn, cuts, intestinal inflammation, skin rashes, skin diseases, urinary complaints
294.	*Codariocalyx motorius*	Whole plant, leaves	Antidote, aphrodisiac, cough, fever, diabetes, menorrhagia, scabies, wound, snake bite, vermifuge
295.	*Coffea arabica*	Seeds	Stimulating and refreshing the body
296.	*Coleus aromaticus*	Leaves	Cold, urinary diseases, sores, ulcers, boils, sprains, swelling and throat pain, headache
297.	*Colocasia esculenta*	Tender leaves; petiole, tuber	Hematinic, otalgia
298.	*Combretum ovalifolium*	Bark	Jaundice, cooling effect, headache, snakebite, urinary complaints
299.	*Commelina benghalensis*	Stem and leaves	Laxative, burns, boils, itches, septic wounds in the breast, bitter, emollient, demulcent, refrigerant, laxative, and beneficial in leprosy, scabies, wounds, to remove poisonous spines, bone fracture

TABLE 6.3 *(Continued)*

Sl.No.	Botanical Name	Part Used	Medicinal uses
300.	*Commelina clavata*	Whole plant	Body pain, migraine and tooth pain
301.	*Commelina coelestes*	Stem	Ring worm, skin diseases
302.	*Commelina erecta*	Whole plant	Scabies
303.	*Commiphora caudata*	Bark, latex	Wounds, rheumatism
304.	*Corallocarpus epigaeus*	Root tuber	Cold, cough in children
305.	*Cordia dichotoma*	Flowers, fruit, seed, stem, bark	Boils, anti-inflammatory, excessive menstruation, kidney stones, dysmenorrhea
306.	*Cordia monoica*	Leaves	Chest pains
307.	*Cordia obliqua* var. *tomentosa*	Bark, kernels	Astringent, ring worm
308.	*Coriandrum sativum*	Leaves, seeds	Anthelmintic, epistaxis, stomachache
309.	*Coronopus didymus*	Aerial parts	Cuts and wounds
310.	*Coscinium fenestratum*	Root, stem	Jaundice, burns, acne
311.	*Crassocephalum crepidioides*	Leaves	Post pregnancy care
312.	*Crataeva magna*	Root, stem bark, leaves	Rheumatism, inflammation, urinary disorder, fever, vomiting, gastric irritation abortion
313.	*Cressa cretica*	Whole plant, leaves, roots	Anthelmintic, aphrodisiac, asthma, bilious, blood purifier, constipation, diabetes, health tonic, jaundice, scabies, sores, stomachic and tonic
314.	*Crinum asiaticum*	Leaves	Eczema
315.	*Crinum defixum*	Bulbs	Skin inflammation due to insect bites or allergy, snake bite
316.	*Crossandra infundibuliformis*	Leaves	Expelling thorn from foot, eczema, scabies
317.	*Crotalaria albida*	Leaves	Carminative
318.	*Crotalaria juncea*	Root, flowers	Measles, antifertility

TABLE 6.3   *(Continued)*

Sl.No.	Botanical Name	Part Used	Medicinal uses
319.	*Crotalaria laburnifolia*	Leaves	Mumps
320.	*Crotalaria laevigata*	Leaves	Edible
321.	*Crotalaria verrucosa*	Whole plant, leaves, flowers	Chest pain, throat complaints, to avoid oedema during pregnancy, skin infection, vomiting and diarrhea
322.	*Croton bonplandianum*	Leaves, roots	Constipation, inflammation and small tumors, arthritic pain
323.	*Croton caudatus*	Leaves, roots	Sprains, diuretic, malaria
324.	*Croton roxburghii*	Roots	Snake bite
325.	*Croton tiglium*	Leaves, seeds	Eczema, alopecia, scabies and ringworm infection
326.	*Croton zeylanicus*	Bark	Stomach ache
327.	*Cryptocoryne retrospiralis*	Whole plant	Edible
328.	*Cryptolepis buchanani*	Root, bark, latex	Paralysis, eczema, herpes, pruritus, scabies, snake bite, cuts
329.	*Cryptostegia grandiflora*	Root, bark	Influenza, refresher, snake bite
330.	*Cucumis prophetarum*	Tender fruits	Eczema
331.	*Cucumis sativus*	Fruit	Urinary diseases
332.	*Cullen coryllifolium*	Seeds	Leucoderma
333.	*Cuminum cyminum*	Seeds	Ear sore, aphthae, ulcer in intestine, gangrene, gingival wounds
334.	*Curculigo orchioides*	Tuber, rhizome, leaves	All skin diseases, biliousness, body pain, bronchitis, gonorrhea, Diabetes, fatigue, leucorrhoea, piles, tinea infection, tonsillitis and wound
335.	*Curcuma amada*	Rhizome	Bone fracture
336.	*Curcuma aromatica*	Rhizome	Relieve pain, insect bite, improves complexion, all types of skin infections and pimples

TABLE 6.3 *(Continued)*

Sl.No.	Botanical Name	Part Used	Medicinal uses
337.	*Curcuma longa*	Rhizome	Ecchymosis, poison, mumps, snake bite, skin diseases, eczema, inter trigo, crack foot, cuts, gangrene, otorrhoea, wounds, bone fracture
338.	*Curcuma pseudomontana*	Rhizome	Wounds and cuts
339.	*Cuscuta reflexa*	Whole plant	Purgative, dandruff, hair fall, flatulence, birth control, haematuria, fever, liver diseases, hepatitis, body ache, constipation, expel worms from the body
340.	*Cyanotis pilosa*	Stem	Eye irritation
341.	*Cyathula prostrata*	Whole Plant	Applied to wounds
342.	*Cyclea peltata*	Whole plant, root, leaves	Joint pain, bodyache, cooling effects, stomachache, eczema, toothache boils, dandruff, intestinal disorder and dysentery.
343.	*Cymbidium aloifolium*	Leaves	Emetic, purgative, ear-ache, otitis, inflammatory conditions, salep used as nutrient and demulcent
344.	*Cymbopogon flexuosus*	Roots	Carminative, flatulence and gastric irritability.
345.	*Cymbopogon caesius*	Whole plant, root	Body fluid, fever, hemicrania.
346.	*Cymbopogon citratus*	Whole plant, root, leaves	Asthma, fever, sprain, toothache, rheumatism, gastric problem
347.	*Cymbopogon confertiflorus*	Leaves	Skin allergy
348.	*Cynodon dactylon*	Whole plant, roots, leaves, fiber	Urinogenital troubles, asthma, diabetes, snake poison, skin allergy, headache, sore nose, chronic dysentery, bleeding piles and irritation of urinary organs, bone fracture, Inflammation, skin burns, skin wounds, sore nose, diuretic, dropsy, secondary syphilis, bleeding from piles, cuts and wounds, astringent, diuretic, anasarca, hysteria, epilepsy, insanity, dysentery, opthalmia, rheumatism, itching, body pain, asthma
349.	*Cynoglossum furcatum*	Roots	Jaundice

TABLE 6.3 *(Continued)*

Sl.No.	Botanical Name	Part Used	Medicinal uses
350.	*Cynoglossum zeylanicum*	Root, leaves	Snake bite, vomiting
351.	*Cyperus pangorei*	Root	Stomach pain
352.	*Cyperus rotundus*	Rhizomes	Dyspepsia, vomiting, cough, bronchitis, indigestion, thirst, worm troubles, induce appetite and improves body health
353.	*Cyphomandra betacea*	Fruits	Diuretic, cough and cold
354.	*Cyrtococcum deccanense*	Whole plant	Body heat
355.	*Cyrtococcum longipes*	Leaves	Scabies
356.	*Cytisus scoparius*	Leaves	Diuretic
357.	*Daemia extensa*	Leaves	Cough
358.	*Dalbergia lanceolaria*	Bark	Jaundice, menorrhagia
359.	*Dalbergia latifolia*	Stem bark	Fever
360.	*Dalberga sissoo*	Tender leaves	Leucorrhoea, skin eruptions
361.	*Datura innoxia*	Leaves, fruit	Mumps, piles, relieve pain and inflammation, dog bite
362.	*Datura metel*	Stem, leaves, flowers, fruit	Asthma, cough, stomachache, wound maggots, pain, dandruff, hair growth, worms in skin, seizure, swelling, rheumatic pain, ear ache, cut, tooth problem
363.	*Datura stramonium*	Leaves, fruit	Dandruff
364.	*Debregeasia longifolia*	Leaves	Skin allergy
365.	*Dendranthema indicum*	Leaves	Boils, cuts
366.	*Dendrobium ovatum*	Leaves	Laxative, emollient, stomachic, liver problem
367.	*Dendrocnide sinuata*	Leaves	Eczema
368.	*Dendrophthoe falcata*	Whole plant, leaves	Aphrodisiac, astringent, narcotic, placental expulsion, inflammation, uterine disorders, small pox, rheumatic complaints, cholesterol reduction, diarrhea, impotency, menstrual troubles, tuberculosis, malarial fever, chicken pox, leucorrhoea

TABLE 6.3 *(Continued)*

Sl.No.	Botanical Name	Part Used	Medicinal uses
369.	*Dendrophthoe falcata* var. *coccinea*	Leaves and flowers	Giddiness
370.	*Derris benthamii*	Leaves and flowers	Giddiness, leukemia, body heat
371.	*Derris scandens*	Leaves	Scabies
372.	*Desmodium gangeticum*	Leaves, root	Vomiting during pregnancy, urinary diseases
373.	*Desmodium laxiflorum*	Root	Dysentery
374.	*Desmodium oojeinensis*	Bark	menorrhagia
375.	*Desmodium repandum*	Whole plant	Antidote for poisons and scabies
376.	*Desmodium triangulare*	Root	Snake bite
377.	*Desmodium triflorum*	Whole plant	Boils, inflammations, cough
378.	*Desmodium umbellatum*	Whole plant	Fever
379.	*Dichrocepala integrifolia*	Leaves, aerial part	Menstrual disorder, immunity in infants
380.	*Dichrostachys cinerea*	Fiber paste	Vomiting
381.	*Dicliptera paniculata*	Leaves	Bone fracture
382.	*Dictyospermum ovalifolium*	Leaves, tender fruits	Throat swelling
383.	*Didymocarpus tomentosa*	Leaves	Skin allergy
384.	*Dillenia pentagyna*	Bark, fruit	Coolant, fever, cough, eczema, measles
385.	*Dimorphocalyx glabellus* var. *lawianus*	Whole plant, leaves, tender fruits	Skin diseases, giddiness
386.	*Dioscorea bulbifera*	Tuber, leaves	Dysentery, burns and boils
387.	*Dioscorea pentaphylla*	Tubers	Piles, asthma, swelling, general good health, bone fracture
388.	*Dioscorea tomentosa*	Tubers	Bowel complaints

TABLE 6.3   *(Continued)*

Sl.No.	Botanical Name	Part Used	Medicinal uses
389.	*Diospyros ebenum*	Heart- wood, stem	Toothache, blood purifier
390.	*Diospyros malabarica*	Leaves, flowers, tender fruits, fruit	Constipation, bone fracture, dysentery, uterine hemorrhage
391.	*Diospyros melanoxylon*	Leaves, fruit	Edible, rheumatism
392.	*Diospyros montana*	Root, stem-bark, leaves, flowers, fruit,	Paralysis, joint pains, body swelling, skin wounds eczema, skin sore, boil, burns, treating bone fracture
393.	*Diospyros vera*	Leaves, flowers, tender fruits	Body pain, blood clotting piles and scabies
394.	*Diploclisia glaucescens*	Roots, leaves	Herpes, inter trigo, pruritus, scabies, snake bite, leprosy
395.	*Diplocyclos palmatus*	Fruit, seed, root/ tuber	Gynaecological complaints, athelets foot disease, body pain, inflammatory sebaceous cysts, snake bite
396.	*Diplopenta odorata*	Stem, leaves, flowers	Gastric, stomach problems
397.	*Dipteracanthus patulus*	Whole plant	Poisonous bites
398.	*Dodonaea viscosa*	Stem, leaves	Astringent, bone fracture, healing wounds, skin cuts, body pain
399.	*Dolichos trilobus*	Whole plant, leaves	Chest pain, headache, induce abortion, fever
400.	*Dorstenia indica*	Leaves and flowers	Gastric complaints
401.	*Dracaena terniflora*	Root	Bed sore
402.	*Drimia indica*	Bulb	Burns, insect bite
403.	*Drymaria cordata*	Leaves	Cuts and wounds
404.	*Drymoglossum heterophyllum*	Leaves	Body heat, white discharge, conceiving
405.	*Drynaria quericifola*	Rhizome	Paralysis, body pain, typhoid, fever, phthisis, dyspepsia, cough, arthralgia, cephalalgia, diarrhea, ulcers, inflammations, migraine, swellings
406.	*Dryopteris cochleata*	Whole plant, rhizome, root	Snake bite, wound, antibacterial, rheumatism, epilepsy, leprosy, amoebic dysentery
407.	*Duranta erecta*	Whole plant	Skin ailments

TABLE 6.3 *(Continued)*

Sl.No.	Botanical Name	Part Used	Medicinal uses
408.	*Dysphania ambrosioides*	Leaves	Fever, giddiness, gynecological complaints, skin ailments, stomachache, intestinal worms and stomach disorders
409.	*Echinops echinatus*	Root, leaves	skin problems, scorpion sting
410.	*Eclipta prostrata*	Whole plant, stem, Leaves	Dandruff, hair growth, toothache, foot cracks, jaundice, asthma, oedema, fever, rheumatism
411.	*Ehretia microphylla*	Leaves	Swelling of leg
412.	*Elaeagnus conferta*	Bark, Fruits	Whitlow, aphthae, ulcer in Intestine
413.	*Elaeocarpus serratus*	Leaves, fruits	Rheumation, antidote to poison, dysentery, diarrhea, skin allergy
414.	*Elaeocarpus serratus* var. *weibelii*	Leaves, flowers, tender fruits	Gastric complaints
415.	*Elatostema lineolatum*	Leaves, flowers, tender fruit	Body pain, nervous weakness, Gastric complaints, giddiness,
416.	*Eleagnus kologa*	Root, leaves, fruit	Heart pain, fever, evil spirits
417.	*Elephantopus scaber*	Whole plant, root leaves	Blood clot, gynecological complaints, skin wounds, dysentery, stomachache, lactation, healing wounds after delivery, urinary problems like urinary blocks, piles, cuts, mouth ulcers, tooth problem
418.	*Elettaria cardamomum*	Leaves, fruits, seed	Cuts, indigestion, vomiting, stomach problems, epilepsy, asthma, rheumatism, headache
419.	*Eleusine coracana*	Root, seed	Fever, cooling effect, abdominal distension
420.	*Embelia basaal*	Leaves, roots	Reducing body heat, jaundice
421.	*Embelia officinalis*	Fruit	Anaemia, jaundice, diabetes, asthma and bronchitis, mouth ulcer, inflammation, dysentery, blood purifier, cough
422.	*Emilia scabra*	Whole plant, leaves	Blood vomiting, constipation, toothache and gum swelling
423.	*Emilia sonchifolia*	Whole plant, leaves	Boils, body pain, dyspepsia, tonsillitis, night blindness, eye soreness, Bowel complaints

TABLE 6.3   *(Continued)*

Sl.No.	Botanical Name	Part Used	Medicinal uses
424.	*Enicostemma axillare*	Whole plant	Blood pressure
425.	*Ensete superbum*	Inflorescence, seed	Kidney stones and painful urination
426.	*Entada rheedii*	Bark, wood, seed	Ulcers, galactagogue, pain, strengthen the joints in infants
427.	*Equisetum ramosissimum*	Stem, leaves	Stomach disorders in children; fracture and dislocation of bones
428.	*Eranthemum roseum*	Root	Scabies
429.	*Erigeron karwinskianus*	Leaves	Cuts and wounds
430.	*Eriolaena hookeriana*	Stem	Skin cuts
431.	*Erythrina indica*	Leaves,	Stops bleeding, helps in fetal development
432.	*Erythrina stricta*	Root, leaves	Mumps
433.	*Erythrina suberosa*	Root	Wounds
434.	*Erythrina variegata*	Stem-bark, leaves	Giddiness, gynecological complaints, paralytic stroke, rheumatism, skin ailments, eczema, malaria, boils, pruritus, scabies, septic due to skin allergies, to remove placenta after delivery, joint pain
435.	*Erythropalum scandens*	Aerial portion/ tender shoots	Rheumatism, air-borne poisons
436.	*Erythroxylon monogynum*	Bark, flowers, tender fruits	Skin disorder, body pain, throat pain
437.	*Eucalyptus camaldulensis*	Leaves	Headache and common cold.
438.	*Eucalyptus globulus*	Leaves, twig	Rheumatism, Headache, migraine
439.	*Eugenia singampattiana*	Leaves	Rheumatism
440.	*Euonymus dichotomous*	Leaves, flowers	Gastric complaints
441.	*Eupatorium glandulosum*	Leaves	Cuts and wounds
442.	*Eupatorium odoratum*	Leaves	Wound healing

TABLE 6.3   *(Continued)*

Sl.No.	Botanical Name	Part Used	Medicinal uses
443.	*Euphorbia antiquorum*	Roots	Stomachic, digestive, wounds, ulcers, deafness, cough and anti-inflammatory
444.	*Euphorbia cyathophora*	Leaves	Galactogogue constipation
445.	*Euphorbia dracunculoides*	Capsules	Warts
446.	*Euphorbia helioscopia*	Latex	Cuts and wounds
447.	*Euphorbia hirta*	Whole plant, latex, shoots, root, leaves, fruit	Abdominal pain, asthma, bone fracture, bowel complaints, bronchial affection, colic, cough, amoebic dysentery, lactation, other respiratory disorders, purifies blood, ringworm, skin diseases, snake bite, stomach ache, stop local bleeding, warts, wound
448.	*Euphorbia indica*	Whole plant, roots, stem, leaves	Fever, giddiness, urinary blockage, weakness, diarrhea, dysentery and leucorrhoea
449.	*Euphorbia ligularia*	Leaves	Eczema, wound, burning
450.	*Euphorbia nivulia*	Leaves, latex, root	Loose motion, stomach complaints in infants, skin disorders, ear disorders, retention of urine, swelling, worm infection
451.	*Euphorbia rosea*	Leaves, seeds	Vermifuge
452.	*Euphorbia rothiana*	Latex, leaves	Boils, vermifuge
453.	*Euphorbia thymifolia*	Whole plant	Ring worm, wounds, asthma, skin diseases.
454.	*Euphorbia tirucalli*	Stem, twig, milky juice	Rheumatism, headache, toothache, earache, cough, asthma, migraine, dog bite, warts
455.	*Euphorbia tortilis*	Mikely juice	Burns
456.	*Evolvulus alsinoides*	Whole plant, root, flower, leaves, latex	Dysentery, fever, leucorrhoea, wounds, brain tonic
457.	*Evolvulus nummularius*	Whole plant	Fever, cold, body heat, hair care
458.	*Excoecaria oppositifolia* var. *crenulata*	Leaves, flowers	Asphyxia, gastric

TABLE 6.3   *(Continued)*

Sl.No.	Botanical Name	Part Used	Medicinal uses
459.	*Fagraea ceilanica*	Leaves, flowers, tender fruits	Blood vomiting, body heat, jaundice, stomach pain
460.	*Ficus auriculata*	Fruit	Leucoderma
461.	*Ficus benghalensis*	Bark, leaves, latex, prop roots	Bonefracture, eczema, all skin diseases, mouth ulcer, to cracked feet, wound, strengthen teeth, dysmenorrhea, jaundice, leucorrhoea, menorrhagia
462.	*Ficus exasperata*	Roots, bark, leaves, fruit	Fertility, eczema, pruritus, ringworm, scabies, promoting fertility
463.	*Ficus glomerata*	Bark, leaves, unripe fruit	Dysentery, aphthome complaints, menorrhagia, haemoptysis
464.	*Ficus hispida*	Tender fruit	Throat problems, stomach pain, ulcers, leukoderma, psoriasis, anemia, hemorrhoids, jaundice
465.	*Ficus microcarpa*	Stem, latex	Cracked feet
466.	*Ficus racemosa*	Bark, root, leaves, latex, flowers, fruits	Aphthae, aphrodisiac, leucorrhoea, blood disorders, burning sensation, fatigue, urine complaints, leprosy, cuts, eczema, all skin diseases, poison, diarrhea, eczema, cracked feet, dysmenorrhea, fertility, menorrhagia
467.	*Ficus religiosa*	Bark, stem, latex, leaves, fruit	Gynaecological complaints, stuttering, eczema, all skin diseases, pruritus, asthma, heel the fissures on the foot, stomachache, indigestion, breast pain, leucorrhoea
468.	*Ficus retusa*	Root bark, bark, leaves	Liver disease, rheumatic headache, wounds and bruises
469.	*Ficus tinctoria*	Bark, leaves	Eczema, all types of skin diseases
470.	*Ficus virens*	Bark	Bone fracture leucorrhoea
471.	*Flickingeria nodosa*		Stimulant, demulcent, tonic, astringent, aphrodisiac, expectorant, asthma, bronchitis, fever, burning sensation, biliousness, blood diseases, ingredient of jeevanti
472.	*Flueggea leucopyros*	Leaves	Boils, cuts, eczema and wounds
473.	*Flueggea virosa*	Root, stem, bark, leaves, flowers, tender fruits	Joint pain, pain around naval, ringworm infection
474.	*Foeniculum vulgare*	Fruits	Gingival wounds, indigestion

TABLE 6.3 *(Continued)*

Sl.No.	Botanical Name	Part Used	Medicinal uses
475.	*Fragaria vesca*	Whole plant	Demulcent, stimulant, headache
476.	*Galinsoga parviflora*	Leaves	Cuts and wounds
477.	*Ganoderma lucidum*	Fruiting body	Mumps
478.	*Garcinia gummi-gutta*	Fruit, seed	Bilious disorders, crack foot
479.	*Gardenia jasminoides*	Root, flower	Eye infections, gastric problem, indigestion, anti-helminthic
480.	*Gardenia resinifera*	Leaves, flowers, tender fruits	Nervous problem, asthma, leukemia
481.	*Garuga pinnata*	Bark, leaves	Stomachache
482.	*Guazuma ulmifolia*	Whole plant	Nervous disorders
483.	*Getonia floribunda*	Fruit, leaves	Ringworm infection, eczema
484.	*Girardinia diversifolia*	Root	Herpes, pruritus, scabies, snake bite, swollen joints and fever
485.	*Givotia moluccana*	Bark	Breathing problems, body heat, dysentery, vomiting, venereal diseases, jaundice, stomach and urinary problem
486.	*Gladiolus dalenii*	Root	Snake bite
487.	*Glochidion ellipticum*	Whole plant	Cold, nose pain
488.	*Glochidion heyneanum*	Stem	Bone fracture
489.	*Glochidion zeylanicum*	Bark, shoot, fruit	Stomachic Itches, coolant
490.	*Gloriosa superba*	Whole plant, tuber, leaves, fruit	Warts, gynecological complaints, scorpion sting, wounds, induce labor pain, pain at the time of child birth, skin diseases, anti-helminthic, laxative, fever, abortion
491	*Glycomis cochinchinensis*	Root	Tumor
492.	*Glycosmis mauritiana*	Whole plant, leaves, flowers, tender fruit, root	Poisonius bites, bone fracture, body pain, itching, white discharge, skin diseases, nasal blockage, purify blood, toothache

TABLE 6.3   *(Continued)*

Sl.No.	Botanical Name	Part Used	Medicinal uses
493.	*Glycosmis pentaphylla*	Root, leaves, flowers, fruit	Gastric complaints, psoriasis, chicken pox, cough, rheumatism, anemia and jaundice
494.	*Glycyrrhiza glabra*	Root	Cuts, ulcers
495.	*Gmelina arborea*	Bark, leaves, flower, fruit	Fever, head and stomach ache, knee pain, bone strengthening after bone setting
496.	*Gnetum ula*	Stem, seed	Jaundice
497.	*Gnidia glauca*	Stem	Dropsy
498.	*Goniothalamus wightii*	Leaves, tender fruits	Rheumatic pain
499	*Gordonia obtusa*	Leaves	Stimulant
500.	*Gossypium hirsutum*	Leaves	Cut, wound
501.	*Grewia abutilifolia*	Root	Stomachache
502.	*Grewia damine*	Root	Boils
503.	*Grewia hirsuta*	Fruit	Bone fracture
504.	*Grewia orbiculata*	Leaves	Ulcer
505.	*Grewia serrulata*	Root	Bones fracture
506.	*Grewia tiliifolia*	Root, stem-bark, bark fibers, leaves, fruit	Boils, gynecological complaints, dandruff, hair fall, hair growth, wounds due to prickly heat
507.	*Guizotia abyssinica*	Root	Anti-inflammation
508.	*Gymnema hirsutum*	Leaves	Diabetes, fertility, inflammation and jaundice
509.	*Gymnema sylvestre*	Leaves	Wounds, diabetes, paralysis, cardiac stimulant fertility, scorpion sting, snake bite, itching
510.	*Gynura nitida*	Leaves, flowers	Blood clot, body pain, mouth ulcer, gastric, piles, rheumatism, wasp poisons
511.	*Hackelochloa granularis*	Whole plant	Skin wounds
512.	*Haldina cordifolia*	Bark, leaves, flowers	Blood-pressure, piles, stomach pain, anti-fungal and jaundice
513.	*Hedychium flavescens*	Rhizome	Fertility, paralysis

TABLE 6.3   *(Continued)*

Sl.No.	Botanical Name	Part Used	Medicinal uses
514.	*Hedyotis corymbosa*	Whole plant	Fertility
515.	*Helichrysum buddleioides*	Leaves	Cuts and wounds
516.	*Helichrysum hookerianum*	Flowers	Skin diseases
517.	*Helicteres isora*	Roots, stem, bark, leaves, fruits, seeds	Asthma, cold, cough, cut, diabetes, dysentery, dyspepsia, ear problem, earache, gastric complaints, general tonic, giddiness, polio, reduce hair fall, skin cuts, snake bite, stomachache, toothache, wounds
518.	*Heliotropium indicum*	Whole plant	Carcinogenic ulcers
519.	*Heliotropium keralense*	Leaves	Scorpion sting
520.	*Heliotropium ovalifolium*	Inflorescence, leaves	Snake bite
521.	*Helminthostachys zeylanica*	Fronds, rhizome	Sciatica, dystery, catarrah, impotency, whooping cough, phthisis, promotes strength and vitality, spermatorrhoea and improving memory power
522.	*Hemidesmus indicus*	Whole plant, root, fruits, latex	Blood purifier, abdominal pain, fever, eczema, leucorrhoea, diabetes and anemia, stomach ache, stomach disorder, leprosy, eye inflammation, abdominal colic, snake bite, wound maggots, skin diseases, burning sensation, leukoderma, menstrual disorder, burns, fever
523.	*Hemidesmus indicus* var. *pubescens*	Whole plant, root	Cold, mouth refresher, eye sight, tooth gums, mouth ulcer, eye complaints, gynecological complaints, tonic, skin diseases, burning sensation, leucoderma
524.	*Hemigraphis colorata*	Leaves	Wound, eczema
525.	*Hemionitis arifolia*	Aerial parts, leaves	Diabetes, joint pain, dog bite, wound
526.	*Heracleum sprengelianum*	Root	Dysentery
527.	*Heterophragma quadriloculare*	Bark	Sores on toe

TABLE 6.3 *(Continued)*

Sl.No.	Botanical Name	Part Used	Medicinal uses
528.	*Hibiscus lobatus*	Leaves	Anorexia
529.	*Hibiscus micranthus*	Roots, leaves	Body swelling
530.	*Hibiscus rosa-sinensis*	Whole plant, root, leaves, flower, bud, fruit	Blood hemorrhage, boils, burns, dysuria, gynecological complaints, hair growth, diarrhea, eczema, dandruff, bleeding, leucorrhoea, inflammations, cold, cough, wounds, relieves breast pain, strengthening of heart, venereal diseases, stomachache
531.	*Hibiscus talbotii*	Root	Menorrhagia
532.	*Hiptage benghalensis*	Leaves and flowers	Ringworm infection
533.	*Holarrhena pubescens*	Root, bark, latex leaves, seed	Menstrual disorder, burns, eczema, pruritus, hemorrhage, leprosy, worms, thirst, pain, diarrhea, boils, stomachache, aphthae, ulcer in intestine, chronic dysentery, lactation, snake bite, skin diseases
534.	*Holigarna arnottiana*	Fruit	Foot crack
535.	*Holigarna grahamii*	Bark	Bone fracture
536.	*Holoptelea integrifolia*	Bark, tender leaves	Ringworm infection, scabies, herpes simplex, malaria
537.	*Holostemma ada-kodien*	Root	Energiser
538.	*Hugonia serrata*	Leaves, Tender fruits	Rheumatism
539.	*Hybanthus enneaspermus*	Whole plant, Leaves, flower, Fruit	Boils, post partum, relapse, sickness digestive problems, kill worms, leucorrhea, improves potency
540	*Hydnocarpus macrocarpa*	*Tender leaves, seed*	Hair problem, leprosy and other skin problems
541.	*Hydnocarpus pentandra*	Root, bark, leaves, seeds	Eczema, pruritus, hemorrhage, leprosy, worms, thirst, pain, boils, diarrhea, dysentery, stomachache, hemorrhage, leprosy, scabies
542.	*Hygrophila auriculata*	Whole plant, root, leaves, seed	Tonic, sexual debility, cough, wounds, swelling, migraine, headache, anti-inflammatory, jaundice, rheumatism, cough, inter trigo

TABLE 6.3 *(Continued)*

Sl.No.	Botanical Name	Part Used	Medicinal uses
543.	*Hymenodictyon orixense*	Bark	Labor pain
544.	*Hymenophyllum javanicum*	Whole plant	Headache
545.	*Hypericum mysurense*	Leaves	Masticatory
546.	*Hypodematium crenatum*	Rhizome, leaves	Fertility, antibacterial
547.	*Hypolepis glandulifera*	Fronds	Boils
548.	*Hyptis suaveolens*	Whole plant, leaves	Blood clot, gastric, ulcer, asthma, headache, fever, measles, sprain, swelling
549.	*Ichnocarpus frutescens*	Roots	Anaemia, kidney stone, headache, fever, cough, diarrhea, wounds between fingers.
550.	*Impatiens chinensis*	Young leaves	Dysentery
551.	*Impatiens fruticosa*	Leaves, flowers	Joint pain, rheumatism
552.	*Impatiens henslowiana*	Whole plant	Kill worms
553.	*Impatiens scabriuscula* var. *rosea*	Whole plant	Giddiness, leukemia
554.	*Impatiens scapiflora*	Whole plant, leaves, flower	Body pain, nerve problem, allergy, chest pain
555.	*Indigofera cassioides*	Flowers	Stomachache
556.	*Indigofera glandulosa*	Seed	General tonic
557.	*Indigofera longiracemosa*	Leaves	Leprosy, leucoderma, scabies, alopecia
558.	*Indigofera uniflora*	Leaves	Skin diseases
559.	*Indigofera tinctoria*	Whole plant, root, leaves	Stomach pain, kidney stone, hair growth, snake poison, rat poison, pain, sprain, diuretic, jaundice, night blindness, joint pain
560.	*Indigofera trita*	Root	Dysentery, general tonic
561.	*Indigofera wightii*	Root	Stomachache
562.	*Ipomoea aquatica*	Leaves	Eye complaints

TABLE 6.3   *(Continued)*

Sl.No.	Botanical Name	Part Used	Medicinal uses
563.	*Ipomoea barlerioides*	Leaves	Body pain
564.	*Ipomoea batatas*	Leaves	Urinary disorders, expelling thorn from foot
565.	*Ipomoea marginata*	Roots	Leucorrhoea, urinary infection
566.	*Ipomoea mauritiana*	Tubers	Lactation
567.	*Ipomoea nil*	Leaves	Sprain
568.	*Ipomoea obscura*	Leaves, root	Foot cracks, skin cuts, sprain, stomachache
569.	*Ipomoea pes-caprae*	Root	Indigestion
570.	*Ipomoea tridentata*	Leaves	Skin allergy
571.	*Isachne kunthiana*	Aerial parts	Coolant, aperient, chicken pox in children
572.	*Isoetes coromandeliana*	Whole plant	Spleen and liver diseases
573.	*Isonandra lanceolata*	Whole plant, leaves, flowers, tender fruit	Blood purifier, giddiness, nasal blockage, rheumatism, stomach ache, throat pain, piles,
574.	*Ixora arborea*	Root, bark	Leucoderma
575.	*Ixora coccinea*	Flower, fruit	Eczema, snake bite, aphthae
576.	*Ixora nigricans*	Leaves, flowers, tender fruit	Giddiness, leukemia, body pain, gastric, headache, kill worms, itching, body hydration
577.	*Jacaranda mimosifolia*	Stem-bark	Gonorrhea
578.	*Janakia arayalpathra*	Tuberous root	Peptic ulcers, related afflictions, coolant, blood purifier
579.	*Jasminum auriculatum*	Root, stem, leaves	Bone fractures
580.	*Jasminum angustifolium*	Leaves, tender twig	Stomach ulcers, good eye sight
581.	*Jasminum bignoniaceum*	Leaves	Skin allergy, paralysis
582.	*Jasminum grandiflorum*	Leaves	Gingivitis, ringworm infection, nasal bleeding, boils, eczema

TABLE 6.3 *(Continued)*

Sl.No.	Botanical Name	Part Used	Medicinal uses
583.	*Jasminum malabaricum*	Root, stem, leaves	Aphthae, diabetes, eye complaints
584.	*Jatropha curcas*	Root, whole plant, young twigs, stem latex, inner bark latex, leaves, fruit, seed	Boils, eye complaints, fever, cuts, wounds, eczema toothache, wormicide, ulcers, tumors, scabies skin diseases, cold and fever (for children), sore throat, cough, whitlow, scabies, tooth brush in toothache, wounds, chronic dysentery, malaria, leucorrhoea
585.	*Jatropha glandulifera*	Leaves, stem	Toothache, tooth gums, bleeding wounds and cuts ulcers
586.	*Jatropha gossypifolia*	Stem bark, leaves, seeds, roots	Stomachic, fever, kidney troubles, liver bladder diseases, diabetes, leprosy, rheumatic pain, promotes menstruation, irregular periods, boils, eczema, itches, foul odor, toothache, wounds and tongue
587.	*Justicia adhatoda*	Whole plant, root stem, leaves	Anti spasmodic, asthma, blood vomiting, body pain, cause abortion, cold, cough, cuts, diarrhea, ear ache, headache, induce menstrual flow, respiratory stimulant, rheumatic pain and scabies
588.	*Justicia beddomei*	Leaves	Asthma
589.	*Justicia gendarussa*	Leaves	Rheumatism
590.	*Justicia tranquebariensis*	Leaves	Intestinal ulcers
591.	*Kaempferia galanga*	Rhizome	Leucoderma, corns, headache, cough, nasal bleeding, pectorial affection
592.	*Kalanchoe laciniata*	Leaves	Applied on wounds, coolant
593.	*Kalanchoe pinnata*	Leaves	Cut, eczema, pruritus, dysentery, cholera, coolant
594.	*Kingiodendron pinnatum*	Resin	Joint pain, fissured feet
595.	*Kleinhovia hospita*	Leaves, flowers	Sprain, rheumatism
596.	*Kleinia grandiflora*	Leaves	Giddiness

TABLE 6.3   *(Continued)*

Sl.No.	Botanical Name	Part Used	Medicinal uses
597.	*Klugia notoniana*	Aerial part, root	Body swelling, polio
598.	*Knoxia sumatrensis*	Leaves, flowers	Asthma, blood clot, body weakness
599.	*Knoxia wightiana*	Whole plant	Digestive problems
600.	*Kyllinga nemoralis*	Wholeplant, rhizome	Asthma, worm infection
601.	*Lobelia excelsa*	Leaves	Masticatory, insecticide
602.	*Lablab purpureus*	Leaves, pod	Irritation, milk secretion
603.	*Lagerstroemia parviflora*	Leaves	Throat problems
604.	*Lannea coromandelica*	Bark	Bone fracture, wounds, stomach pain
605.	*Lantana camara*	Whole plant, leaves, flowers, fruit	Burns, coolant, cuts, fever, insect bite, worms, loose motion, malaria, rheumatism, ringworms, tetanus, tonic, abdominal viscera, wounds
606.	*Lantana camara* var. *aculeata*	Leaves	Fever, headache
607.	*Laportia terminalis*	Root	Diabetes
608.	*Launaea procumbens*	Leaves	Abdominal colic
609.	*Lawsonia inermis*	Leaves, flowers	Athelets foot disease, sore throat, bronchitis, burns, cracked heels, felon, hair growth, headache, immunity in children, insomnia, leprosy, mouth ulcers, rheumatism, ringworm infection, skin diseases
610.	*Leea asiatica*	Root	Burns
611.	*Leea indica*	Root, tender shoots, stem, leaves, flowers	Body hydration, rheumatism, snake bite, cuts, wounds, diarrhea
612.	*Leea macrophylla*	Root, tender stem	Expel worms
613.	*Lepianthes umbellata*	Leaves	Wasp bites
614.	*Lepidagathis cuspidata*	Root	Tooth problem

TABLE 6.3 *(Continued)*

Sl.No.	Botanical Name	Part Used	Medicinal uses
615.	*Leucas aspera*	Whole plant, root, leaves, flowers	Anthelmintic, bronchial disease, chickenpox, dog-bite, cold, fever, color blindness, cough in children, dhobis itch, dysentery, dysuria, epistaxys, eye complaints, fever, headache, mumps, piles, poison, ringworm, skin allergy, skin cuts, skin diseases, smallpox, snake bite, toothache, worm infestation, wounds and typhoid
616.	*Leucaena leucocephala*	Bark	Bone fracture
617.	*Leucas cephalotes*	Whole plant	Headache, jaundice, itching, cold, dermatitis
618.	*Leucas ciliata*	Whole plant	Dysentery
619.	*Leucas indica*	Whole plant	Scabies
620.	*Leucas vestita*	Leaves, flowers	Throat, skin infection
621.	*Leucostegia immerse*	Rhizome	Boils, constipation, antibacterial
622.	*Ligustrum perrottetii*	Tender leaves, flowers	Gastric, cough, blood purifier, weakness, paralysis and skin diseases
623.	*Limonia acidissima*	Fruit	Dyspepsia, breathing troubles
624.	*Lindernia caespitosa*	Whole plant	Body sprain, gastric
625.	*Lippia nodiflora*	Whole plant	Dandruff, skin infection on scalp, leucorrhoea
626.	*Litsea deccanensis*	Leaves, Flowers	Nasal blockage
627.	*Lobelia nicotianifolia*	Whole plant, root, stem bark, leaves, seed	Boils, aches, toothache, wounds, tineapedis, antispasmodic, scorpion sting, Antiseptic and asthma
628.	*Loranthus* sp.	Whole plant	Whitlow
629.	*Loranthus longiflorus*	Bark	Astringent, wounds, menstrual troubles, asthma and mania
630.	*Ludwigia adscendens*	Aerial part	Wounds
631.	*Ludwigia octovalvis*	Leaves	Skin ailments
632.	*Luffa acutangula*	Fruit, seeds	All skin diseases, snake bite
633.	*Luffa cylindrica*	Leaves	Hemorrhoids, leprosy

TABLE 6.3   *(Continued)*

Sl.No.	Botanical Name	Part Used	Medicinal uses
634.	*Luisia zeylanica*	Pseudobulb	Emollient, boils, abscesses, tumors
635.	*Lygodium flexuosum*	Rhizome	Rheumatism, sprains, scabies, ulcers, eczema, coughs, gonorrhea, herpes, jaundice, fever
636.	*Lygodium microphyllum*	Leaves	Dysentery, skin diseases, swelling, hiccough
637.	*Macaranga indica*	Gum	Antiseptic, applied on the sores
638.	*Macaranga peltata*	Bark, leaves, gum/latex	Boils, cuts, sores, antiseptic
639.	*Macrotyloma uniflorum*	Leaves, seed	Tooth problem
640.	*Madhuca indica*	Bark, leaves	Aphthae, itching, bed sore, ulcers in intestine
641.	*Madhuca insignis*	Fruit, seed	Relieves body and burning sensation of eyes
642.	*Madhuca longifolia*	Bark, leaves, seed	Skin diseases, delivery pain, Bone fracture, foot cracks, jaundice, joint pain, muscular pain
643.	*Medinilla beddomei*	Leaves	Fever
644.	*Maerua oblongifolia*	Leaves	Asthma
645.	*Maesa indica*	Fruit, whole plant	Body heat, skin infection
646.	*Magnolia champaca*	Root, bark, flower, fruits, seed	Malaria, rheumatism, ophthalmia, headache, stomach problem, flatulence, menstrual disorders, abscesses, gonorrhea, eye pain, cracked foot, leech repellent,
647.	*Malachra capitata*	Whole plant, root	Wounds, urinary diseases
648.	*Mallotus atrovirens*	Leaves, flowers, tender fruits	Stomachache, ulcer
649.	*Mallotus aureopunctatus*	Leaves, flowers	Giddiness, itching, leukemia, white discharge, gastric
650.	*Mallotus beddomei*	Leaves, flowers	Skin diseases
651.	*Mallotus philippensis*	Whole plant, bark, leaves, flower, tender fruits, seed	Expelling tapeworms, nervous weakness, rheumatism, throat pain, blood discharge, gastric, piles, mouth ulcers, antiseptic skin affection, muscular pain, dysentery, stomachache

TABLE 6.3 *(Continued)*

Sl.No.	Botanical Name	Part Used	Medicinal uses
652.	*Mallotus resinosus*	Whole plant, leaves, flowers, tender fruits	Blood purify, body heat, syphilis, mouth ulcer, giddiness, vomiting
653.	*Mallotus tetracoccus*	Leaves	Fever
654.	*Malvastrum coromandelianum*	Roots, leaves	Menstrual pain, wounds
655.	*Mammea suriga*	Bark, fruit bud	Eczema, wound, tick bite
656.	*Mangifera indica*	Root bark, stem bark, leaves, fruit, seed	Bleeding, cut, dysentery, diabetes, fertility, antihelminthic, pruritus, scabies, septic due to skin allergies, rheumatism, dysentery, stomachache, wound, body pain, delivery pain
657.	*Manilkara hexandra*	Bark	Hip pains
658.	*Maranta arundinacea*	Rhizome, leaves	Asthma, infants loose motion, urinary complaint
659.	*Maranta virgata*	Rhizome	Stomach pain
660.	*Marattia fraxinea*	Whole plant	Ancylostomiasis
661.	*Marselia minuta*	Whole plant	Cough, spastic condition of leg and muscles, psychopathy, opthalmia, strangury, diarrhea, leprosy, skin diseases, hemorrhoids, dyspepsia, fever
662.	*Martynia annua*	Whole plant, root, tender fruit, seed	Cold, deworming, snake posion, skin disease, tooth problem
663.	*Medinilla beddomei*	Leaves	Body heat, giddiness
664.	*Melia azedarach*	Leaves, stem, root bark, leaves	Stomachache, body pain, improve fertility
665.	*Melia dubia*	Bark, berries	Leprosy, intestinal worms, stomachache
666.	*Meliosma simplicifolia*	Wood and leaves	Skin allergy
667.	*Melothria maderaspatana*	Leaves	Cold, cough and fever
668.	*Memecylon lushingtonii*	Leaves, flowers, tender fruits	Leg sores, rheumatism, blood discharge
669.	*Memecylon rivulare*	Leaves	Vermifuge

TABLE 6.3   *(Continued)*

Sl.No.	Botanical Name	Part Used	Medicinal uses
670.	*Memecylon terminale*	Leaves, flowers	Deworming
671.	*Memecylon umbellatum*	Leaves	Snake bite
672.	*Mentha arvensis*	Whole plant	Paralysis
673.	*Merremia hastata*	Whole plant	Hairgrowth
674.	*Merremia tridendata*	Entire plant, leaves	Dandruff, hair growth, leprosy, various skin infections in children
675.	*Mesua ferrea*	Root, bark, Leaves, flowers, flower buds, fruits, seed	Cold, cough, blood vomiting, blood discharge, giddiness
676.	*Mesua ferrea* subsp. *pulchella*	Leaves, flowers	Gastric, nervous, sprain
677.	*Mesua ferrea* subsp. *pulchella var. coromandeliana*	Flowers, seeds	Sores, rheumatism, stomachache
678.	*Meteoromyrtus wynaadensis*	Whole plant, leaves	Eye, skin disease, throat infection
679.	*Meyna laxiflora*	Leaves	Fever, mumps
680.	*Meyna spinosa*	Seed, stem	Headache, scabies
681.	*Microcos paniculata*	Leaves, flowers	Cough, cold
682.	*Micromeria biflora*	Whole plant	Post-natal treatment
683.	*Microsorum punctatum*	Leaves	Wounds, purgative, diuretic
684.	*Microstachys chamaelea*	Leaves	Diarrhea
685.	*Miliusa eriocarpa*	Leaves, tender fruits	Piles
686.	*Mimosa pudica*	Whole plant, root, leaves, flowers	Asthma, boils, cold, constipation, cracks on foot, dermatitis, diabetes, dysentery, eczema, gingival wounds, goiter, gynecological complaints, hair complaints, headache, inflammations, kidney problem, laxative, liver disease, piles, pimples, psoriasis, ringworm, scorpion sting, skin cuts, skin infection and wounds

TABLE 6.3 *(Continued)*

Sl.No.	Botanical Name	Part Used	Medicinal uses
687.	*Mimusops elengi*	Flower, fruit, seed	Body glow
688.	*Mirabilis jalapa*	Root, flowers, leaves, seeds	Boils, otorrhoea, wounds, spasmolytic, cysts, inflammatory viral infections
689.	*Mitragyna parvifolia*	Fruit, stem-bark	Rheumatism, snake bite, menorrhagia
690.	*Mitrephora heyneana*	Leaves, flowers	Body heat, giddiness
691.	*Mollugo cerviana*	Whole plant	Cold and cough
692.	*Mollugo nudicaulis*	Whole plant, leaves	Fever, cough, headache, stomachache, venereal diseases
693.	*Mollugo pentaphylla*	Whole plant	Giddiness, coolant
694.	*Momordica charantia*	Leaves, fruit	Anaemia, boils, cut, chest pain, diabetes, gynecological complaints, leprosy, skin diseases, dhobis itch, ringworm, vermifuge, asthma, wheezing and wounds
695.	*Momordica dioica*	Root, fruit	Boils, stomachache
696.	*Morinda pubescens*	Leaves, bark	Rheumatism, indigestion, wounds, jaundice
697.	*Morinda umbellata*	Leaves, flowers	Urinary blockage
698.	*Moringa concanensis*	Bark, leaves, fruits	Bowel disorders, fever, de-worming, dysentery, blood purifier, paralysis, loose motion, vomiting, stomach pain
699.	*Moringa oleifera*	Stem bark, young leaves, leaves, flower, fruits	Aphrodisiac, gynecological complaints, toothache, inflammation, piles, sprain, constipation, anemia, rheumatic swelling, reduce blood pressure, diabetes mellitus, sprain and bone fracture
700.	*Morus alba*	Aerial part	Diabetes and vermifuge
701.	*Moullava spicata*	Root	Skin ailments
702.	*Mucuna atropurpurea*	Leaves, fruit	Swelling and itching
703.	*Mucuna pruriens* var. *pruriens*	Seed	Scorpion sting, impotency, urinary diseases, intestinal worms, cut

TABLE 6.3 *(Continued)*

Sl.No.	Botanical Name	Part Used	Medicinal uses
704.	*Mukia maderaspatana*	Leaves, fruit	Giddiness paronychia, cold and cough, reduce bile problem, body strength
705.	*Mundulea sericea*	Leaves, flowers	Breathing, giddiness, earache
706.	*Murdannia esculenta*	Whole plant	Blood purify, vomiting
707.	*Murdannia lanceolata*	Whole plant	Cool body
708.	*Murraya koenigii*	Root, bark, leaves, fruit	Sickness, gynecological complaints, epilepsy, malaria, stomachache, hair growth, delay graying of hair, morning sickness, stomach upsets, insect bites, dog bite, skin eruption, de-worming, eye pain, eye coolant, dandruff and hair loss
709.	*Murraya paniculata*	Leaves, fruit	Gastric problems, hemicrania, oedema, rheumatic pain
710.	*Musa paradisiaca*	Sucker, stem, leaves, fruit	Boils, gingival wounds, gynecological complaints, menstrual complaints, worms, kidney stones, renal colic, excess hair loss
711.	*Mussaenda frondosa*	Leaves	Gangrene, dandruff, eye diseases
712.	*Mussaenda glabrata*	Whole plant, root, stem, leaves, flowers	Scorpion sting, heal wounds, toothache, cooling effects, eye complaints, hair growth, dandruff, skin allergy, inflammation
713.	*Myriactis wightii*	Leaves	Wound healing
714.	*Myristica fragrans*	Leaves, fruits	Aphthae, blood vomiting, ulcers in intestine, scabies
715.	*Myxopyrum serrulatum*	Leaves	Skin rashes, scabies, ringworm infections
716.	*Naravelia zeylanica*	Whole plant, root, stem, leaves	Sores in buccal cavity, headache, wounds
717.	*Naregamia alata*	Whole plant, rhizome	Eczema, pruritus, scabies, dysentery, contagious skin diseases
718.	*Naringi crenulata*	Leaves	Aphthae, gangrene, leg pains
719.	*Neanotis decipiens*	Stem, leaves, flowers	Leg pain, rheumatism
720.	*Neanotis indica* var. *affinis*	Whole plant, leaves, flowers	Gastric, edema

TABLE 6.3 *(Continued)*

Sl.No.	Botanical Name	Part Used	Medicinal uses
721.	*Nelumbo nucifera*	Tuber, leaves, flower	Dropsy, boils, piles
722.	*Neolamarckia cadamba*	Bark	Measles, snake bite, diabetes, cough, musculoskeletal disorder, fever, anesthetic
723.	*Neolitsea scrobiculata*	Leaves, flowers, tender flowers, tender fruits	Gastric, nasal blockage, back itching, body pain, sprain
724.	*Nephrolepis cordifolia*	Rhizome, pinnae	Antibacterial, antifungal, antitussive, cough, rheumatism, chest congestion, nose blockage, loss of appetites, styptic, coughs, wounds, jaundice
725.	*Nerium indicum*	Root	Leprosy
726.	*Nerium oleander*	Latex, leaves	Muscle pain, speech
727.	*Nervilia aragoana*	Whole plant, leaves	Mental diseases, cough, vomiting, wound
728.	*Netholaena standleyi*	Whole plant	Post natal problems
729.	*Nicandra physalodes*	Leaves	Cuts and wounds
730.	*Nicotiana tabacum*	Leaves	Skin cut, toothache.
731.	*Nothapodytes foetida*	Bark, root, leaves	Fever, dog-bite and snake bite
732.	*Nothopegia heyneana*	Leaves, flowers, tender fruits	Urinary problems, body heat, blood pressure, sprain, body pain, itching, bone fracture
733.	*Nyctanthes arbor-tristis*	Root	Bone fracture
734.	*Ocimum americanum*	Whole plant, fruit, leaves	Eye and hair complaints insect bite, dandruff, dark pigmentation on skin, dyspepsia, dental care, cough, fever, chest pain and indigestion
735.	*Ocimum basilicum*	Leaves, flowers, seed	Body pain, cold, coolant, headache, bone fracture
736.	*Ocimum gratissimum*	Leaves, flowers	Body pain, gastric problem, infantile cough, cold, catarrh, dysentery, skin diseases, muscle pain

TABLE 6.3 *(Continued)*

Sl.No.	Botanical Name	Part Used	Medicinal uses
737.	*Ocimum tenuiflorum*	Whole plant, leaves	Antiviral, cold, cough, cut, dermatitis, dyspepsia, ear pain, fever, gingival wounds, headache, insect bite, jaundice, lymphogranuloma venereum, mumps, poison, rheumatism, ringworm infection, scabies, skin diseases, toothache
738.	*Odontosoria chinensis*	Leaves	Chronic enteritis
739.	*Oleandera musifera*	Stipe, rhizome	Emmenagogue, snake bite, anthelmintic
740.	*Oldenlandia auriculata*	Whole plant	Cholera
741.	*Oldenlandia puberula*	Leaves	Eye pain and infection
742.	*Oldenlandia umbellata*	Whole plant	Leucoderma
743.	*Oligochaeta ramosa*	Whole plant, leaves, stem	Rheumatism, sprain, cut wounds, cold, cough
744.	*Ophioglosum gramineum*	Fronds, rhizome	Angina, contusions, wounds, hemorrhages, boils
745.	*Ophioglossum reticulatum*	Whole plant	Inflammation, wounds, contusions and hemorrhages
746.	*Ophiorrhiza mungos*	Whole plant	Skin infection, cancer, ulcer, venomous snake bites and mad dog bites
747.	*Opuntia dillenii*	Fruit, sucker, stem	Whitlow, wounds
748.	*Oreocnide integrifolia*	Leaves, flowers	Rheumatism
749.	*Orophea uniflora*	Whole plant	Post delivery problems
750.	*Oroxylum indicum*	Bark	Coolant, jaundice, joint pain, muscular pain, diarrhea, fever, ulcer
751.	*Orthosiphon thymiflorus*	Whole plant, leaves	Body pain, skin cuts
752.	*Osbeckia aspera*	Leaves, flowers	Body pain, joint pain, gastric, piles, blood clot, rheumatism, skin diseases, blood discharge
753.	*Osbeckia minor*	Leaves, flowers	Cough
754.	*Osbeckia zeylanica*	Whole plant, leaves, flowers	Cold, cough, itching, body pain, chest pain

TABLE 6.3   *(Continued)*

Sl.No.	Botanical Name	Part Used	Medicinal uses
755.	*Osmunda hugeliana*	Fronds	Rickets, rheumatism, intestinal problems
756.	*Osyris lanceolata*	Leaves	Scorpion sting
757.	*Oxalis corniculata*	Whole plant, leaves, fruits	Dysentery, diarrhea, headache dyspepsia, hemorrhoids, syphilis, pregnancy, piles, fever, wart
758.	*Oxalis latifolia*	Whole plant, tuber, leaves	Paralysis
759.	*Pachygone ovata*	Seeds	Snake bite
760.	*Pandanus fascicularis*	Floral bracts, young twig, leaves	Earache, jaundice
761.	*Panicum vulgare*	Seed	Body energiser
762.	*Paracalyx scariosus*	Root	Stomach pain
763.	*Paramignya monophylla*	Root	Boils
764.	*Parthenium hysterophorus*	Leaves	Cuts and burns
765.	*Paspalum scrobiculatum*	Root, leaves	Pain during delivery
766.	*Passiflora edulis*	Flower, fruit	Headache
767.	*Passiflora foetida*	Leaves, fruit	Cold, cough, breathing problem, diabetes, eczema
768.	*Pavetta blanda*	Leaves, flowers	Urinary disorder, rheumatism
769.	*Pavetta hispidula*	Leaves, tender fruits	Giddiness, leukemia
770.	*Pavetta indica*	Roots, leaves, fruits	Skin allergy, fertility, eczema
771.	*Pavonia odorata*	Whole plant	Fever and cough
772.	*Pennisetum pedicellatum*	Root	To prevent abortion
773.	*Peperomia pellucida*	Leaves, stem	Wounds
774.	*Pergularia daemia*	Roots, leaves	Fever, hair complaints, itching, eczema, cough, chest pain, joint pain, worms, gas problem, wounds, urinary complaints
775.	*Persea macrantha*	Stem/Bark	Bone fracture

TABLE 6.3   *(Continued)*

Sl.No.	Botanical Name	Part Used	Medicinal uses
776.	*Petiveria alliacea*	Root	Cold and fever
777.	*Phaseolus* spp.	Leaves	Dysentery
778.	*Phlebodium aureum*	Rhizome	Cough, fever, sudorific
779.	*Phoenix loureroi* var. *humilis*	Leaves	Dog bite
780.	*Phoenix loureroi*	Stem, tender leaves, fruits	Body heat
781.	*Phoenix sylvestris*	Fruit	Aphrodisiac, body strength
782.	*Pholidota imbricata*	Pseudobulbs	Spine problems and inflammation
783.	*Phyllanthus amarus*	Whole plant, root, leaves	Jaundice, skin ailments, anemia, dandruff, chronic dysentery, fever
784.	*Phyllanthus baillonianus*	Whole plant	Jaundice, cure blood, piles
785.	*Phyllanthus debilis*	Leaves	Jaundice
786.	*Phyllanthus emblica*	Root bark, leaves, fruit	Anaemia, asthma, bleeding, body strength, bronchitis, coolant, cardiac disorders, cough, dandruff, diabetes, diarrhea, dysentery, dyspepsia, grayness of hairs, gynecological complaints, hemorrhage disorders, inflammation, jaundice, leprosy, mouth ulcers, peptic ulcers, skin diseases, snake bite, stomachache, tonic, oral problems, worms in gums and teeth, worms indigestive tract
787.	*Phyllanthus fraternus*	Whole plant, leaves	Dizziness, jaundice, ulcers, diarrhea, dysentery, intermittent fever, urino-genital diseases, scabies, wounds
788.	*Phyllanthus gardnerianus*	Whole plant	Giddiness, headache, body pain, cold, fever
789.	*Phyllanthus maderaspatensis*	Leaves, seeds	Infusion in headache, carminative, diuretic
790.	*Phyllanthus polyphyllus*	Whole plant	Rheumatism
791.	*Phyllanthus reticulatus*	Whole plant, leaves, fruits	Burning sensation, fever, sores, diarrhea, headache and skin eruptions
792.	*Phyllanthus rheedei*	Whole plant	Stomach disorders

TABLE 6.3 *(Continued)*

Sl.No.	Botanical Name	Part Used	Medicinal uses
793.	*Phyllanthus singampattiana*	Leaves	Cold and cough
794.	*Phyllanthus urinaria*	Whole plant, leaves	Substitute for *Phyllanthus amarus*
795.	*Phyllanthus virgatus*	Whole plant	Bleeding, diabetes, renal infections
796.	*Phyllocephalum phyllolaenum*	Stem	Throat problems
797.	*Physalis angulata*	Whole plant, leaves	Dog-bite, psoriasis, inflammation
798.	*Physalis minima*	Whole plant, leaves, fruit	Constipation, ulcers, cough, venereal diseases, galactagogue
799.	*Physalis peruviana*	Roots, fruit	Vomiting
800.	*Piliostigma foveolata*	Bark	Stomach pain
801.	*Pimpinella heyneana*	Whole plant, seed	Gum swellings, stomach pain
802.	*Piper attenuatum*	Seed, leaves, tender fruit	Cough, gastric
803.	*Piper betle*	Leaves	Cuts, dandruff, itching, skin irritation
804.	*Piper longum*	Fruit, root, stem	Diarrhea, cholera, scarlatina, chronic malaria, viral hepatitis, otorrhoea, bronchitis, cough & cold, fever
805.	*Piper nigrum*	Roots, leaves fruit, seeds	Aphthae, arthritic diseases; bed sore, body pain, boils, bronchitis, cholera, cold, coma, cough, crack foot, cuts, dyspepsia, eczema, fever, dandruff, flatulence, gangrene, gingival wounds, hair complaints, indigestion, malarial fever, otorrhoea, paraplegia, piles, pruritus, psoriasis, rheumatic pain, rubefacient, scabies, skin diseases, snake bites, sore throat, ulcers in intestine, vertigo
806.	*Piper umbellatum*	Leaves	Insect bites
807.	*Pithecellobium dulce*	Bark	Scorpion sting
808.	*Pityrogramma calomelanos*	Frond, rhizome	Renal disorders, hypertensions, fever, cough, boils in mouth and nose, asthma, cold, chest congestion
809.	*Pittosporum tetraspermum*	Root bark	Snake bite

TABLE 6.3   *(Continued)*

Sl.No.	Botanical Name	Part Used	Medicinal uses
810.	*Plantago erosa*	Leaves	Muscle pain
811.	*Plantago ovata*	Seed	To prevent abortion
812.	*Plectranthes malabarica*	Aerial part	Fever
813.	*Plectranthus amboinicus*	Whole plant, leaves	Cold, cuts, cough, fever, headache, bed sore, giddiness, wound healing, nasal congestion
814.	*Plectranthus mollis*	Root	Tonic
815.	*Pleiospermium alatum*	Bark, leaves	Chest pains, shoulder, rheumatism
816.	*Pleopeltis macrocarpa*	Fronds, rhizome	Cold, sore throat, itches, cough
817.	*Pluchea indica*	Flower and seeds	Coolant
818.	*Plumbago indica*	Root	Eczema, appetizer, stomach pain, vitality, skin diseases
819.	*Plumbago zeylanica*	Root, bark, latex	Skin allergy, itching, rheumatism, wounds, eczema, snake bite, sore-wounds, breast cancer, jaundice, eye diseases, poisonous insect bites, joint pain, measles, stomach pain
820	*Plumeria acuminata*	Root	Birth control
821.	*Plumeria rubra*	Flower, fruit	Gonorrhea, anorexia, spasm scabies
822.	*Poeciloneuron pauciflorum*	Bark	Infectious diseases, mental disorders, exorcism activities
823.	*Pogonatherum crinitum*	Whole plant	Urinary blockage, gastric
824.	*Pogostemon benghalensis*	Leaves	Wounds, fever, stomach ache
825.	*Polyalthia longifolia*	Stem bark	Gynaecological complaints, fever, skin, diabetes, wounds
826.	*Polygala chinensis*	Root	Poisonous bites
827.	*Polygala elongata*	Root	Fever
828.	*Polygala javana*	Leaves	To reduce pain after breast feeding is stopped

TABLE 6.3 *(Continued)*

Sl.No.	Botanical Name	Part Used	Medicinal uses
829.	*Polygala sibirica*	Whole plant	Urinary infection
830.	*Polygonum barbatum*	Root	Coolant
831.	*Polygonum chinensis*	Whole plant, Leaves, fruit, young stem	Aphthae, paralysis, giddiness, quenching thirst
832.	*Polygonum glabrum*	Root, young shoot	Piles, jaundice, constipation
833.	*Polygonum molli folium*	Root	Vomiting, fever, tuberculosis
834.	*Polygonum ple-beium* var.*indica*	Leaves	All skin diseases
835.	*Polystichum squarrosum*	Sporophylls	Antibacterial
836.	*Pongamia pinnata*	Roots, stem bark, heart wood, leaves, flowers, seed	Bodyache, bone fracture, cooling effects, cuts, diabetes, digestion troubles, eczema, giddiness, herpes, itching, piles, prevention of conception, pruritus, psoriasis, body pain after delivery, ringworm infection, scabies, scratches of elephants, skin cuts, skin diseases, snake bite, wounds
837.	*Pothos scandens*	Whole plant	Bone fracture
838.	*Pouzolzia auriculata*	Whole plant	Cold, headache, chest pain, body pain, vomiting
839.	*Pouzolzia bennettiana*	Stem bark	Hair nourisher, coolant to the eyes
840.	*Pouzolzia indica*	Whole plant	Sprains
841.	*Pouzolzia wightii*	Whole plant	Sores
842.	*Premna latifolia*	Stems, leaves, flowers, tender fruits	Rheumatism
843.	*Premna serratifolia*	Leaves	Wound healing, boils
844.	*Premna tomentosa*	Bark, leaves	Paralysis, antirheumatic agent, dropsy, stomach disorders, diarrhea
845.	*Prosopis cineraria*	Bark, leaves	Snake bite, poisonous bite
846.	*Prunus persica*	Leaves twig	Stomach pain

TABLE 6.3   *(Continued)*

Sl.No.	Botanical Name	Part Used	Medicinal uses
847.	*Pseudarthria viscida*	Leaves	Lactation
848.	*Psidium guajava*	Root, leaves, fruits, tender fruits	Constipation, diarrhea, dysentery, wound healing, oedema, uterine hemorrhage, vomiting, snake bite, food poisoning, renal diseases, ulcers, loose motion
849.	*Psilotum nudum*	Spores	Diarrhea, purgative, antibacterial
850.	*Psychotria bisulcata*	Whole plant	Gastric problem
851	*Psychotria dalzellii*	Root	Pruritus, scorpion sting
852.	*Psychotria flavida*	Leaves, flowers, root	Throat pain, whooping cough, snake bite
853.	*Psychotria nilgiriensis*	Tender fruits	Rheumatism
854.	*Psychotria nudiflora*	Whole plant	Cure syphilis, rheumatism, giddiness, leukemia
855.	*Psychotria subintegra*	Leaves, flowers	Joint pain
856.	*Psydrax didoccos*	Bark, leaves	All type of fevers
857.	*Psydrax umbellatum*	Whole plant	Giddiness, leukemia, gastric complaints
858.	*Pteris cretica*	Fronds	Antibacterial, wounds
859.	*Pteris quadriaurita*	Rhizome	Pus, boils
860.	*Pteris vittata*	Whole plant	Demulcent, hypotensive, tonic, antiviral, antibacterial
861.	*Pterocarpus marsupium*	Stem bark, stem, heart wood, sap, resin	Paralysis, fertility, chest pain, diabetes, fever, dysentery, stomachache, body pain, diabetes, tooth gums, toothache, muscular relaxation, ulcers, joint pain, jaundice
862.	*Pterocarpus marsupium* var. *acuminatus*	Gum, wood	Menorrhagia, blood purifier, skin diseases
863.	*Pterocarpus santalinus*	Heartwood	Poisonous affections
864.	*Pterolobium*	*hexapetalumn* Leaves	Body pain after delivery

TABLE 6.3 *(Continued)*

Sl.No.	Botanical Name	Part Used	Medicinal uses
865.	*Pterospermum canescens*	Leaves	Fracture and inflammation
866.	*Punica granatum*	Stem bark, leaves, bud, flower, fruits, seeds	Anaemia, bronchitis, antipyretic, vomiting, stomach disorders, dysuria, dysentery, worm infections, diarrhea
867.	*Putranjiva roxburghii*	Leaves, seeds	Cold, fever, abortion, sterility, burning sensation
868.	*Pyrrosia heterophylla*	Whole plant	Swelling, sprains, pain
869.	*Pyrossia lanceolata*	Fronds	Cold, sore throat, itch guard
870.	*Pyrossia strictus*	Aerial part	Asthma, hernia
871.	*Pyrus communis*	Leaves	Dysentery
872.	*Quercus infectoria*	Gall	Dental problems
873.	*Radermachera xylocarpa*	Stem bark, fruit	Skin burns, bones fracture, jaundice, menorrhagia, snake bite, dysentery, stomach pain, urinary complaints
874.	*Raphidophora pertusa*	Whole Plant	Swellings in groin joints
875.	*Rauvolfia densiflora*	Leaves and flower	Rheumatic complaints
876.	*Rauvolfia serpentina*	Roots, leaves, flowers, stem	Blood pressure, expel worm from the body, gastric, headache, herpes, insect bite, intestinal disorders, mental disorders, mind tension, otorrhoea, poisonous bite, pruritus, psoriasis, rheumatism, scabies, scorpion sting, snake bite, sore, stomachache, toothache
877.	*Rauvolfia tetraphylla*	Whole plant	Skin diseases, snake bite
878.	*Rauvolfia verticillata*	Leaves, flowers	Rheumatism, sprain
879.	*Remusatia vivipara*	Root, leaves	Alexipharmic, cytotoxic, anti-tumor, wounds, breast tumor, itching & soil borne diseases (skin diseases), leucoderma, athelets foot disease, ringworm infection
880.	*Rhaphidophora pertusa*	Tuber	Skin diseases, intestinal ulcers, body inflammation, otorrhoea
882.	*Rhinacanthus nasutus*	Whole plant, leaves	Burn, blood disorder, poisoning, ring worm infection

TABLE 6.3   *(Continued)*

Sl.No.	Botanical Name	Part Used	Medicinal uses
883.	*Rhodomyrtus*	Stem, fruit	Dental diseases
884.	*Rhynchoglossum notonianum*	Stem and leaves	Bite scar against the poison
885.	*Rhynchosia rufescens*	Leaves	Abortion
886.	*Rhynchostylis retusa*	Leaves, fruit	Ear sore, emollient, throat inflammation, ear pain
887.	*Richardia scabra*	Leaves	Hey fever
888.	*Ricinus communis*	Roots, twig, leaves, seed	Dysentery, clear black scars on the face, coolant, jaundice, cuts, dysentery, toothache, constipation, muscle and joint pain, dog bite, abortifacient, headache, pain reliever
889.	*Rivea hypocrateriformis*	Leaves	Fever, diarrhea, stomach upset
890.	*Rivina humilis*	Leaves	Insect bites
891.	*Rostellularia procumbens*	Tender stem, leaves, flowers	Blood clot, chest pain, fever
892.	*Rotula aquatica*	Whole plant, root	Urinary disorders
893.	*Rubia cordifolia*	Whole plant, root, tender shoot, Stem, leaves, fruits	Antiseptic, bronchitis, diabetes, dysentery, snake bite, epilepsy, gynecological complaints, healing the injury, renal lithiasis, rheumatism, ringworm infection, scorpion sting, giddiness, skin ailments
894.	*Rubus ellipticus*	Leaves, fruits	Easy digestion, paralysis
895.	*Rubus moluccanus*	Leaves, fruits	Easy digestion, paralysis
896.	*Rubus recemosus*	Leaves, fruits	Promotion digestion, paralysis
897.	*Rumex nepalensis*	Root, leaves	Jaundice
898.	*Rungia apiculata*	Whole plant	Cold, gastric, body pain
899.	*Rungia pectinata*	Whole plant	Jaundice
900.	*Rungia repens*	Whole plant	Scorpion sting, cold, hydration, rheumatism, throat pain
901.	*Ruta graveolens*	Leaves, flowers	Cough, dyspnoea, epilepsy, fever, antiseptic, intestinal worm, rheumatic pain
902.	*Saccharum officinarum*	Stem	Jaundice
903.	*Sageraea laurifolia*	Leaves	Rheumatism

TABLE 6.3 (Continued)

Sl.No.	Botanical Name	Part Used	Medicinal uses
904.	*Salacia chinensis*	Root	Snake bite
905.	*Salacia fruticosa*	Root	Psoriasis
906.	*Salacia oblonga*	Flowers, tender fruits	Body pain, arthritis
907.	*Saline gallica*	Leaves	Headache
908.	*Salvinia molesta*	Whole plant	Antifungal agent
909.	*Sansevieria roxburghiana*	Leaves	Earache, rheumatic pain
910.	*Santalum album*	Heart wood, stem, leaves	Dysuria, ear ailment, headache, mumps, rheumatism, pimples, herpes simplex, coolant, prickly heat, skin rashes and allergy
911.	*Sapindus emarginatus*	Root, bark, fruit	Substitute for soap, used in tooth problems, snake bite
912.	*Sapindus mukorossi*	Fruits	Rheumatic pain
913.	*Sapindus trifoliatus*	Root, leaves, fruit	Pruritus, fever, mumps, snake bite
914.	*Saraca asoca*	Root bark, bark, flowers	Bed sore, bleeding, irregular menstruation, menorrhea, leucorrhoea, dysmenorrhea, eczema and scabies
915.	*Sarcococca saliga*	Leaves, fruits	Render pest (veterinary)
916.	*Sarcostemma acidum*	Latex	Heat boils in children
917.	*Sarcostemma brunonianum*	Root	Vomiting.
918.	*Sauropus androgynus*	Whole plant, leaves, flowers, tender fruits	Giddiness, vomiting, fever, bladder, constipation, ulcers, lactation
919.	*Sauropus quadrangularis*	Leaves	Tonsils
920.	*Schumannianthus virgatus*	Rhizome	Skin disease
921.	*Scilla hyacinthina*	Bulb	Body pain
922.	*Scleria lithosperma*	Rhizome	Eczema, leucoderma and scabies

TABLE 6.3 *(Continued)*

Sl.No.	Botanical Name	Part Used	Medicinal uses
923.	*Scoparia dulcis*	Whole plant, root, stem, leaves	Gastric, liver complaints, skin ailments, itching, inflammation, urinary diseases, poisonous bite, kidney stone
924.	*Scurrula cordifolia*	Leaves, flowers	Nervous problems
925.	*Scurrula parasitica*	Whole plant	Internal bleeding
926.	*Scutellaria discolor*	Wholeplant	Lactation
927.	*Scutellaria oblonga*	Leaves	Ear Pain
928.	*Scutellaria violacea*	Whole plant	Pimples
929.	*Scutellaria wightiana*	Whole plant	Measles, Boils
930.	*Sechium edule*	Leaves, fruits	Constipation
931.	*Selaginella delicatula*	Whole plant, fronds	Antibacterial, wounds
932.	*Selaginella involvens*	Whole plant	Rejuvenate, prolapse of rectum, cough, bleeding piles, gravel amenorrhea and antibacterial, poisonous bites
933.	*Selaginella radicata*	Fronds	Antibacterial
934.	*Semecarpus anacardium*	Bark, fruit, seed	Corn, foot cracks, headache, rheumatic pain, dysentery, wounds
935.	*Senecio candicans*	Leaves	Jaundice, eye problem, fertility
936.	*Senecio corymbosus*	Leaves	Wound healing
937.	*Senna alata*	Leaves, stem bark	Ring worm infection, abortion, alopecia, boils
938.	*Senna auriculata*	Leaves, flowers	Fracture, colic pain, coolant, stomachache, itching, skin irritation
939.	*Senna kleinii*	Whole plant	Eczema
940.	*Senna occidentalis*	Leaves, roots, whole plant, tender leaves, flowers, tender pods	Asthma, cough, pain during delivery, fever, vomiting, body swellings, eye problem, dog-bite, mumps, stomach disorder in children

TABLE 6.3 *(Continued)*

Sl.No.	Botanical Name	Part Used	Medicinal uses
941.	*Senna tora*	Whole plant, tender leaves, leaves, seed	Cuts and wounds; ringworms, skin disease. Fungal infection, eye pain, jaundice, thorn from foot, headache, muscle pain, dysentery, athelets foot disease
942.	*Sesamum indicum*	Seeds	Piles, urinary complaints, burns and hair growth, skin irritation, itching, pain reliever
943.	*Sesbania grandiflora*	Leaves, flower	Dizziness, coolant
944.	*Setaria italica*	whole plant, buds, grains	Promoting vigor, rheumatism, bone fracture
945.	*Sida acuta*	Whole plant, root, leaves	Labor pain, dysmenorrhea, menstruation, heat boils, vomiting, jaundice, boils, skin cuts, leucorrhoea, dandruff, rheumatism, breathing problems and cough
946.	*Sida alnifolia*	Whole plant, root	Normal delivery, arthritis, rheumatism, tuberculosis, strengthen the body
947.	*Sida cordata*	Whole plant, leaves	Urinary problem, boils, coolant, wounds
948.	*Sida cordifolia*	Root, leaves	Skin cuts, fever, coolant, wounds
949.	*Sida rhombifolia*	Root, stem, leaves	Bodyache, hair complaints, heart diseases, oral care
950.	*Sida spinosa*	Leaves	Hair complaints
951.	*Siegesbeckia orientalis*	Leaves	Skin allergy, fertility
952.	*Sirhookera lanceolata*	Aerial part	Antidote for poisonous bites
953.	*Smilax ovalifolia*	Rhizome, root	Venereal diseases, rheumatism, arthritis, urinary complaints
954.	*Smilax zeylanica*	Root, fruit	Venereal diseases, swelling, skin diseases, antidote, piles, rheumatism, arthritis, urinary complaints
955.	*Smithia conferta*	Whole plant, stem, leaves	Fertility, muscular pain
956.	*Smithia gracilis*	Roots, leaves, flowers	Leukemia, piles

TABLE 6.3   *(Continued)*

Sl.No.	Botanical Name	Part Used	Medicinal uses
957.	*Solanum americanum*	Roots, stem, leaves, fruit	Anaemia during pregnancy, tonsillitis, fungal infection, induce lactation, gingivitis, earache, stomach ache, gynecological complaints, poultice, cancerous sores, ulcers, boils, leucoderma, tuberculosis
958.	*Solanum ferox*	Fruit and seed	Skin ailments, toothache
959.	*Solanum melongena*	Leaves	Cuts
960.	*Solanum rudepannum*	Leaves	Cough and fever
961.	*Solanum sisymbrifolium*	Fruit	Vermifuge, antifertility
962.	*Solanum surattense*	Root, fruit, leaves, seeds	Tooth problem, vomiting, tooth paste, cold, cough, fever, chest pain, rheumatism, viral fever, asthma and cough
963.	*Solanum torvum*	Root, fruit/unripe and seed	Liver and spleen ailment, cracks in the feet, cough, sore throat, asthma, vermifuge
964.	*Solanum trilobatum*	Whole plant, flower, leaves	Chronic bronchitis, asthma, cold, cough, sputum expulsion, constipation, fertility
965.	*Solanum violaceum*	Root, seeds	Poison, cough and bronchial diseases
966.	*Solanum virginianum*	Whole plant	Cough, toothache and swelling
967.	*Solena amplexicaulis*	Fruit and seed	Ear complaints
968.	*Sonerila tinnevelliensis*	Leaves	Rheumatism
969.	*Sonchus oleraceus*	Leaves	Skin diseases
970.	*Sophora glauca*	Leaves	Hair nourisher, cooling effect to the eyes
971.	*Soymida febrifuga*	Bark	Bone fracture, snake bite, dysentery
972.	*Spathodea campanulata*	Bark	Malaria
973.	*Spergularia arvensis*	Aerial part	Digestive

TABLE 6.3 (Continued)

Sl.No.	Botanical Name	Part Used	Medicinal uses
974.	Spermacoce hispida	Whole plant, root, leaves, flowers, tender fruits	Blood clot, body pain, rheumatism, cooling effects, sprain, expel gall stones, stomach pain
975.	Spermacoce latifolia	Whole plant	Ringworm infection
976.	Spermacoce pusilla	Root, stem, flowers	Body pain
977.	Sphaeranthus indicus	Whole plant, leaves	Scabies, painful asthma, giddiness, skin irritation, cut, bones fracture
978.	Sphaerostephanos unitus	Rhizome	Antibacterial agent
979.	Spilanthes acmella	Leaves, flowers	Dental pain, throat infections and toothe ache
980.	Spilanthes paniculata	Inflorescence	Stuttering speech, throat infection
981.	Spinacia oleracea	Leaves	Cooling effect
982.	Spondius pinnata	Bark	Boils
983.	Stachytarpheta indica	Bark, leaves	Diarrhea, dysentery, ophthalmic problem, ulcers
984.	Stachytarpheta jamaicensis	Whole plant	Dysentery
985.	Stemodia viscosa	Leaves	Fever, headache
986.	Stenochlaena palustris	Rhizomes, fronds, leaves	Fever, skin disease, throat, gastric ulcers, burns
987.	Stenosiphonium russellianum	Stem, leaves, flowers	Blood purifier
988.	Stephania japonica	Rhizome/ root	Stomach pain, dental diseases. cough, fever, diarrhea
989.	Sterculia foetida	Seed	Constipation
990.	Sterculia urens	Bark, gum, leaves	Throat problems, contraction of uterus, cooling effect, foot cracks, general tonic, vomiting, delivery pain, bone fracture
991.	Sterculia villosa	Root	Bone fracture
992.	Stereospermum colais	Bark	Stomach pain
993.	Streblus asper	Leaves	Muscular knots in stomach, giddiness
994.	Strobilanthus kunthianus	Stem, bark	Masticatory

TABLE 6.3 *(Continued)*

Sl.No.	Botanical Name	Part Used	Medicinal uses
995.	*Strychnos colubrina*	Fruit, root, fresh leaves	Bruises, diarrhea, pains in the joints, tumors.
996.	*Strychnos nux-vomica*	Root, stem bark, leaves, seeds, fruit	Body pain, boils, cold, cut, depression, dysentery, eye diseases, fever, gangrene, hair complaints, headache, heart diseases, gynecological complaints, stomach pain, migraine headaches, nervous conditions, paralysis, rheumatism, snake bite, wound
997.	*Strychnos potatorum*	Whole plant, bark	Faintness, piles, urinary and kidney problems
998.	*Suregada lanceolata*	Whole plant	Skin diseases, blood vomiting, piles, weakness, toothache, worms
999.	*Symplocos cochinchinensis* subsp. *laurina*	Stem bark	For fairness, discoloration
1000.	*Symplocos macrocarpa* subsp. *kanarana*	Leaves, flowers, tender fruits	Rheumatism
1001.	*Syzygium aromaticum*	Bark, leaves, flower bud, tender fruits	Aphthae, ulcers in intestine, crack in foot, gangrene, gingival wounds, nausea, decaying teeth, tooth ache, dandruff, pruritus, scabies, eczema, bed sore
1002.	*Syzygium cumini*	Bark, leaves, flower bud, fruit, seed	Tooth problems, rheumatism, body pain, headache, diabetes, dysentery, stomachache, aphthae, ulcers in intestine, leucorrhoea, diabetes, liver diseases indigestion
1003.	*Syzygium mundagam*	Bark, stem, leaves, flowers, tender fruit, seed	Body fluid, body pain, breathing problems
1004.	*Syzygium rubicundum*	Leaves, flowers	Body pain, gastric problem
1005.	*Syzygium zeylanicum*	Leaves, flowers, fruit	Urinary disorder
1006.	*Tabernaemontana divaricata*	Leaves, flowers	Boils, cuts, eye disease
1007.	*Tabernaemontana gamblei*	Leaves, flowers	Rheumatism
1008.	*Tabernaemontana heyneana*	Stem, bark, root, flower	Cuts, fever, headache, herpes, pruritus, scabies, skin sore, scorpion sting, snake bite, improve vision, burning sensation of sore eye, toothache

TABLE 6.3 *(Continued)*

Sl.No.	Botanical Name	Part Used	Medicinal uses
1009.	*Tacca leontopetaloides*	Tuber	Labor pain
1010.	*Tagetes erecta*	Leaves	Boils, wounds
1011.	*Tamarindus indica*	Stem-bark, leaves, fruit, seed	Cut, eye pain, stomachache, scabies, inflammations, rheumatism, herpes simplex, bleeding piles, to relieve pain of scorpion sting, coolant, indigestion, treating bone fracture
1012.	*Tamarix ericoides*	Leaves	Cough
1013.	*Taxillus cuneatus*	Bark, leaves	Skin allergy,
1014.	*Taxillus tomentosus*	Bark, leaves	De-worming
1015.	*Tecomella undulata*	Bark	Bone fracture
1016.	*Tectaria coadunata*	Rhizome	Asthma, bronchitis, stings of honeybee, diarrhea in children, stomach trouble
1017.	*Tectaria wightii*	Rhizome	anthelmintic
1018.	*Tectona grandis*	Burn, wood, stem bark, tender leaves	All types of skin diseases, inter trigo, gynecological complaints, skin burns, eczema, hair loss, dandruff, measles, dysentery
1019.	*Tephrosia purpurea*	Leaves, root bark	Bowel disorders, STD, asthma, Bowel disorders in children, gastric problems, joint pain, general tonic
1020.	*Tephrosia tinctoria*	Root	Dyspepsia
1021.	*Terminalia arjuna*	Leaves	Earache, heart troubles
1022.	*Terminalia bellirica*	Stem bark, root, tender leaves, seed, fruit	Allergy, diabetes, dropsy, piles, diarrhea, skin burns, snake bite, dysentery, dropsy, piles, diuretic, asthma, jaundice, scorpion sting
1023.	*Terminalia catappa*	Bark, young leaves	Astringent, diuretic, potent cardio tonic, leprosy, scabies, skin diseases
1024.	*Terminalia chebula*	Stem bark, leaves, fruit, seeds	As an ingredient in triphala, asthma, bleeding gums, bodyache, cold, cough, digestive disorders, dysentery, eczema, fever, headache, inter trigo, intestinal ulcers, mumps, otorrhoea, chronic ulcers and wounds, cuts, skin diseases, snoring sound of children, sterility in males, stomachache, aphrodisiac, toothache

TABLE 6.3   *(Continued)*

Sl.No.	Botanical Name	Part Used	Medicinal uses
1025.	*Terminalia crenulata*	Stem sap, stem bark	Chest pain, dysentery, stomachache, rheumatic pain, stomach pain
1026.	*Terminalia cuneata*	Bark	Itching, bone fracture
1027.	*Terminalia paniculata*	Bark, flowers	Parotitis, cholera
1028.	*Tetrastigma sulcatum*	Leaves, stem	Headache
1029.	*Thespesia populnea*	Tender shoot, leaves	Abscess, scorpion sting, scabies
1030.	*Thottea siliquosa*	Roots and leaves	Scabies, eczema, sedative for treating snake bite, ringworm infection
1031.	*Thunbergia fragrans*	Leaves	Dyspepsia, wounds
1032.	*Thunbergia fragrans* var. *laevis*	Leaves	Dizziness
1033.	*Thunbergia laevis*	Stem	Vesicles
1034.	*Tinospora cordifolia*	Whole plant, stem, leaves	Diarrhea, dysentery, fever, urinary disorders, dyspepsia, anti-inflammation, influenza, mouth ulcers, snake bite, boils, scabies, joint pain, felon, insomnia, rheumatism and bone fracture
1035.	*Tinospora crispa*	Stem, fruit	General weakness, fever, inflammation, rheumatism, jaundice
1036.	*Tinospora sinensis*	Stem, leaves	Bone fracture, for bone strengthening
1037.	*Toddalia asiatica*	Whole plant, root, stem, stem bark, leaves, flowers, fruits, seeds	Skin diseases(itching), intestinal problems, mumps, vomiting, snake bite, for all skin diseases, kidney stones, painful urination, genito-urinary disorders, boils, eczema, wounds, gastric problems, rheumatic pain
1038.	*Tolypanthus lageniferus*	Aerial part	Menstrual disorders, check excess bleeding, rheumatism, antidote for poisonous bites
1039.	*Trachyspermum ammi*	Seed	Aphthae, ulcers in intestine
1040.	*Tragia involucrata*	Whole plant, roots	Dandruff, skin eruption, venereal, piles
1041.	*Trema orientalis*	Whole plant	Antidote for poisons

TABLE 6.3   *(Continued)*

Sl.No.	Botanical Name	Part Used	Medicinal uses
1042.	*Trianthema portulacastrum*	Leaves, root	Immunity
1043.	*Tribulus terrestris*	Fruits, seeds	Urinary infection and irritation, fever, leucorrhoea, headache, cracked heels
1044.	*Trichilia connaroides*	Bark	Bone fracture, nervous problems
1045.	*Trichodesma indicum*	Leaves	Skin cuts
1046.	*Trichodesma sedgwickianum*	Root	Anti-inflammatory, superficial injuries on skin
1047.	*Tricholepis glaberrima*	Whole plant, stem	Throat problems, skin diseases
1048.	*Trichopus zeylanicus*	Leaves, fruits	Good health, vitality, bleeding piles, asthma, venereal diseases
1049.	*Tridax procumbens*	Whole plant, leaves, root	Cuts, wounds, giddiness, nervous disloca-tion, tiredness, itching, scabies, skin infec-tions, boils, skin cuts, wounds, injuries, eczema, shoulder pain, swellings, stomach pain, bone fracture
1050.	*Trignospora caudipinna*	Rhizome	Fever
1051.	*Triticum aestivum*	Seed	Boils
1052.	*Triumfetta annua*	Whole plant	Body heat
1053.	*Triumfetta pilosa*	Root	Dysentery
1054.	*Triumfetta rhomboidea*	Leaves	Foot cracks
1055.	*Triumfetta rotundifolia*	Leaves	Dysentery
1056.	*Tylophora indica*	leaves	Respiratory problems, bronchial asthma, gynecological complaints, jaundice, snake bite, bronchitis
1057.	*Typha domingensis*	Fruit	Skin cuts
1058.	*Uraria picta*	Leaves, fruits	Joint pain
1059.	*Urena lobata* subsp. *lobata* var. *viminea*	Leaves	Wiping away wound maggots
1060.	*Urena lobata* subsp. *lobata*	Root	Fever, urinary infections, rheumatism, tooth problem, Bone fracture

TABLE 6.3 *(Continued)*

Sl.No.	Botanical Name	Part Used	Medicinal uses
1061.	*Urginea indica*	Bulb	Painful corns on the lower surface of the feet
1062.	*Utleria salicifolia*	Tuber	Indigestion and gas troubles
1063.	*Utricularia graminifolia*	Whole plant	Itching hands and legs
1064.	*Uvaria narum*	Stem, flowers, tender fruits	Rheumatism
1065	*Vaccinium neilgherrense*	Whole plant	Body pain, cough, joint pain, nervous problem
1066.	*Vanda spathulata*	Root	Frenzy, asthma
1067.	*Vanda tessellata*	Leaves, root	Alexiteric, antipyretic, sexual stimulant, ear-ache, sprains, lumbago, back pain, fever, otitis, nervous disorders, rheumatism, bronchitis, inflammation, hiccups, piles, boils on the scalp, joint pain, stomach pain
1068.	*Vanda testacea*	Leaves	Asthma, malaria, rheumatism, nervous disorders
1069.	*Ventilago denticulata*	Bark	Bone fracture
1070.	*Ventilago madraspatana*	Stem bark, bark	Dermatitis, tinea infection, joint pains
1071.	*Verbascum thapsus*	Leaves	Skin allergy, cuts and wounds
1072.	*Verbena bonariensis*	Leaves	Cuts and wounds
1073.	*Vernonia anthelmintica*	Seed	All types of skin diseases, eczema, pruritus, scabies, bed sore, dandruff, crack foot, cut and gangrene, stomachache
1074.	*Vernonia cinerea*	Whole plant, root, leaves	Eczema, boils, wounds, fever, filaria, leucorrhoea, tonsillitis, eye diseases, jaundice, diarrhea and stomachache
1075.	*Vetiveria zizanioides*	Root	Dandruff, skin diseases, used as hair tonic, hair growth of hair, swelling in the body, summer boils, headache, to prevent abortion
1076.	*Vicoa indica*	Leaves	Headache
1077.	*Vigna radiata*	Whole plant, flower	Boils, backache
1078.	*Viscum angulatum*	Whole plant, stem	Bone Fracture, headache, skin cuts, inflammation

TABLE 6.3 *(Continued)*

Sl.No.	Botanical Name	Part Used	Medicinal uses
1079.	*Viscum articulatum*	Whole plant, stem	Cooling, alexipharmic, aphrodisiac, fever, blood diseases, ulcers, epilepsy, biliousness
1080.	*Viscum monoicum*	Whole plant, stem	Jaundice, fever, typhoid, stomach disorders, fungal infection, hip inflammation
1081.	*Viscum ramosissimum*	Leaves, flowers	Asthma, gastric
1082.	*Vitex altissima*	Stem bark, leaves, tender fruits	Giddiness, leukemia, fever, rheumatic swellings and chest pains.
1083	*Vitex leucoxylon*	Leaves, flowers, tender fruits	Weakness, piles
1084.	*Vitex negundo*	Whole plant, root, leaves	Anticancer, body pain, bone fracture, diuretic, expectorant, expelling lice, eye disease, fever, giddiness, gingivitis, headache, insecticidal, joint pain, muscular pain, poison bites, rheumatism, ringworm infection, snake bite, tonic, tooth ache, vermifuge, whitlow, prickly heat, wounds with maggots
1085.	*Vitex trifolia*	Leaves	Hair fall
1086.	*Vitis vinifera*	Fruit	Itching
1087.	*Vittaria elongata*	Leaves	Rheumatism
1088.	*Walsura trifoliolata*	Leaves, tender flowers	Cancer, fits
1089.	*Wattakaka volubilis*	Root, stem, bark, fruits	Dysentery, boils, abscesses, paronchia, throat infection in cattle, inter trigo, urinary troubles, bone fracture
1090.	*Wedelia calendulacea*	Leaves	Hair loss, graying of hair
1091.	*Wedelia urticifolia*	Whole plant	Jaundice
1092.	*Wendlandia thyrsoidea*	Root, tender leaves	Skin ailments, conjunctivitis
1093.	*Wendlandia tinctoria*	Leaves	Knee pain
1094.	*Withania somnifera*	Roots	Asthma, aphrodisiac, sprain, wounds, ulcers, fever, cough and leucoderma, weight loss, weakness

TABLE 6.3   *(Continued)*

Sl.No.	Botanical Name	Part Used	Medicinal uses
1095.	*Woodfordia fruticosa*	Bark, root	Jaundice, snake bite, to strengthen teeth, itching
1096.	*Wrightia arborea*	Stem bark	Stomachache
1097.	*Wrightia tinctoria*	Latex, stem bark, leaves, fruits	Chest pain, to expel worm in children, boils, skin wounds, psoriasis, eczema, ringworm infection, mouth ulcers, fever, lactation, snake bite, dysentery, stomach pain
1098.	*Xanthium indicum*	Roots, leaves, flowers	Dog bite, edema, ulcer
1099.	*Xanthophyllum flavescens*	Whole plant	Throat pain, swelling, skin diseases, itching, chest pain
1100.	*Xylia xylocarpa*	Bark	Leprosy
1101.	*Zanthoxylum rhetsa*	Spines	Galactogogue and bone fracture
1102.	*Zea mays*	Seeds (grains)	Food, body strength
1103.	*Zehneria maysorensis*	Whole plant, aerial part, leaves, fruits	Skin allergy, swelling body pain, blood purifier
1104.	*Zingiber cernuum*	Rhizome	Stomachache
1105.	*Zingiber officinale*	Rhizome	Aphthae, cough, gas problems, ulcers in intestine, throat problems
1106.	*Zingiber roseum*	Rhizome	Stomach ulcers
1107.	*Zingiber zerumbet*	Rhizome	Leucoderma, cold, gastroenteritis, digestion, whooping cough, fever.
1108.	*Ziziphus abyssinica*	Bark	Anti-cancerous, antibiotics
1109.	*Ziziphus mauritiana* var. *fruticosa*	Leaves, stem-bark	Fracture, stomachache
1110.	*Ziziphus mauritiana*	Bark	Dysentery, diarrhea, removing intestinal worms, cough
1111.	*Ziziphus oenoplia*	Bark, leaves, fruit	Scabies, diarrhea, cough, wounds
1112.	*Ziziphus rugosa*	Roots, stem-bark, leaves, fruit,	Fever, toothache, boils, dysentery, to set bone fracture, menorrhagia, scabies
		Bark, flower	Menorrhagea, diarrhea, scabies and ringworm infection
1113.	*Ziziphus xylopyrus*	Whole plant, leaves, stem bark	Stomach problems, eczema, boils

Plate 1: Threatened Medicinal Plants of the Western Ghats

a) *Aristolochia tagala*; b) *Canarium strictum*; c) *Cinnamomum wightii*
d) *Coscinium fenestratum*; e) *Garcinia gummi-gutta*; f) *Janakia arayalpathra*
g) *Kingiodendron pinnatum*; h) *Nothopodytes foetida*; i) *Oroxylum indicum*
j) *Pterocarpus santalinus*; k) *Rauvolfia serpentina* & l) *Utleria salicifolia*

Plate 2: Some Interesting Medicinal Plants from the Western Ghats

a) *Adenia hondala - tuber*; b) *Adenia hondala - fruit*; c) *Alseodaphne semecarpifolia*;

d) *Ancistrocladus heyneanus*; e) *Aristolochia krisagathra*; f) *Azanza lampas*

g) *Baccaurea courtallensis*; h) *Barringtonia racemosa*; i) *Berberis tinctoria*;

j) *Breynia retusa*; K) *Cadaba trifoliata*

Plate 3: Some Interesting Medicinal Plants from the Western Ghats

a) *Calotropis procera*; b) *Capparis zeylanica*; c) *Chasalia curviflora*
d) *Cinnamomum sulphuratum*; e) *Clerodendrum serratum*; f) *Cassia tora*
g) *Corollacarpus epigaeus*; h) *Dichrostachys cinerea*; i) *Diospyros malabarica*

Plate 4: Some Interesting Medicinal Plants from the Western Ghats

a) *Echinops echinatus*; b) *Embelia basal*; c) *Gordonia obtusa*

d) *Gymnena sylvestre*; e) *Justicia adhatoda* ; f) *Kleinia grandiflora*

g) *Limonia acidissima*; h) *Maerua oblongifolia*; i) *Mesua ferrea*; j) *Meyna laxiflora*

Plate 5: Some Interesting Medicinal Plants from the Western Ghats

a) *Mitragyna parvifolia*; b) *Mullohua spicata*; c) *Phoenix loureirii*
d) *Phyllanthus singampattiana*; e) *Pleiospermium alatum*; f) *Pongamia pinnata*
g) *Premna serratifolia*; h) *Radermachera xylocarpa*

Plate 6: Some Interesting Medicinal Plants from the Western Ghats

a) *Schumannianthus virgatus*; b) *Smilax zeylanica*; c) *Solanum trilobatum*
d) *Soymida febrifuga*; e) *Stereospermum colais*; f) *Syzygium cumini*
g) *Tabernaemontana heyneana*; h) *Terminalia crenulata*; i) *Vitex leucoxylon*
j) *Wattakaka volubilis*; k)*Woodfordia fruiticosa*

KEYWORDS

- **Edible**
- **Ethnobotany**
- **Ethnomedicine**
- **Medicinal Plants**
- **Tribe**
- **Western Ghats**

REFERENCES

Abraham, Z. (1981). Ethnobotany of the Thodas, the Kotas and the Irulas of the Nilgiris. In: Glimpses of Indian Ethnobotany. Jain, S.K. (ed.) Oxford & IBH Publishing Co., New Delhi, pp. 308.

Achar, S.G., Rajkumar, N. & Shivanna, M.B. (2010). Ethno-medico-botanical knowledge of Khare-Vokkaliga community in Uttar Kannada District of Karnataka, India. *J Compliment Integr Med. 7,* 1–18.

Ajesh, T.P. & Kumuthakalavalli, R. (2013). Botanical ethnography of muthuvans from the Idukki District of Kerala. *Int. J. Pl. Anim. Environ. Sci., 3,* 67–75.

Ajesh, T.P., Krishnaraj, M.V., Kumuthakalavalli, R. & Prabu, M. (2012). Ethno-gynecological observations on Leguminosae (nom. alt: Fabaceae) among Mannan Tribes of Kerala. *Plant Archives, 12(2),* 1115–1119.

Anitha, B., Mohan, V.R., Athiperumalsami, T. & Sutha, S. (2008). Ethnomedicinal plants used by the Kanikkars of Tirunelveli District, Tamil Nadu, India to treat skin diseases. *Ethnobotanical Leaflets 12,* 171–180.

Anonymous (1994). Ethnobotany in India: A status report. All India coordinated research project in ethnobiology. Min. Environment and Forests, Government of India, New Delhi.

Augustine, J., Sreejesh, K.R. & Bijeshmon, P.P. (2010). Ethnogynecological uses of plants prevalent among the tribes of Periyar Tiger Reserve, Western Ghats. *Indian J. Trad. Knowl. 9(1),* 73–76.

Ayyanar, M. & Ignacimuthu, S. (2005). Traditional knowledge of Kani tribals in Kouthalai of Tirunelveli hills, Tamil Nadu, India. *J. Ethnopharmacol. 102,* 246–255.

Ayyanar, M. & Ignacimuthu, S. (2005). Medicinal plants used by the tribals of Tirunelveli hills, Tamil Nadu to treat poisonous bites and skin diseases. *Indian J. Trad. Knowl. 4(3),* 229–236.

Ayyanar, M. & Ignacimuthu, S. (2009). Herbal medicines for wound healing among tribal people in Southern India: Ethnobotanical and scientific evidences. *Intern. J. Applied Res. Nat. Prod., 2(3),* 29–42.

Ayyanar, M. & Ignacimuthu, S. (2010). Plants used for non-medicinal purposes by the tribal people in Kalakad Mundanthurai Tiger Reserve, Southern India. *Indian J. Trad. Knowl. 9(3),* 515–518.

Ayyanar, M. & Ignacimuthu, S. (2011). Ethnobotanical survey of medicinal plants commonly used by Kani tribals in Tirunelveli hills of Western Ghats, India. *J. Ethnopharmacol. 134*, 851–864.

Ayyanar, M., Sankarasivaraman, K. & Ignacimuthu, S. (2008). Traditional herbal medicines used for the treatment of diabetes among two major tribal groups in South Tamil Nadu, India, *Ethnobotanical Leaflets 12*, 276–280.

Ayyanar, M., Sankarasivaraman, K., Ignacimuthu, S. & Sekar, T. (2010). Plant species with ethno botanical importance other than medicinal in Theni district of Tamil Nadu, southern India. *Asian J. Exp. Biol. Sci. 1(4)*, 765–771.

Balasubramanian, P. & Prasad, S.N. (1996). Medicinal plants among the Irulas of Attapady and Boluvampatty forest in the Nilgiri Biosphere Reserve. *J. Econ. Taxon. Bot. Addl. Ser. 12*, 253–259.

Balasubramanian, P., Rajasekaran, A. & Prasad, S.N. (1997). Folk medicine of the Irulas of Coimbatore forests. *Ancient Science of Life 16 (3)*, 1–4.

Balasubramanian, P., Rajasekaran, A. & Prasad, S.N. (2010). Notes on the distribution and ethnobotany of some medicinal orchids in Nilgiri Biosphere. *Zoos' Print Journal 15(11)*, 368.

Benjamin, A. & Manickam, V.S. (2007). Medicinal pteridophytes from the Western Ghats. *Indian J. Trad. Knowl. 6(4)*, 611–618.

Bhambare, P.B. (1995). Some antivenom medicinal plants from tribals of Dhule district (Maharashtra). *JAST 1(1)*, 36–37.

Bhandary, M.J. & Chandrashekar, K.R. (2011). Herbal therapy for herpes in the ethno-medicine of Coastal Karnataka. *Indian J. Trad. Knowl. 10(3)*, 528–532.

Bhandary, M.J., Chandrashekar, K.R. & Kaveriappa, K.M. (1995). Medical ethnobotany of the Siddis of Uttar Kannada district, Karnataka, India. *J. Ethnopharmacol. 47*, 149–56.

Bhandary, M.J., Chandrashekar, K.R. & Kaveriappa, K.M. (1996). Ethnobotany of Gawlis of Uttar Kannada district, Karnataka. *J. Econ. Taxon. Bot. 12*, 244–249.

Bhat, P., Hegde, G. & Hegde, G.R. (2012). Ethnomedicinal practices in different communities of Uttar Kannada district of Karnataka for treatment of wounds. *J. Ethnopharmacol., 143*, 501–514.

Bhat, P., Hegde, G.R., Hegde G. & Mulgund, G.S. (2014). Ethnomedicinal plants to cure skin diseases – An account of the traditional knowledge in the coastal parts of Central Western Ghats, Karnataka, India. *J. Ethnopharmacol., 151*, 493–502.

Bodding, Rev. D.P. (1925). Studies in Santal Medicine and Connected Folklore: The Santals and Disease. *Mem. Asiat. Soc. Bengal 10*, 1–132.

Bodding, Rev. D.P. (1927). Studies in Santal Medicine and Connected Folklore-II. Santal Medicine. *Mem. Asiat. Soc. Bengal 10*, 133–426.

Bodding, Rev., D.P. (1940). Studies in Santal medicine and Connected Folklore. *Mem. Asiat. Soc. Bengal. 10*, 427–502.

Bose, M.F.J.N., Aron, S. & Mahalingam, P. (2014). An ethnobotanical study of medicinal plants used by the Paliyars aboriginal community in Virudhunagar district, Tamil Nadu, India. *Indian J. Trad. Knowl. 13(3)*, 613–618.

De Britto, J. & Mahesh, R. (2007). Exploration of Kani tribal botanical knowledge in Agasthiayamalai Biosphere Reserve – South India. *Ethnobotanical Leaflets, 11*, 258–265.

Deepthy, R. & Remashree, A.B. (2014). Ethnobotanical studies on medicinal plants used for skin diseases in Malabar region of Kerala. *Intern. J. Herbal Medicine 2(1)*, 92–99.

Deokule S.S. (2006). Ethno-Medicinal Plants of Baramati Region of Pune District, Maharashtra (India). *J. Econ. Taxon. Bot. 30 (Suppl.)*, 59–66.

Deokule, S.S. & Mokat, D.N. (2004). Ethno-medico-botanical survey of Ratnagiri district of Maharashtra. *J. Econ. Taxon. Bot. 28 (3)*, 19–23.

Desale, M.K., Bhamane, P.B., Sawant, P.S., Patil, S.R. & Kamble, S.Y. (2013). Medicinal plants used by the rural people of Taluka Purandhar, district Pune, Maharashtra. *Indian J. Trad. Knowl. 12(2)*, 334–338.

Dey, A. & De, J.N. (2011). Ethnobotanical aspects of *Rauvolfia serpentina* (L.) Benth. *ex* Kurz. in India, Nepal and Bangladesh. *J. Medicinal Plants Res.*, *5(2)*, 144–150.

Divya, V.V., Karthick, N. & Umamaheswari, S. (2013). Ethnopharmacological studies on the Medicinal plants used by Kani tribes of Thachamalai Hills, Kanyakumari, Tamil Nadu, India. *Intern. J. Advanced Biol. Res. 3(3)*, 384–393.

Ganesan, S., Suresh, N. & Kesaven, L. (2004). Ethnomedicinal survey of lower Palani Hills of Tamil Nadu. *Indian J. Trad. Knowl. 3(3)*, 299–304.

Garg, A. Mahendra Darokar, P., Sundaresan, V., Uzma Faridi, Suaib Luqman, Rajkumar, S. & Suman Khanuja, P.S. (2007). Anticancer activity of some medicinal plants from high altitude evergreen elements of Indian Western Ghats. *J. Res. Educ. Indian Med. 8(3)*, 1–6.

Gayake, D.N., Awasarkar, U.D. & Sharma, P.P. (2013). Indigenous traditional medicinal plant resources from Ahmednagar district, Maharashtra. *Asian J. Biomedical & Pharmaceutical Sci.*, *3(22)*, 1–5.

Gireesha, J. & Raju, N.S. (2013). Ethnobotanical study of medicinal plants in BR Hills region of Western Ghats, Karnataka. *Asian J. Plant Sci. Res.*, *3(5)*, 36–40.

Harsha, V.H., Hebbar, S.S., Hegde, G.R. & Shripathi, V. (2002). Ethnomedical knowledge of plants used by Kunabi Tribe of Karnataka in India. *Fitoterapia. 73*, 281–287.

Harshberger, J.W. (1896). The purpose of ethnobotany. *Bot. Gaz. 21*, 146–154.

Hebbar, S.S., Hegde, H.V., Shripathi, V. & Hegde, G.R. (2004). Ethnomedicine of Dharwad district in Karnataka, India – plants used in oral health care. *J. Ethnopharmacol. 94*, 261–266.

Hegde, H.V., Hebbar, S.S., Shripathi, V. & Hegde, G.R. (2003). Ethnomedicobotany of Uttar Kannada District in Karnataka, India – plants in treatment of skin diseases. *J. Ethnopharmacol. 84*, 37–40.

Henry, A.N., Hosagoudar, V.B. & Ravikumar, K. (1996). Ethno-Medico-Botany of the Southern Western Ghats of India. In: S.K. Jain (Ed.). Ethnobiology in Human Welfare, pp. 173–180.

Hosagoudar, V.B. & Henry, A.N. (1996a). Ethnobotany of Kadars, Malasars and Muthuvans of the Anamalais in Coimbatore district, Tamil Nadu, India. *J. Econ. Taxon. Bot. 12*, 260–267.

Hosagoudar, V.B. & Henry, A.N. (1996b). Ethnobotany of tribes Irular, Kurumban and Paniyan of Nilgiris in Tamil Nadu, southern India. *J. Econ. Taxon. Bot. 12*, 272–283.

Hosamani, P.A., Lakshman, H.C., Sandeepkumar, K., Kulkarni, S.S. & Gadi, S.B. (2012). Documentation of ethnobotanical medicinal plants growing in rock crevices of river Kali in Dandeli Wild Life Sanctuary. *Life Sciences Leaflets, 3*, 36–39.

Ignacimuthu, S., Ayyanar, M. & Sankara Sivaraman, K. (2006). Ethnobotanical investigations among tribes in Madurai District of Tamil Nadu (India). *J. Ethnobiol. Ethnomed.*, *2*, 25, 7 Pages.

Jain, S.K. (1991). Dictionary of Indian folk medicine and ethnobotany. Deep publications, Delhi.

Jain, S.K. & Mudgal, V. (1999). A Handbook of Ethnobotany. Bishen Singh Mahendra Pal Singh, Dehra Dun.

Jain, S.K., Sinha, B.K. & Gupta, R.C. (1991). Notable plants in ethnomedine. Deepti Publications, Delhi.

Janaki Ammal, E.K. (1956). Introduction to subsistence economy of India. In: William, L.T. Jr. (ed.) Man's role in changing the face of the earth. University of Chicago Press, Chicago, pp. 324–335.

Jayakumar, G., Ajithabai, M.D., Sreedevi, S., Viswanathan, P.K. & Remeshkumar, B. (2010). Ethnobotanical survey of the plants used in the treatment of diabetes. *Indian J. Trad. Knowl., 9(1),* 100–104.

Jegan, G., Kamalraj, P. & Muthuchelian, K. (2008). Medicinal plants in Tropical Evergreen forest of Pachakumachi Hill, Cumbum Valley, Western Ghats, India. *Ethnobotanical Leaflets 12,* 254–260.

Jenisha, S.R. & Jeeva, S. (2014). Traditional remedies used by the inhabitants of Keezhakrishnanputhoor – A coastal village of Kanyakumari district, Tamilnadu, India. *Med Aromat Plants, 3,* 50–54.

Jeyaprakash, K., Ayyanar, M., Geetha, K.N. & Sekar, T. (2011). Traditional uses of medicinal plants among the tribal people in Theni District (Western Ghats), Southern India. *Asian Pacific J. Tropical Biomed., 1(1),* S20–S25.

John, D. (1984). One hundred useful raw drugs of the Kani tribes of Trivandrum forest division, Kerala, India. *Int. J. Crude Drug. Res. 22(1),* 17–39.

Joseph, J.K., Antony, V.T. (2008). Ethnobotanical investigations in the Genus *Momordica* L. in the Southern Western Ghats of India. *Genet. Resour. Crop Evol., 55,* 713–721.

Joseph, J.M., Thomas, B., Rajendran, A. & Prabhu Kumar, K.M. (2015). Medicinal Chasmophytes of Urumbikkara Hills, Idukki district, Kerala, India. *Asian J. Pharmaceutical Science & Technology 5(1),* 11–17.

Jothi, G.J., Benniamin, A. & Manickam, V.S. (2008). Glimpses of Tribal Botanical Knowledge of Tirunelveli Hills, Western Ghats, India. *Ethnobotanical Leaflets 12,* 118–126.

Kadam, N.V., Khatale, S. & Kochar, S. (2013). Medicinal plants used in Ghoti tribal region of Nasik district, Maharashtra. *PhTechMed, 2(5),* 348–352.

Kalaiselvan, M. & Gopalan, R. (2014). Ethnobotanical studies on selected wild medicinal plants used by Irula tribes of Bolampatty valley, Nilgiri Biosphere Reserve (NBR), Southern Western Ghats, India. *Asian J. Pharm. Clin. Res., 7, Suppl 1,* 22–26.

Kamble, S.Y., More, T.N., Patil, S.R., Pawar, S.G., Ram Bindurani & Bodhankar, S.L. (2008). Plants used by the Tribes of Northwest Maharashtra for the treatment of gastrointestinal disorders. *Indian J. Trad. Knowl. 7(2),* 321–325.

Kamble, S.Y., More, T.N., Patil, S.R., Singh, E.A. & Pawar, S.G. (2009). Ethnobotany of Thakar tribes of Maharashtra. *J. Econ. Taxon. Bot. 33(Suppl.),* 95–122.

Kamble, S.Y., Patil, S.R., Sawant, P.S., Sawant, S.P. & Singh, E.A. (2009). Traditional medicines used by the tribes for the treatment of upper respiratory tract disorders. *J. Econ. Taxon. Bot. 33(3),* 682–687.

Kamble, S.Y., Patil, S.R., Sawant, P.S., Sawant, S., Pawar, S.G. & Singh, E.A. (2010). Studies on plants used in traditional medicine by Bhilla tribe of Maharashtra. *Indian J. Trad. Knowl. 9(3),* 591–598.

Keshava Murthy, K.R. & Yoganarasimhan, S.N. (1990). Flora of Coorg (Kodagu), Karnataka, India, with data on medicinal plants and chemical constituents. Vimsat Publishers, Bangalore.

Khairnar, D.N. (2006). Medico-ethnological studies and conservation of Medicinal plants of north Sahyadri. *Asian J. Microbiol. Biotech. Environ. Sci. 8(3)*, 535–539.

Kholkunte, S.D. (2008). Database on ethnomedicinal plants of Western Ghats. *Final report 05-07-2005 to 30-06-2008, ICMR* New Delhi.

Khyade, M.S., Takate, Y.A. & Divekar, M.V. (2011). Plants used as an antidote against Snake bite in Akole Taluka of Ahmednagar District (MS), India. *J. Natural Remedies. 11(2)*, 182–192.

Khyade, M.S., Wani, P.S., Awasarkar, U.D. & Petkar, A.S. (2008). Ethnomedicinal Plants used in the treatment of toothache by Tribal's of Akole, Ahmednagar (MS) *Enrich Environment Multidisciplinary Intern. Res. J., 1 (1–3)*, 76–80.

Kola, I. & Landis, J. (2004). Can the pharmaceutical industry reduce attrition rates? *Nat. Rev. Drug Discov. 3*, 711–715.

Kothari, M.J. & Moorthy, S. (1996) Ethnobiology in Human welfare in Raigad district in Maharashtra state, India. In: Jain, S.K. (ed.). Ethnobiology in Human Welfare. pp. 403–407.

Kshirsagar, R.D. & Singh, N.P. (2000a). Less known ethnomedicinal uses of plants reported by Jenu Kuruba tribe of Mysore district, southern India. *Ethnobotany 12*, 118–22.

Kshirsagar, R.D. & Singh, N.P. (2000b). Less known ethnomedicinal uses of plants in Coorg district of Karnataka state, southern India. *Ethnobotany. 12*, 12–6.

Kshirsagar, R.D. & Singh, N.P. (2001). Some less known ethnomedicinal uses from Mysore and Coorg districts, Karnataka State, India. *J. Ethnopharmacol. 75*, 231–8.

Kulkarni, C.G. & Deshpande, A. (2011). Folk therapies of Katkaris from Maharashtra. *Indian J. Trad. Knowl. 10(3)*, 554–558.

Kumar, G.S. & Manickam, V.S. (2008). Ethnobotanical utilization of *Poecilineuron pauciflorum* Bedd. by the Kani Tribes of Agasthiamalai, Western Ghats, Tamil Nadu, India. *Ethnobotanical Leaflets, 12*, 719–22.

Kumar, J.I.N., Soni, H. & Kumar, R.N. (2005). Aesthetic values of selected floral elements of Khatana and Waghai forests of Dangs, Western Ghats. *Indian J. Trad. Knowl. 4(3)*, 275–286.

Kumar, J.I.N., Kumar, R.N., Patil, N. & Soni, H. (2007). Studies on plant species used by tribal communities of Saputara and Purna forests, Dangs district, Gujarat. *Indian J. Trad. Knowl. 6(2)*, 368–374.

Kumar, J.I.N., Soni, H. & Kumar, R.N. (2004). Ethnobotanical values of certain plant species of Dang Forest, extreme northern parts of Western Ghats, south Gujarat, India. *J. Curr. Biosc. 2(1)*, 63–74.

Kumar P.S., Tambhekar, N. & Nayak, S.U. (2014). Ethnobotanical survey of drugs used in South India for respiratory disorders: A Meta-Analysis. *Photon, 121*, 787–792.

Kumar, S.S., Samydurai, P. & Nagarajan, N. (2014). Indigenous knowledge on some medicinal pteridophytic plant species among the Malasar Tribe's in Valparai hills, Western Ghats of Tamilnadu. *American J. Ethnomedicine, 1(3)*, 164–173.

Mahesh, T. & Shivanna, M.B. (2004). Ethno-botanical wealth of Bhadra wild life Sanctuary in Karnataka. *Indian J. Trad. Knowl. 3(1)*, 37–50.

Mahishi, P., Srinivasa, B.H. & Shivanna, M.B. (2005). Medicinal plant wealth of local communities in some villages in Shimoga District of Karnataka, India. *J. Ethnopharmacol., 98*, 307–312.

Malhotra, K.C., Gokhale, Y., Chatterjee, S. & Srivastava, S. (2001). Cultural and ecological dimensions of Sacred groves in India. Indian National Science Academy, New Delhi & Indira Gandhi Rashtriya Manav Sangrahalaya, Bhopal.

Mali, P.R. (2012). Ethnobotanical studies of Peth and Trimbakeshwar district Nashik, Maharashtra, India. *Trends in Life Sciences 1(4)*, 35–37.

Manikandan, P.N.A. (2005). Folk herbal medicine: A survey on the Paniya tribes of Mundakunnu village of the Nilgiri Hills, South India. *Ancient Science of Life, 25(1)*, 21–27.

Manilal, K.S. (1981). Ethnobotany of the rice's of Malabar In: S.K. Jain (Ed.). Glimpses of Indian Ethnobotany. Oxford & IBH, New Delhi, pp. 297–307.

Maridass, M. & Victor, B. (2008). Ethnobotanical Uses of *Cinnamomum* Species, Tamil Nadu, India. *Ethnobotanical Leaflets 12*, 150–155.

Mathew, S.P., Mohandas, A., Shareef, S.M. & Nair, G.M. (2006). Biocultural diversity of the Endemic 'Wild Jack Tree' on the Malabar Coast of South India. *Ethnobotany Research & Applications.* 25–40 (http://hdl.handle.net/10125/235).

Mohan, V.R., Rajesh, A., Athiperumalsami, T. & Sutha, S. (2008). Ethnomedicinal Plants of the Tirunelveli District, Tamil Nadu, India. *Ethnobotanical Leaflets 12*, 79–95.

Muthukumarasamy, S., Mohan, V.R., Kumaresan, S. & Chelladurai, V. (2004). Traditional medicinal practices of Palliyar tribe of Srivilliputhur in antenatal and post – natal care of mother and child. *Natural Product Radiance. 3(6)*, 422–426.

Nanjunda, D.C. (2010). Ethno-medico-botanical investigation of Jenu Kuruba ethnic group of Karnataka State, India. *Bangladesh J. Med. Sci., 9(3)*, 161–169.

Narayanan, M.K.R., Mithunlal, S., Sujanapal, P., Anil Kumar, N., Sivadasan, M., Alfarhan, A.H. & Alatar, A.A. (2011). Ethnobotanically important trees and their uses by Kattunaikka tribe in Wayanad Wildlife Sanctuary, Kerala, India. *J. Med. Plants Res. 5(4)*, 604–612.

Navaneethan, P., Sunil Nautiyal, Kalaivani, T. & Rajasekaran, C. (2011). Cross-cultural ethnobotany and conservation of medicinal and aromatic plants in the Nilgiris, Western Ghats: A case study. *Medicinal Plants, 3(1)*, 27–45.

Noorunnisa Begum, S., Ravikumar, K., Vijaya Sankar, R. & Ved, D.K. (2004). Profile of medicinal plants diversity in Tamil Nadu MPCAs. In: Abstracts of XIV Annual Conference of IAAT and National Seminar on New Frontiers in Plant Taxonomy and Biodiversity Conservation. TBGRI, Thiruvananthapuram. p. 97.

Noorunnisa Begum, S., Ravikumar, K., Vijaya Sankar, R. & Ved, D.K. (2005). Profile of medicinal plants diversity in Kerala MPCAs: In: N. Sasidharan et al. (Eds.). Medicinal Plants of Kerala: Conservation & Beneficiation. Compendium on the focal Theme of 17th Kerala Science Congress. pp. 23–28.

Noorunisa Begum, S., Ravikumar, K. & Ved, D.K. (2014). *"Asoka"* – A Culturally and traditionally important medicinal plant: Its' Present Day Market Demand and Conservation Measures in India. *Current Sci. 107(1)*, 26–28.

Oak, G., Kurve, P., Kurve S. & Pejaver, M. (2015). Ethno-botanical studies of edible plants used by tribal women of Thane District. *J. Medicinal Plant Studies, 3(2)*, 90–94.

Pal, D.C. & Jain, S.K. (1998). Tribal medicine. Naya Prokash, Calcutta.

Palekar, R.P. (1993). Ethno-medical traditions of Thakur tribals of Karjat, Maharashtra. *Anc. Sci. Life 12*, 388–393.

Pandiarajan, G., Govindaraj, R., Makesh Kumar, B. & Sankarasivaraman, K. (2011). A survey of medicinally important plants in around Srivilliputtur Taluk, Virudhunagar District, Tamil Nadu. *Intern. J. Pharm. Res., Dev.* – Online (IJPRD)/PUB/ARTI/VOL-3/ISSUE-3/ April/009, 77–86.

Parinitha, M., Harish, G.U., Vivek, N.C., Mahesh, T. & Shivanna, M.B. (2004). Ethnobotanical wealth of Bhadra Wild Life Sanctuary in Karnataka. *Indian J. Trad. Knowl., 3(1)*, 37–50.

Patil, H.M. & Bhaskar, V.V. (2006). Medicinal Knowledge system of tribals of Nandurbar district, Maharashtra. *Indian J. Trad. Knowl., 5(3)*, 327–330.

Patil, K.J. & Patil, S.V. (2012). Biodiversity of vulnerable and endangered plants from Jalgaon district of North Maharashtra. *Asian J. Pharm. Life Sci., 2(2)*, 144–150.

Patil, M.V. & Patil, D.A. (2005). Ethnomedicinal practices of Nasik district, Maharashtra. *Indian J. Trad. Knowl., 4(3)*, 287–290.

Patil, S.I. & Patil, D.A. (2007). Ethnomedicinal plants of Dhule district, Maharashtra. *Nat. Prod. Rad. 6(2)*, 148–151.

Paul, T. & Prajapati, M.M. (2014). Ethno-therapeutic remedies for Bone Fracture, Dang District. Gujarat, India. *J. Pharm. Biol. Sci. 9(2)*, 45–50.

Paulsamy, S., Vijayakumar, K.K., Murugesan, M., Padmavathy, S. & Senthilkumar, P. (2007). Ecological status of medicinal and other economically important plants in the Shola under-stories of Nilgiris, the Western Ghats. *Nat. Prod. Rad., 6(1)*, 55–61.

Patwardhan, B. (2005). Ethnopharmacology and drug discovery. *J. Ethnopharmacol., 100*, 50–52.

Patwardhan, B., Vaidya, A.D.B. & Chorghade, M. (2004). Ayurveda and natural products drug discovery. *Curr. Sci. 86*, 789–799.

Pesek, T.J., Helton, L.R., Reminick, R., Kannan, D. & Nair, M. (2008). Healing Traditions of Southern India and the conservation of culture and biodiversity: A preliminary study. *Ethnobot. Res., Applications 6*, 471–479.

Poornima, G., Manasa, M., Rudrappa, D. & Prashith Kekuda, T.R. (2012). Medicinal plants used by herbal healers in Narasipura and Manchale villages of Sagara taluk, Karnataka, India. *Sci. Technol. Arts Res. J., 1(2)*, 12–17.

Powers, S. (1874). Aboriginal Botany. *Proc. Calif. Acad. Sci. 5*, 373–379.

Pradheeps, M. & Poyyamoli, G. (2013). Ethnobotany and utilization of plant resources in Irula villages (Sigur plateau, Nilgiri Biosphere Reserve, India). *J. Med. Plants Res. 7(6)*, 267–276.

Prakash, B.N. & Unnikrishnan, P.M. (2013). Ethnomedical survey of herbs for the management of malaria in Karnataka, India. *Ethnobot Res Applications. 11*, 289–98.

Prakash, J.W., Raja, R.D.A., Anderson, N.A., Williams, C., Regini, G.S., Bensar, K., Rajeev, R., Kruba, S., Jeeva, S. & Das, S.S.M. (2008). Ethnomedicinal plants used by Kani tribes of Agasthiyarmalai biosphere reserve, Southern Western Ghats. *Indian J. Trad. Knowl., 7(3)*, 410–413.

Prakasha, H.M., Krishnappa, M., Krishnamurthy, Y.L. & Poornima, S.V. (2010). Folk medicine of NR Pura taluk in Chikmagalur district of Karnataka. *Indian J. Trad. Knowl. 9 (1)*, 55–60.

Prasad, P.N. & Abraham, Z. (1984). Ethnobotany of Nyadis of Kerala. *J. Econ. Taxon. Bot. 5*, 1.

Prasad, P.N., Jabadhas, J.W. & Janaki Ammal, E.K. (1987). Medicinal plants used by the Kanikkars of South India. *J. Econ. Taxon. Bot. 11*, 149–155.

Prasad, P.N., Devi, V.N.M., Syndia, L.A.M., Rajakohila, M. & Ariharan, V.N. (2012). Ethnobotanical studies on Thozhukanni and Azhukanni among the Kanikkars of South India. *Int. J. Pharm. Sci. Rev. Res., 14(2)*, 135–138.

Pushpakarani, R. & Natarajan, S. (2014). Ethnomedicines used by Kaniyakaran tribes in Kaniyakumari district, Southern Western Ghats of Tamil Nadu, India. *J. Applied Pharmaceutical Sci. 4(2)*, 56–60.

Pushpangadan, P., Rajasekharan, S., Ratheesh Kumar, P.K., Jawahar, C.R. & Saradmmal, L. (1988). Aroyogyappacha (*Trichopus zeylanicus* Gaertn). The Ginseng of Kani Tribes of Agasthyar Hill (Kerala) for evergreen health and vitality. *Ancient Science of Life. 7,* 13–16.

Rajakumar, N. & Shivanna, M.B. (2010). Traditional Herbal Medicinal Knowledge in Sagar Taluk of Shimoga District, Karnataka, India. *Indian J. Nat. Prod. Res., 1(1),* 102–108.

Rajan, S., Baburaj, D.S., Sethuraman, M. & Parimala, S. (2001). Stem and stem bark used medicinally by the tribals Irulas and Paniyas of Nilgiri District, Tamil Nadu. *J. Natural Remedies. 1/1,* 49–54.

Rajan, S., Jayendran, M. & Sethuraman, M. (2003). Medico-ethnobotany: A study on the Kattunayaka Tribe of Nilgiri Hills, Tamil Nadu. *J. Natural Remedies. 3/1,* 68–72.

Rajendran, A. & Henry, A.N. (1994). Plants used by the tribe Kadar in Anamalai hills of Tamil Nadu. *Ethnobotany 6,* 19–24.

Rajendran, A., Rama Rao, N., Ravikumar, K. & Henry, A.N. (1997). Some medicinal orchids of southern India. *Ancient Science of Life, 17(1),* 10–14.

Rajendran, A., Ravikumar, K. & Henry A.N. (2000). Plant Genetic Resources and Knowledge of Traditional Medicine in Tamil Nadu. *Ancient Science of Life. 20(1&2),* 25–28.

Rajendran, A., Ravikumar, K. & Henry, A.N. (2002). Some useful rare and endemic plants of the southern Western Ghats. *J. Econ. Tax. Bot. 26(1),* 181–184.

Rajendran, S.M., Chandra Sekar, K. & Sundaresan, V. (2002). Ethnomedicinal lore of Valaya tribals in Seithur Hills of Virudunagar district, Tamil Nadu, India. *Indian J. Trad. Knowl. 1(1),* 59–71.

Rajith, N.P. & Ramachandran, V.S. (2010). Ethnomedicines of Kurichyas, Kannur district, Western Ghats, Kerala. *Indian J. Natural Prod. Resources. 1(2),* 249–253.

Rajput, A.P. & Yadav, S.S. (1998). Medico-Botanical and phytochemical studies on medicinal plants of Dhule and Nandurbar districts of Maharashtra state. *J. Phytol. Res. 13(2),* 161.

Ramachandran, V.S. (1987). Further notes on the ethnobotany of Cannanore district, Kerala. *J. Econ.. Taxon. Bot. 11,* 47–50.

Ramachandran, V.S., Joseph, S. & Aruna, R. (2009). Ethnobotanical studies from Amaravathy Range of Indira Gandhi Wildlife Sanctuary, Western Ghats, Coimbatore District, Southern India. *Ethnobotanical Leaflets, 13,* 1069–1087.

Ramachandran, V.S. & Manian, S. (1982). Ethnobotanical notes on the Irulars, Puliyars and Koravas of Coimbatore district, Tamil Nadu. *Indian Bot. Reptr. 8(2),* 85–91.

Rmachandran, V.S. & Nair, N.C. (1981). Ethnobotanical observations on Irulars of Tamil Nadu, India. *J. Econ. Taxon. Bot. 2,* 183–190.

Ramachandran, V.S., Selvalakshmi, S. & Betty, T. (2014). Floral diversity of Karian shola MPCA, Coimbatore district, Tamilnadu, with special emphasis on the conservation of RET and Endemic plants of Anamalai hills. *Elixir Appl. Botany, 66,* 20653–20655.

Ramana, P., Nayak, R. & Sunti, A. (2011). Plants used for oral care in Sirsi region of Uttar Kannada district of Karnataka. *My Forest, 47(4),* 295–301.

Rani, S.L., Devi, V.K., Soris, P.T., Maruthupandian, A. & Mohan, V.R. (2011). Ethnomedicinal plants used by Kanikkars of Agasthiarmalai Biosphere Reserve, Western Ghats. *J. Ecobiotech., 3(7),* 16–25.

Rasingam, L. (2012). Ethnobotanical studies on the wild edible plants of Irula tribes of Pillur Valley, Coimbatore district, Tamil Nadu, India. *Asian Pacific J. Tropical Biomed.,* S1493–S1497.

Revathi, P. & Parimelazhagan, T. (2010). Traditional knowledge on medicinal plants used by the Irula Tribe of Hasanur hills, Erode district, Tamil Nadu, India. *Ethnobotanical Leaflets*, *14*, 136–60.

Revathi, P., Parimelazhagan, T. & Manian, S. (2013). Ethnomedicinal plants and novel formulations used by Hooralis tribe in Sathyamangalam forests, Western Ghats of Tamil Nadu, India. *J. Med. Plants Res., 7(28)*, 2083–2097.

Rothe, S.P. (2003). Ethnomedicinal plants from Katepurna wildlife sanctuary of Akola district. *Indian J. Trad. Knowl., 2(1)*, 378–382.

Roxburgh, W. (1832). Flora Indica. 3 vols. In: W. Thackers, Serampore.

Sabnis, S.D. & Bedi, S.J. (1983). Ethnobotanical studies in Dadra-Nagar Haveli and Daman. *Indian J. For. 6(1)*, 65–69.

Sajeev, K.K. & Sasidharan, N. (1997). Ethnobotanical observations on the tribals of Chinnar Wildlife Sanctuary. *Ancient Science of Life. 16(4)*, 284–292.

Salave, A.P., Reddy, P.G. & Diwakar, P.G. (2011). Some unreported ethnobotanical uses from Karanji Ghat areas of Pathardi Tahasil in Ahmednagar district (M.S.) India. *Intern. J. Applied Biol. Pharmaceutical Tech. 2(4)*, 240–245.

Samy, R.P., Thwin, M.M., Gopalakrishnakone, P. & Ignacimuthu, S. (2008). Ethnobotanical survey of folk plants for the treatment of snake bites in Southern part of Tamilnadu, India. *J. Ethnopharmacol., 115(2)*, 302–312.

Sarvalingam, A., Rajendran, A. & Aravindhan, V. (2011). Curative Climbers of Maruthamalai Hills in the Southern Western Ghats of Tamil Nadu, India. *Int. J. Med. Arom. Plants. 1(3)*, 326–332.

Sathyavathi, R. & Janardhanan, K.J. (2011). Folklore medicinal practices of Badaga community in Nilgiri Biosphere Reserve, Tamilnadu, India. *Intern. J. Pharma. Res., Dev.* – Online Publication ijprd/pub/arti/vol-3/issue-2/April/007.

Shalini, C.B., Chidambaram Pillai, S. & Mohan, V.R. (2014a). Ethnomedicinal plants used by the Kanikkars of Southern Western Ghats. *Int. J. Pharm. Sci. Rev. Res., 28(2)*, 101–107.

Shalini, C.B., Chidambaram Pillai, S. & Mohan, V.R. (2014b). Ethnomedicinal plants used for the treatment of skin-related ailments by the Kanikkars, an indigenous tribe inhabiting Southern Western Ghats. *Intern. J. Advanced Res. 2(8)*, 368–377.

Sharma, P.P. & Singh, N.P. (2001), Ethnobotany of Dadra Nagar Haveli and Daman (Union Territory). Director BSI, Calcutta.

Shanavaskhan, A.E., Sivadasan, M., Alfarhan, A.H. & Thomas, J. (2012). Ethnomedicinal aspects of Angiospermic epiphytes and parasites of Kerala, India. *Indian J. Trad. Knowl. 11(1)*, 250–258.

Sharma, T. (2014). India's Western Ghats: Biodiversity and Medicinal Plants. Retrieved from http://www.eoearth.org/view/article/53df77740cf2541de6d027c5-10/06/2015.

Sharmila, S., Kalaichelvi, K. & Abirami, P. (2015). Ethnopharmacobotanical informations of some herbaceous medicinal plants used by Toda tribes of Thiashola, Manjoor, Nilgiris, Western Ghats, Tamilnadu, India. *Int. J. Pharmaceut. Sci. Res., 6(1)*, 315–320.

Sharmila, S., Kalaichelvi, K., Rajeswari, M. & Anjanadevi, N. (2014). Studies on the folklore medicinal uses of some indigenous plants among the tribes of Thiashola, Manjoor, Nilgiris South Division, Western Ghats. *Intern. J. Pharmaceutical Sci. Res. 4(3)*, 14–22.

Shiddamallayya, N., Yasmeen, A. & Gopakumar, K. (2010). Medico-botanical survey of Kumar Parvatha Kukkke Subramanya, Mangalore, Karnataka. *Indian J. Trad. Knowl. 9(1)*, 96–99.

Shivanna, M.B. & Rajakumar, N. (2010). Ethno-medico-botanical knowledge of rural folk in Bhadravathi taluk of Shimoga district, Karnataka. *Indian J. Trad. Knowl., 9(1)*, 158–162.

Silja, V.P., Samitha Varma, K. & Mohanan, K.V. (2008). Ethnomedicinal plant knowledge of the Mullu Kuruma tribe of Wayanad district, Kerala. *Indian J. Trad. Knowl. 7(4),* 604–612.

Simon, S.M., Norman, T.S.J., Suesh, K. & Ramachandran, V. (2011). Ethnobotanical knowledge on single drug remedies from Idukki district, Kerala for the treatment of some chronic diseases. *Intern. J. Res. Ayurveda & Pharmacy 2(2),* 531–534.

Singh, E.A., Kamble, S.Y., Bipinraj, N.K. & Jagtap, S.D. (2012). Medicinal plants used by the Thakar tribes of Raigad district, Maharashtra for the treatment of snake-bite and scorpion-sting. *Intern. J. Phytotherapy Res., 2(2),* 26–35.

Sivakumar, A. & Murugesan, M. (2005). Ethnobotanical Studies on the Wild Edible Plants used by the Tribals of Anaimalai hills, the Western Ghats. *Ancient Science of Life, 25(2),* 69–73.

Sivasankari, B., Pitchaimani, S. & Marimuthu Anandharaj, M. (2013). A study on traditional medicinal plants of Uthapuram, Madurai District, Tamilnadu, South India. *Asian Pac J. Trop Biomed. 3(12),* 975–979.

Smitha Kumar, P., Abdul Latheef, K. & Remashree, A.B. (2014). Ethnobotanical survey of Diuretic and Antilithiatic medicinal plants used by the traditional practitioners of Palakkad District. *Intern. J. Herbal Medicine, 2(2),* 52–56.

Soejarto, D.D. & Fonf, H.H.S. et al. (2005). Ethnobotany/ethnopharmacology and mass bioprospecting issues on intellectual property and benefit-sharing. *J. Ethnopharmacol. 100,* 15–22.

Soman, G. (2014). Some ethnomedicinal plants of Panhala taluka used as Anti Allergics. *Bull. Env. Pharmacol. Life Sci., 3(6),* 128–131.

Soman, S.G. (2011). Diversity of Ethnomedicinal plants used by Tribals of Karjat Taluka in Maharashtra, India. *Indian J. Applied & Pure Bio. 26(1),* 75–78.

Sripathi, S.K. & Sankari, U. (2010). Ethnobotanical documentation of a few medicinal plants in the Agasthiayamalai region of Tirunelveli district, India. *Ethnobotanical Leaflets, 14,* 173–81.

Subramanian, A., Mohan, V.R., Kalidass, C. & Maruthupandian, A. (2010). Ethno-medico botany of the Valaiyans of Madurai District, Western Ghats, Tamil Nadu. *J. Econ. Taxon. Bot. 34(2),* 363–379.

Sukumaran, S.A., Jeeva, S., Raj, A.D.S. & Kannan, D. (2008). Floristic diversity, conservation status and economic value of miniature Sacred Groves in Kanyakumari district, Tamil Nadu, Southern Peninsular India. *Turk. J. Bot., 32,* 185–199.

Sukumaran, S. & Raj, A.D.S. (2008). Rare and Endemic Plants in the Sacred Groves of Kanyakumari District in Tamil Nadu. *Indian J. Forestry, 31(4),* 611–616.

Suresh, M., Ayyanar, M., Amalraj, L. & Mehalingam, P. (2012). Ethnomedicinal plants used to treat skin diseases in Pothigai hills of Western Ghats, Tirunelveli district, Tamil Nadu, India. *J. Biosci. Res., 3(1),* 112–121.

Sutha, S., Mohan, V.R., Kumaresan, S., Murugan, C. & Athiperumalsami (2010). Ethnomedicinal plants used by the tribals of Kalakad-Mundanthurai Tiger Reserve (KMTR), Western Ghats, Tamil Nadu for the treatment of rheumatism. *Indian J. Trad. Knowl., 9(3),* 502–509.

Thomas, B., Arumugam, R., Veerasamy, A. & Ramamoorthy, A. (2014). Ethnomedicinal plants used for the treatment of cuts and wounds by Kuruma tribes, Wayanadu districts of Kerala, India. *Asian Pacific J. Tropical Biomedicine. 4,* Supplement 1, S488–S491.

Thomas, B. & Rajendran, A. (2013). Less known ethnomedicinal plants used by Kurichar Tribe of Wayanad District, Southern Western Ghats Kerala, India. *Bot. Res. Intern., 6(2)*, 32–35.

Thomas, J. & De Britto, A.J. (1999). An ethnobotanical survey of Naduvil Panchayat in Kannur district Kerala. *Ancient Science of Life. 18 (3&4)*, 279–283.

Udayan, P.S., Begum, S.N., Mudappa, A. & Singh, A. (2003). Medicinal Plants Diversity of Karnataka MPCAs. *J. Econ. Taxon. Bot., 27(3)*, 635–639.

Udayan, P.S., George, S., Tushar, K.V. & Balachandran, I. (2005). Medicinal plants used by the Kaadar tribes of Sholayar forest Thrissur district, Kerala. *Indian J. Trad. Knowl. 4(2)*, 159–163.

Umapriya, T., Rajendran, A., Aravindha, V., Thomas, B. & Maharajan, M. (2011). Ethnobotany of Irular tribe in Palamalai Hills, Coimbatore, Tamil Nadu. *Indian J. Naural Prod. Res. 2(2)*, 250–255.

Upadhya, V., Hegde, H.V., Bhat, S., Hurkadale, P.J., Kholkute, S.D. & Hegde, G.R. (2012). Ethnomedicinal plants used to treat bone fracture from North-Central Western Ghats of India. *J. Ethnopharmacol., 142*, 557–562.

Vijayalakshmi, N., Anbazhagan, M. & Arumugam, K. (2014). Studies on ethno-medicinal plants used by the Irulas tribe of Thirumurthi Hill of Western Ghats, Tamil Nadu, India. *Intern. J. Res. Plant Sci. 4(1)*, 8–12.

Vijayan, A., Liju, V.B., Reena John, J.V., Parthipan, B. & Renuka, C. (2006). Traditional remedies of Kani tribes of Kottoor reserve forest, Agasthyavanam, Thiruvananthapuram, Kerala. *Indian J. Trad. Knowl., 6(4)*, 589–594.

Vikneshwaran, D., Viji, M. & Raja Lakshmi, K. (2008). Ethnomedicinal plants survey and documentation related to Paliyar community. *Ethnobotanical Leaflets, 12*, 1108–15.

Viswanathan, M.B., Premkumar, E.H. & Ramesh, N. (2001). Ethnomedicines of Kanis in Kalakkad Mundanthurai Tiger Reserve, Tamil Nadu. *J. Econ. Taxon. Bot. 11*, 149–155.

Viswanathan, M.B., Prem Kumar, E.H. & Ramesh, N. (2006). Ethnobotany of the Kanis (Kalakkad Mundanthurai Tiger Reserve in Tirunelveli District, Tamil Nadu, India). Bishen Singh Mahendra Pal Singh, Dehradun.

Warrier, N. & Ganapathy (2001). Some Important Medicinal Plants of the Western Ghats, India – A Profile. Arya Vaidyasala (Kerala), MAPPA & IDRC (Canada).

Xavier, T.F., Kannan, M., Lija, L., Auxillia, A., Rose, A.K. & Senthilkumar, S. (2014). Ethnobotanical study of Kani tribes in Thoduhills of Kerala, South India. *J. Ethnopharmacol., 152*, 78–90.

Yabesh, J.E., Prabhu, S. & Vijayakumar, S. (2014). An ethnobotanical study of medicinal plants used by traditional healers in Silent Valley of Kerala, India. *J. Ethnopharmacol. 154*, 774–789.

Yesodharan, K. & Sujana, K.A. (2007). Ethnomedicinal knowledge among Malamalasar tribe of Parambikulam wildlife sanctuary, Kerala. *Indian J. Trad. Knowl. 6(3)*, 481–485.

Yoganarasimhan, S.N. (1996). Medicinal plants of India, Karnataka, Volume 1. Interline Publishers Pvt. Ltd.

CHAPTER 7

CONTEMPORARY RELEVANCE OF ETHNO-VETERINARY PRACTICES AND A REVIEW OF ETHNO-VETERINARY MEDICINAL PLANTS OF WESTERN GHATS

M. N. B. NAIR[1] and N. PUNNIAMURTHY[2]

[1]Trans-disciplinary University, Veterinary Ayurveda Group, School of Health Sciences, 74/2, Jarakabandekaval, Attur Post, Yelahanka, Bangalore, India, E-mail: nair.mnb@frlht.org

[2]Veterinary University Training and Research Centre, Pillayarpatty, Thanjavur–Trichirapally National Highway (Vallam Post), Near RTO's Office, Thanjavur – 613403, India, E-mail: thanjavurvutrc@tanuvas.org.in

CONTENTS

ABSTRACT

The Veterinary science in India has a documented history of around 5000 years. The veterinary and animal husbandry practices are mentioned in *Rigveda* (2000–1400 BC) and *Atharvaveda*.

India is the world's largest milk producer. The increase in the advocacy of exotic breed for higher milk production has led to many problems. There is high incidence of disease in cross-breed animals and indiscriminate use of antibiotics and other veterinary medicine in dairy animals leading to high veterinary drug residues in the various animal products. People use milk, meat, eggs and other dairy products as their food. The widespread use of antimicrobials and poor infection control practices in livestock management lead to the antibiotic residue in the milk and other animal products, and encourages spread of antimicrobial resistance (AMR). This could cause health hazards to consumers. By 2050 the death by AMR is estimated to be 10 million. The unique Ethno-veterinary heritage of India, documented, rapidly assessed and subsequently mainstreamed into the livestock management among veterinarians and farmers as the first line of treatment for management of animal health conditions. This could lead reduction of use of antibiotics and associated AMR. Nearly 10% of the native plant species, including those having medicinal properties in the Western Ghats are part of the endangered species list. 281 species from Western Ghats are listed with their specific use for animal health.

7.1   INTRODUCTION

Livestock rearing is considered as a supplementary occupation to many farmers and also a source of additional income for those engaged in agricultural operations in India. Decline in the animal husbandry budget for veterinary services has led to the scanty veterinary services provided by the government to the poor in the rural areas (Anonymous, 2004). Prevention, control and eradication of diseases among domesticated animals are major concern as diseases in animals will lead to economic loses and possible transmission of the causative agents to humans. Veterinary services have a crucial role in controlling highly contagious diseases and zoonotic infections, which have implications for human health as well as that of livestock. Mainstreaming EVP for livestock keeping is essential for reducing the antibiotic use and for sustainable livestock production.

People use milk, meat, eggs and other dairy products as their food. Indiscriminate use of antibiotics and other chemical veterinary medicine in dairy animals cause high veterinary drug residues in the animal products leading to Antimicrobial resistance (Hill and McLaughlin, 2006). Food safety and Standard act 2006 of India, Chapter 4 Item 21 indicates that pesticide, veterinary drug residues, antibiotic residues and microbial counts (1) "No article of food shall contain insecticide or pesticide residues veterinary drug residues, antibiotic residues, solvent residues, pharmacologically active substances and microbial counts in excess of such tolerance limit as may be specified by regulations." However, there is no effective government regulation to control antibiotic use in human beings and domestic animals in India. The widespread use of antimicrobials and poor infection control practices in livestock management lead to the antibiotic residue in the milk and other animal products, and encourages spread of antimicrobial resistance (AMR).

The projection of deaths by 2050 due to AMR per year will be 10 million compared to other major causes of death (Cancer 8.2 million), (Asia: 4,730,000 every year, Africa: 4150,00 every ear, Europe: 390,000, North America: 319,000, Latin America: 392,00). By 2050, world can expect to lose between 60 and 100 trillion USD due to AMR. 100,000–200,000 tons of antibiotics are used worldwide. US Production is 22.7 million kilograms and 70% of antibiotics dispensed are given to healthy livestock to prevent infections and/or promote growth. This practice is outlawed in EU countries (Md. Nadeem Fairoze 2015 – Presented in the International conference at TDU April 30–May 1st, 2015).

TDU and TANUVAS had documented Ethno-veterinary practices from 24 locations in 10 states, rapidly assessed them and established that 353 out of 460 formulations documented are safe and efficacious. A total of 140 medicinal plants were used in these formulations. The unique Ethno-veterinary heritage of India, assessed subsequently mainstreamed into the livestock management among veterinarians and farmers. The safe and efficacious formulations used as the first line of treatment for management of animal health conditions subsequently leading to the reduction of antibiotics. An observational study indicate 97% efficacy of EVP for mastitis, 99% in Foot and mouth disease, 100% in enteritis, udder pox, pododermitis, udder oedema, arthritis, pesticide/mimosine poisoning and downer (Table 7.1). An intervention impact analysis indicates a trend in reduction of antibiotic residue in the milk after training the farmers and veterinarians to use EVP for 15 clinical conditions in animals in selected areas. The goal of judicious use of antibiotic has to be achieved through creating awareness among the veterinary doctors, dairy cooperatives and people, by training them to use these effective herbal alternatives to antibiotics. The use of antibiotics and other chemical veterinary drugs have in fact been banned for animal health care in many countries and the world is looking for safer herbal alternatives.

TABLE 7.1 Efficacy of Herbal Medicine –Field Study

No	Conditions	Number treated with EVM alone	% of complete cure	EVM +Veterinary drug	% of complete cure
1	Mastitis	314	97	130	97
2	FMD	829	99	63	98
3	Enteritis	81	100	0	
4	Udderpox	66	100	0	
5	Pododermitis	5	100	0	
6	Udder edima	10	100	0	
7	Arthritis	10	100	0	
8	Pesticide/Mimosine poisoning	4	100	0	
9	Downer	4	100	0	
10	Post-partum complication	30	99	0	

Ethno-veterinary practices were used by humans from ancient times to take care of the health of domesticated, which are recorded in the river valley civilizations. Egyptian had used knowledge of more than 250 medicinal plants and 120 mineral salts (Swarup and Patra, 2005). The veterinary science in India has a documented history of around 5000 years. The veterinary and animal husbandry practices are mentioned in *Rigveda* (2000–1400 BC) and *Atharvaveda*. A detailed historical perspective ethno-veterinary practices is available in the publication by Nair and Unnikrishnan (2010).

In India the veterinary medical knowledge can be classified into codified traditions and folk medicine. In the 'codified' traditions, the medical knowledge is documented and presented in thousands of medical manuscripts, such as *Mrugayurveda* and *Hastyayurveda*. The medical literature from all these traditions is in both classical and regional script and languages. There is no 'exhaustive' catalog of the corpus of medical literature available in any of these medical traditions.

The folk traditions are oral and are passed on from one generation to the other by word of mouth. They are dynamic, innovative, and evolving. These oral or folk medical traditions are extremely diverse, since they are rooted in natural resources located in so many different ecosystems and community. Ethno-veterinary practices comprise of belief, knowledge, practices and skills pertaining to health care and management of livestock. They form an integral part of the family and play an important social, religious and economic role. As the ethno-veterinary practices are eco-system and ethnic-community specific, the characteristics, sophistication, and intensity of these practices differ greatly among individuals, societies, and regions. There are local healers and livestock raisers (both settled and nomadic) who are knowledgeable and experienced in traditional veterinary health care and are very popular in their communities. EVP has great potential to address current challenges faced by veterinary medicine as they are safe, efficacious and create no adverse effects in the animals. The Ethno-veterinary traditions can take care of wide range of ailments. However, they are facing the threat of rapid erosion. According to Anthropological Survey of India (ASI), there are 4635 ethnic communities in India. In principle, each of these communities could be having their own oral medical traditions for human and animals that have been evolving across the time and space. The urgent revival of these traditional veterinary practices is a high priority in the light of the constraints of modern medicine and the benefits of these practices.

An annotated bibliography on the Indian ethno-veterinary research was published by Ramdas and Ghtoge (2004). The Indian Council of Agricultural Research in the year 2000 collected 595 veterinary traditions from different sources and recorded (Swarup and Patra, 2005). About 48 of them were recommended for scientific validation and some have shown therapeutic and ameliorative potential. In another work on 'identification and evaluation of medicinal plants for control of parasitic diseases in livestock,' 158 plants have been cataloged and 50 have been evaluated for anti-parasitic activity (Anonymous, 2004). Ethno-veterinary practices on eye diseases, helminthiasis, repeat breeding, corneal opacity, simple anorexia, lacerated wounds, mastitis and the common health problems in cattle and their management were described by Nair (2005) and Nair and Unnikrishnan (2010).

Ethno-veterinary research should put more emphasis on first, documentation of the knowledge in new unexploited areas, and secondly, on learning procedures and methods as used in tradition knowledge. Studies on Pharmacognosy in ethno-veterinary medicine seem to be very limited. The reason for this is that since many pharmacognosy studies have been done on plants used in human medicine and several medicinal plants used for humans and animals are often overlap, the pharmacognosy of many ethno-veterinary plants does not need to be repeated for animals. For validation of the ethno-veterinary medicine all factors including the prevailing culture and belief should be considered (Mathias, 2006).

7.2   THE WESTERN GHATS OR SAHYADRI

The Western Ghats or Sahyadri are a unique mountain range that runs almost parallel to the western coast of the India. Western Ghats are amongst the 34 biodiversity hot-spots identified in the world. The Western Ghats are region of immense global importance for the conservation of biological diversity and also contain areas of high geological, cultural and esthetic values. Besides the Western Ghats contains numerous medicinal plants, it also houses large number of economically important genetic resources of the wild relatives of grains, fruits and spices plants (Conservation International: Biodiversity Hotspots – Western Ghats and Sri Lanka (March, 2011) accessed http://www.arkive.org/eco-regions/western-ghats/image-H436-10/06/2015. Western Ghats directly and indirectly supports the livelihoods of over 200 million people through ecosystem services. In addition to rich biodiversity, the Western Ghats is home to diverse social, religious, and linguistic groups.

Nearly 10% of the native plant species, including those having medicinal properties, in the Western Ghats are part of the endangered species list. As the part of protection of rich wildlife 13 national parks two biosphere reserves and many wildlife sanctuaries are established across the mountain range. The Nilgiri Biosphere Reserve is the largest protected area in the Western Ghats. Spread across three different states this reserve covers an area of around 5500 sq.km. The most important national parks in the Western Ghats are Bandipur National Park, Karnataka, Kudremukh National Park, Karnataka, Chandoli National Park, Maharashtra, Eravikulam National Park, Kerala, Grass Hills National Park, Tamilnadu, Karian Shola National Park, Tamilnadu, and Silent Valley National Park, Kerala (http://en.wikipedia.org/wiki/Western_Ghats).

There are many reports on the use of ethnoveterinary plants in Western Ghats (Prasad et al., 2014; Nair, 2005; Nair and Punniamurthy, 2010). Kiruba et al. (2006) enumerated 34 ethnoveterinary plants of Cape Comorin, Tamilnadu.

Veterinary wisdom for the Western Ghats was elucidated by Rajagopalan and Harinarayan (2001). Mini and Sivadasan (2007) described the use of 39 species of flowering plants by Kurichya tribe of Wayanad district of Kerala for the treatment of diseases of domestic animals, such as cattle, dogs and poultry. Ashok et al. (2012) reported that 21 plants belonging to 15 families are used for EVP by Dhangar, Laman and Vanjaris tribes from Karanji Ghat in Ahmednagar, Maharashtra. Shimoga district has 52 species belonging to 38 families which are used for veterinary healthcare (Rajkumar and Shivanna, 2012). There are 281 species of Ethno-veterinary medicinal Plants reported from Western Ghats (Table 7.2) (Kiruba et al., 2006; Kholkunte, 2008; Deokule and Mokat, 2004; Ashok et al., 2012; Alagesaboopathi, 2015; Harsha et al., 2005; Rajkumar and Shivanna, 2012; Raveesha and Sudhama 2015; Nair and Unnikrishnan 2010; Renjini et al., 2015; Prasad et al. 2014). 198 species are reported to be used for ethno-veterinary purpose without mentioning the specific disease conditions for which they are used (Somkuwar et al., 2015). An Ethno-veterinary survey was carried out on the indigenous knowledge of tribes and folk medicine practitioners of Mallenahalli village, Chikmagalur Taluk, Karnataka. The study indicates that 52 medicinal plant species belonging to 32 families have been used to treat against anthrax, foot and mouth diseases, bloat, conjunctivitis, dysentery, fractures, snake bite, rot tail, Kasanoor forest disease (Raveesha and Sudhama, 2015).

Ethno-veterinary medicinal uses of plants from Agasthiamalai Biosphere Reserve (KMTR), Tirunelveli District, Tamilnadu was given by Kalidass et al. (2009). Clinical trials using ferns *Actiniopteris radiata, Acrostichum aureum* and *Hemionitis arifolia* for anthelmintic property on naturally infected sheep against *Haemonchus contortus* are proved to be effective (Rajesh et al., 2015). 25 formulations from 39 plant species belonging to 30 families used to treat 21 diseases conditions of domestic animals are reported from Uttar Kannada district of Western Ghats. The method of preparation, dose of each plant along with its botanical name, family and local names are reported (Harsha et al., 2005). The petroleum ether extract of leaves of *Tetrastigma leucostaphylum* (Dennst.) Alston showed acaricidal activity against *Rhipicephalus (Boophilus) annulatus* (Krishna et al., 2014). The acaricidal properties of *Acorus calamus, Aloe vera* and *Allium sativum*) have also been reported (Prathipa et al., 2014). Herbal medicines create suitable environment for the natural wound healing process (Gopalakrishnan, 2013).

TABLE 7.2   Ethnoveterimedicinal Plants of Western Ghats, Parts Used and Ethnoveterinary Uses

No	Botanical Name	Family	Plant Parts	Uses
1.	*Abelmoschus manihot* (L.) Medik.	Malvaceae	Fruit	Blood dysentery
2.	*Abrus precatorius* L.	Fabaceae	Root	Dysentery, wounds
3.	*Abutilon hirtum* (Lam.) Sweet	Malvaceae	Leaf	Enteritis, Bloat fetal membrane
4.	*Abutilon indicum* (Link) Sweet	Malvaceae	Leaves	Dysentery
5.	*Acacia concinna* (Willd.) DC.	Fabaceae	Fruits	Throat cancer
6.	*Acacia ferruginea* DC.	Fabaceae	Bark	Retention of placenta
7.	*Acacia nilotica* (L.) Willd. ex Delile	Fabaceae	Leaf, tender pods	Mouth ulcer in oxen and buffaloes; for lactation
8.	*Acalypha indica* L. (Figure 7.1b)	Euphorbiaceae	Entire plant, leaf	Constipation, skin diseases
9.	*Achyranthes aspera* L.	Amaranthaceae	Leaves, entire plant	Stomach pain, TRP, eye troubles, intestinal worms, wounds
10.	*Achyranthes bidetata* Blume	Amaranthaceae	Roots, spike and leaves	Asthma, antidote, contraceptive and night blindness
11.	*Acmella calva* (DC.) R. K. Jansen	Asteraceae	Inflorescence	Tooth ache
12.	*Acorus calamus* L.	Araceae	Rhizome	Skin ailment.
13.	*Adhatoda zeylanica* Medik.	Acanthaceae	Leaves	Wound, poison bite
14.	*Adhatoda vasica* Nees	Acanthaceae	Leaves	Fever, cough, skin diseases, ectoparasites
15.	*Aegle marmelos* Corr.	Rutaceae	Leaf, fruit	Wounds and ulcers of cattle, dysentery, fetal membrane, urinary disorders
16.	*Ageratina adenophora* (Spreng.) R. M. King & H. Rob.	Asteraceae	Leaves	Wound healing
17.	*Agave sisalana* Perrine.	Agavaceae	Leaf, root	Ear disease, wounds
18.	*Agrostis peninsularis* Hook.f.	Poaceae	Aerial parts	Fodder

TABLE 7.2 *(Continued)*

No	Botanical Name	Family	Plant Parts	Uses
19.	*Alangium salvifolium* Lam.	Alangiaceae	Root bark, leaves, stem bark	Bloat, fever (*Jvara*), intestinal disorders, opacity of cornea, fever, madness, to treat rinderpest in cattle
20.	*Albizia lebbeck* (L.) Benth.	Fabaceae	Leaf and stem bark	Anthrax, to repel lice and wasp bite
21.	*Allium cepa* L.	Liliaceae	Bulb, leaf	Foot & mouth disease, enteritis, Kasanoor Forest disease, unknown insect bite
22.	*Allium sativum* L.	Liliaceae	Bulb	Anthrax, Kasanoor Forest disease, throat cancer, Three day sickness, digestion, wounds
23.	*Aloe vera* (L.) Burm.f. (Syn. *Aloe barbadensis* Mill.)	Liliaceae	Leaf	Ticks, wound healing, swelling, reproduction, deworming, septicemias, skin diseases
24.	*Alpinia galanga* Willd.	Zingiberaceae	Rhizome	Impaction, cough, fever
25.	*Alpinia officinarum* Hance	Zingiberaceae	Rhizome	Anorexia, digestion
26.	*Alseodaphne semecarpifolia* Nees	Lauraceae	Stem bark	Rinderpest disease, dysentery
27.	*Anacardium occidentale* L.	Anacardiaceae	Bark	Broken horn
28.	*Andrographis alata* Nees	Acanthaceae	Whole plant	Snake bite, diarrhea, insect bite and scorpion stinger, skin ailments
29.	*Andrographis paniculata* Nees (Figure 2b)	Acanthaceae	Whole plant	Fever and cough
30.	*Annona squamosa* L.	Annonaceae	Leaves, seeds, fruit	Lice infestations, Maggot infestation, tick, wound, applied on the wounds of cattle to kill worms
31.	*Anogeissus latifolia* (Roxb. ex DC.) Wall. ex Guill. & Perr.	Combretaceae	Stem	Dysentery

TABLE 7.2 *(Continued)*

No	Botanical Name	Family	Plant Parts	Uses
32.	*Aporusa lindleyana* (Wight) Baillon	Euphorbiaceae	Leaves	Dog bite
33.	*Arachis hypogaea* L.	Fabaceae	Seeds, leaf	Fodder, protein content
34.	*Argyreia cuneata* Willd. ex Ker-Gawl.	Convolvulaceae	Leaves	Foetal membrane
35.	*Aristolochia bracteolata* L.	Aristolochiaceae	Leaves	Intestinal worms, poison bite, kill ticks, ephemeral fever
36.	*Aristolochia indica* L.	Aristolochiaceae	Leaves, stem	Anorexia, bloat, insect bite
37.	*Artemisia nilagirica* (Clarke) Pamp.	Asteraceae	Whole plant	Remove ectoparasites
38.	*Artocarpus lakoocha* Roxb.	Moraceae	Leaves	Bone fracture
39.	*Asparagus adscedens* Roxb.	Liliaceae	Root tuber	Enhances lactation period with superior quality
40.	*Asparagus fysoni* J. F. Macbr.	Asparagaceae	Tubers	Tonic
41.	*Asparagus racemosus* Willd.	Asparagaceae	Tubers	Stimulant, arthritis in cattle, mastitis, remove ectoparasites, lactation
42.	*Azadirachta indica* A. Juss.	Meliaceae	Bark, fruit, leaf, oil	Ulcer, wounds, fly infestation, Kasanoor Forest disease, arthritis, remove ectoparasites, odema, antiviral, antiseptic
43.	*Azima tetracantha* Lam.	Salvadoraceae	Leaves, branches	Anthrax, throat cancer FMD
44.	*Bambusa arundinadea* (Retz.) Roxb.	Poaceae	Bark	Retention of placenta, internal wounds, wound
45.	*Bauhinia purpurea* L.	Fabaceae	Leaf and stem	Better tonic for healthy growth
46.	*Bauhinia vaiegata* L.	Fabaceae	Bark	Wounds in foot and mouth; Foot and Mouth Disease
47.	*Bidens biternata* (Lour.) Merr. & Sherff.	Asteraceae	Whole plant	Wound healing

TABLE 7.2 *(Continued)*

No	Botanical Name	Family	Plant Parts	Uses
48.	*Bidens pilosa* L.	Asteraceae	Leaves	Antiseptic and cough
49.	*Biophytum sensitivum* (L.) DC.	Oxalidaceae	Leaves	Uterine vaginal prolapse
50.	*Bombax ceiba* L.	Bombacaceae	Bark	Excess bleeding after delivery
51.	*Borassus flabellifer* L.	Arecaceae	Inflorescence	Dysentery
52.	*Brassica hirta* Moench.	Brassicaceae	Seeds	Stomach disorder
53.	*Brassica juncea* (L.) Czern.	Brassicaceae	Seeds	Digestion, bloat
54.	*Bridelia scandens* (Roxb.) Willd.	Euphorbiaceae	Stem bark	Odema of jaws
55.	*Butea monosperma* (Lam.) Taub.	Fabaceae	Seeds	Intestinal worms.
56.	*Calotropis gigantea* (L.) R. Br. ex Schult.	Asclepiadaceae	Leaf, flower	Poison bite, wounds
57.	*Calotropis procera* (Aiton) W.T. Aiton	Asclepiadaceae	Root, latex	Snake bite, wound healing
58.	*Canthium parviflorum* Lam.	Rubiaceae	Leaf paste	Fractures
59.	*Cannabis sativa* L.	Cannabinaceae	Leaves	Conjunctivitis
60.	*Capsicum annum* L.	Solanaceae	Fruit	Kasanoor Forest disease
61.	*Capsicum frutescens* L.	Solanaceae	Fruits	Dysentery, Hemorrhagic, Hemorrhagic septicaemia, pyrexia, acidic indigestion
62.	*Cardamine africana* L.	Brassicaceae	Leaves, flowers	Psoriasis
63.	*Cardiospermum halicacabum* L.	Sapindaceae	Entire plant	Fits, anorexia
64.	*Carex longipes* D. Don ex Tilloch. & Taylor.	Cyperaceae	Flowers	Wound healing, analgesic
65.	*Careya arborea* Roxb.	Barringtoniaceae	Leaves	Dislocation of bones, corneal opacity, foot and mouth disease

TABLE 7.2   *(Continued)*

No	Botanical Name	Family	Plant Parts	Uses
66.	*Carica papaya* L.	Caricaceae	Fruit	FMD
67.	*Carissa spinarum* L.	Apocynaceae	Root	Wound
68.	*Carum carvi* L.	Apiaceae	Leaf	Health, fever, cough
69.	*Caryota urens* L.	Arecaceae	Roots	Skin allergy
70.	*Cassia absus* L.	Fabaceae	Leaf paste	Wound healing
71.	*Cassia alata* (L.) Roxb.	Fabaceae	Leaf	Skin conditions
72.	*Cassia auriculata* L.	Fabaceae	Flower, leaf	Enteritis, wounds
73.	*Cassia fistula* L.	Fabaceae	Leaves	Render pest, dog bite
74.	*Cassia italica* Mill.	Fabaceae	bark	Constipation
75.	*Cassia obtusa* Roxb.	Fabaceae	Whole plant	Laxative
76.	*Cassia occidentalis* (L.) Link	Fabaceae	Leaf, fruit	Bone fracture, wound healing, skin diseases
77.	*Cassia tora* L.	Fabaceae	Seed paste	Ringworm, skin diseases
78.	*Catharanthus roseus* G. Don	Apocynaceae	Leaf	For healing of wounds due to dog bite
79.	*Catunaregam spinosa* (Thunb.) Tirveng. (Syn. *Randia dumetorum* Lam.)	Rubiaceae	Leaves	Dislocated bones
80.	*Cayratia pedata* (Lam.) Gagnep. var. *glabra* Gamble	Vitaceae	Whole plant	Antiseptic, cancer, ulcer and refrigerant
81.	*Centella asiatica* Urban	Apiaceae	Whole plant	Antiseptic, cancer, ulcer and refrigerant
82.	*Centratherum anthelminticum* (L.) Kuntze	Asteraceae	Seeds	Fever and cough
83.	*Chlorophytum borivilianum* (Roxb.) Baker	Liliaceae	Root-tuber	To increase healthy growth of horses
84.	*Chloroxylon swietenia* DC.	Flindersiaceae	Leaves	Wound healing
85.	*Chromolaena odorata* (L.) King & Robinson	Asteraceae	Leaves	Wound healing

TABLE 7.2   *(Continued)*

No	Botanical Name	Family	Plant Parts	Uses
86.	*Cissampelos pareira* L.	Menispermaceae	Stem and leaf	Scorpion sting, healthy growth with higher vitality in horses
87.	*Cissus discolor* Bl.	Vitaceae	Whole plant	Heal fracture
88.	*Cissus quadrangularis* L. (Figure 1e)	Vitaceae	Entire plant	Enteritis, bloat, stomach disorders, removal of placenta
89.	*Cissus trilobata* L.	Vitaceae	Leaves	To treat fibrous tissue formed on the neck.
90.	*Citrullus colocynthis* (L.) Schrad.	Cucurbitaceae	Fruit, root	Digestive disorder, bloat, enteritis, horn itching
91.	*Citrullus lanatus* (Thunb.) Matsum. & Nakai	Cucurbitaceae	Fruit	Heat, fever
92.	*Citrus aurantifolia* (Christm.) Swingle	Rutaceae	Fruit	Mastitis, stomach disorders, heat
93.	*Citrus aurantium* L.	Rutaceae	Fruit	Odema of jaws
94.	*Citrus medica* L.	Rutaceae	Fruit, leaf	Fever, cough
95.	*Cleistanthus collinus* (Roxb.) Benth. ex Hook. f.	Euphorbiaceae	Stem bark	Stem bark made into a paste and is applied on the sores of cattle.
96.	*Clematis roylei* Rehder.	Ranunculaceae	Aerial parts	Cold
97.	*Cleome viscosa* L.	Capparaceae	Leaf juice	Wound healing
98.	*Clerodendron phlomides* L.	Verbenaceae	Leaf	Urinary disease
99.	*Coccinia grandis* (L.) Voigt	Curcurbitaceae	Leaves, fruit	Enterotoxemia, dislocation, fever, cough, ear and eye disease
100.	*Cochlospermum religiosum* (L.) Alston	Cochlospermaceae	Stem	To kill ticks
101.	*Cocos nucifera* L.	Arecaceae	Flower, fruit, young leaves, tender fruits, seed oil	Anestrus and subestrus enteritis, wound healing, dysentery, skin diseases
102.	*Coffea arabica* L.	Rubiaceae	Seeds	Foetal membrane, wound healing, pyrexia, fever

TABLE 7.2 *(Continued)*

No	Botanical Name	Family	Plant Parts	Uses
103.	*Coleus amboinicus* Lour.	Lamiaceae	Leaves	Intestinal worms (round worms)
104.	*Corallocarpus epigaeus* (Rottl. & Willd.) C.B. Clarke	Cucurbitaceae	Root tuber	The tuber paste along with water is given for colic pain to cattle; destroy and expel tape worms
105.	*Coriandrum sativum* L.	Apiaceae	Leaf, seeds	Kasanoor Forest disease fever, cough
106.	*Crinum viviparam* (Lam.) Hemadri	Amaryllidaceae	Leaves	Black quarter
107.	*Crocus sativus* L.	Iridaceae	Fruit, leaf	Urinary disorder
108.	*Crossandra undulaefolia* (L.) Nees	Acanthaceae		Dislocated bones
109.	*Cryptolepis buchanani* Roem. & Schult.	Asclepiadaceae	Leaves	Snake bite, leaf paste in doses of 200 g once a day for 7–10 days to cattle as galactagogue and for enhancing the lactation in cattle.
110.	*Cucurbita maxima* Duch.	Cucurbitaceae	Fruit stalk	Dengue fever
111.	*Cuminum cyminum* L.	Apiaceae	Seeds	Bloat, anorexia, enteritis, acidic indigestion
112.	*Curculigo orchioides* Gaertn.	Hypoxidaceae	Tubers	For impaction
113.	*Curcuma amada* Roxb.	Zingiberaceae	Rhizome	Digestion, ulcer
114.	*Curcuma aromatica* Salisb.	Zingiberaceae	Rhizome	Wounds
115.	*Curcuma longa* L.	Zingiberaceae	Rhizome	Intestinal worms, foot and mouth disease, wound healing, Raniket disease, flow pox
116.	*Cyanotis arachnoidea* C.B. Clarke	Commelinaceae	Whole plant	Rheumatic
117.	*Cyclea peltata* (Burm.f.) Hook.f. & Thoms.	Menispermaceae	Whole plant	Pyrexia

TABLE 7.2 *(Continued)*

No	Botanical Name	Family	Plant Parts	Uses
118.	*Cynodon dactylon* (L.) Pers.	Poaceae	Grass	Fever and cough
119.	*Cynoglossum zeylanicum* Thunb.	Boraginaceae	Roots	Jaundice
120.	*Cyrtococcum deccanense* Bor	Poaceae	Aerial parts	Fodder
121.	*Datura metel* L.	Solanaceae	Fruit, leaf	Swellings, bites, enteritis
122.	*Dendrocalamus strictus* Nees	Poaceae	Leaves	Dysentery
123.	*Dendropthoe falcata* (L.f.) Ettingsh	Loranthaceae	Whole plant	Rinder pest disease
124.	*Desmodium scalpe* DC.	Fabaceae	Aerial parts	Diarrhea, dysentery, diuretic, astringent
125.	*Dichrocephala integrifolia* Kuntze	Asteraceae	Tender shoots, flower buds	Wounds and cuts
126.	*Diospyros montana* Roxb.	Ebenaceae	Leaves	Uterine Vaginal Prolapse
127.	*Dodonaea viscosa* Lam.	Sapindaceae	Leaves	Leg fever, bone fracture
128.	*Dolichos lablab* (L.) Sweet	Fabaceae	Seed, fruit	Anorexia, health
129.	*Drymaria cordata* (L.) Roemer ex Schult.	Caryophyllaceae	Whole plant	Headache
130.	*Eclipta alba* (Figure 1c)	Asteraceae	Leaf, root	Horn polish, injury
131.	*Elephantopus scaber* L.	Asteraceae	Whole plant	Increases lactation, avoids weakness in cattle, to control loose motions.
132.	*Elettaria cardamomum* (L.) Maton	Zingiberaceae	Whole plant	Acidic indigestion, Pyrexia

TABLE 7.2   *(Continued)*

No	Botanical Name	Family	Plant Parts	Uses
133.	*Eleucina coracana* Gaertn.	Poaceae	Seed, flour	Horn fractures, fractures, health, milk production, dysentery, fetal membrane
134.	*Eragrostis nigra* Nees ex Steud.	Poaceae	Aerial parts	Fodder
135.	*Erigernon karvinskianus* DC.	Asteraceae	Whole plants	Skin diseases
136.	*Ervatamia coronaria* (Jacq.) Stapf (Syn. *Tabernaemontana divaricata* R.Br. ex Roem. & Schult.)	Apocynaceae	Flower	Fever and cough
137.	*Ervatamia heyneana* (Wall.) T. Cooke	Apocynaceae	Bark	Dysentery and diarrhea, snake bite/scorpion sting
138.	*Erythrina indica* Lam.	Fabaceae	Bark, leaf	Heamorrhagic enteritis, delayed parturition
139.	*Erythrina suberosa* Roxb.	Fabaceae	Seed	Rabies
140.	*Eucalyptus tereticornis* Smith	Myrtaceae	Oil	Remove ectoparasites
141.	*Euphobia cyathophora* Murray	Euphorbiaceae	Entire plant	Retention of placenta
142.	*Euphorbia hirta* L.	Euphorbiaceae	Stem, leaf, latex	Hemorrhagic enteritis, wounds
143.	*Euphorbia rothiana* Spreng.	Euphorbiaceae	Latex	Boils and acne
144.	*Euphorbia tirucalli* L.	Euphorbiaceae	Stem, latex	Black quarters
145.	*Ferula asafoetida* L.	Apiaceae	Bark	Digestion, ear disease, eye disease, fits
146.	*Ficus benghalensis* L.	Moraceae	Fruit, bark, leaf	Wounds
147.	*Ficus glomerata* Roxb.	Moraceae	Bark	Tympanites
148.	*Ficus religiosa* L.	Moraceae	Leaf, bark	Fever, retention of placenta
149.	*Ficus tinctoria* G.Forst.	Moraceae	Leaves	Uterine Vaginal Prolapse, cough

TABLE 7.2 *(Continued)*

No	Botanical Name	Family	Plant Parts	Uses
150.	*Foeniculum vulgare* Mill.	Apiaceae	Seeds	Digestion, anorexia
151.	*Gaultheria fragrantissima* Wall.	Ericaceae	Leaves	Arthritis
152.	*Girardinia diversifolia* (Link) Friis	Urticaceae	Roots, leaves	Stimulant, headache, swollen joints and fever
153.	*Glycine max* (L.) Merr.	Fabaceae	Seeds	To induce milk secretion
154.	*Glycosmis pentaphylla* (Retz.) DC.	Rutaceae	Whole plant	Acidic indigestion, pyrexia
155.	*Gnaphalium indicum* DC.	Asteraceae	Whole plant	Fever
156.	*Gossypium hirsutum* L.	Malvaceae	Seeds	Health, to induce milk secretion
157.	*Grewia tiliaefolia* Vahl.	Tiliaceae	Bark	Root bark paste is applied as plaster on dislocated joints to recovery of cattle.
158.	*Gymnema sylvestre* (Retz.) R. Br.	Asclepiadaceae	Bark, leaves	Fever, ephemeral fever, opacity of cornea
159.	*Helichrysum hookeriana* Wight & Arn.	Asteraceae	Flowers	Skin diseases
160.	*Hibiscus rosa-sinensis* L.	Malvaceae	Leaves	Weakness, dislocated bones
161.	*Holoptelea integrifolia* (Roxb.) Planch.	Ulmaceae	Leaves	Throat problem (swelling)
162.	*Hydrocotyle javanica* Thunb.	Apiaceae	Leaves	Blood purifier
163.	*Hypochaeris glabra* L.	Asteraceae	Roots and leaves	Tonic, astringent and diuretic
164.	*Indigofera tinctoria* L.	Fabaceae	Whole plant	Rabies in dogs
165.	*Ionidum suffruticosum* Ging	Violaceae	Leaf	Nostril congestion
166.	*Isachne kunthiana* (Steud.) Miq.	Poaceae	Aerial parts	Fodder

TABLE 7.2 *(Continued)*

No	Botanical Name	Family	Plant Parts	Uses
167.	*Ixora coccinea* L.	Rubiaceae	Leaves	Wound
168.	*Jatropha curcas* L.	Euphorbiaceae	Stem bark	Dysentery
169.	*Jatropha podogrina* L.	Euphorbiaceae	Seed	For curing loose motion
170.	*Lagerstroemia micro-carpa* Wight	Lythraceae	Bark, fruits, leaves	Foot and mouth disease
171.	*Launaea pinnatifida* Cass.	Asteraceae	Whole plant	Antifungal
172.	*Leonotis nepetaefolia* (L.) R.Br.	Lamiaceae	Root, fruit	Mastitis (inflammation). to achieve successful conception in goats and sheep
173.	*Leucas aspera* (Willd.) Link (Figure 1f)	Lamiaceae	Leaves	Conjunctivitis, Kasanoor Forest disease, snake bite, Tick & lice
174.	*Leucas marrubioides* Desf.	Lamiaceae	Stem	wound
175.	*Limonia acidissima* L.	Rutaceae	Fruit and leaves	Bade Siro's
176.	*Lippia nodiflora* (L.) Greene	Verbenaceae	Leaf	Digestion, anorexia
177.	*Litsea glutinosa* (Lour.) Robins.	Lauraceae	Bark	Bone fracture.
178.	*Lobelia nicotianaefo-lia* Roth ex Roem. & Schult.	Lobeliaceae	Leaves, plant	Wound, maggot wounds, abscess
179.	*Luffa acutangula* L.var. *amara* Roxb.	Cucurbitaceae	Leaf	Mastitis, wounds
180.	*Lycianthes bigemi-nata* (Nees) Bitter	Solanaceae	Fruits, leaves	Ulcer
181.	*Machilus macrantha* Nees	Lauraceae	Bark	Bone fractures
182.	*Madhuca indica* (J.Konig) J.F.Macbr.	Sapotaceae	Flower, leaf	Nostril congestion
183.	*Mangifera indica* L.	Anacardiaceae	Seeds, leaf, fruit, stem bark	To induce milk secretion, foot and mouth disease, healing fracture, dysentery

TABLE 7.2 *(Continued)*

No	Botanical Name	Family	Plant Parts	Uses
184.	*Manilkara hexandra* (Roxb.) Dubard.	Sapotaceae	Stem Bark	To treat throat diseases in cattle
185.	*Martynia annua* L.	Zygophyllaceae	Leaves	Leaf paste is applied on the wounds and sores of cattle
186.	*Melia azedarach* L.	Meliaceae	Bark, fruit, leaf	Intestinal worms, to induce milk secretion
187.	*Melia composita* Willd.	Meliaceae	Bark, leaf, fruit	Fever, cough, wounds
188.	*Melochia corchorifolia* L.	Meliaceae	Leaf	Mastitis
189.	*Melothria maderaspatna* (L.) Cogn.	Cucurbitaceae	Leaf	Anorexia, cough, fever
190.	*Mimosa pudica* L. (Figure 1a)	Mimosaceae	Leaf, whole plant	Wound healing, tick and lice
191.	*Momordica charantia* L.	Cucurbitaceae	Fruit, leaf, roots	Anorexia, internal worms, wound healing
192.	*Momordica dioica* Roxb. ex Willd.	Cucurbitaceae	Fruit, roots	Cures mouth ulcer, mastitis
193.	*Moringa oleifera* Lam.	Moringaceae	Bark	Black quarter, fever and cough
194.	*Mucuna pruriens* (L.) DC.	Fabaceae	Whole plant	Remove ectoparasites
195.	*Mulu jaaruvudu*			Dislocated bones
196.	*Murraya koenigii* (L.) Sprengel	Rutaceae	Leaves	Heat, fever, pyrexia
197.	*Musa paradisiaca* L.	Musaceae	Young leaves and roots	Reduce heat, Enteritis, mouth disease, hemorrhagic enteritis in cattle
198.	*Mussaenda frondosa* L.	Rubiaceae	Root	Poisonous bites in cattle
199.	*Neanotis indica* (DC.) Lewis	Rubiaceae	Aerial parts	Fever
200.	*Nicotiana tabacum* L.	Solanaceae	Leaf	Tick and lice
201.	*Nothopodytes foetida* (Wight) Sleumer.	Icacinaceae	Root	Poisonous bites

TABLE 7.2 *(Continued)*

No	Botanical Name	Family	Plant Parts	Uses
202.	*Ocimum basilicum* L. (Figure 1d)	Lamiaceae	Entire plant	Ear, eye disease
203.	*Ocimum teuniflorum* L.	Lamiaceae	Entire leaf, root, fruit plant	Snake bite, cough, ulcer lungs disorder, pyrexia
204.	*Oplismenus compositus* (L.) P. Beauv.	Poaceae	Aerial parts	Fodder
205.	*Oroxylum indicum* (L.) Bent. ex Kurz Figure 2 c	Bignoniaceae	Bark	Black quarter, paralysis, wounds
206.	*Oryza sativa* L.	Poaceae	Paddy husk, whole plant	Dysentery, fetal membrane, fractures, remove ectoparasites, odema of jaws
207.	*Oxalis latifolia* Kunth	Oxalidaceae	Whole plant	Astringent, antiseptic and anemia
208.	*Parochetus communis* Buch.— Ham. ex D. Don	Fabaceae	Flowers, leaves	Antimicrobial
209.	*Passiflora edulis* Sims	Passifloraceae	Fruits	Edible
210.	*Pedalium murex* L.	Pedaliaceae	Whole plant, leaf, stem	Anestrous & sub estrus, retention of placenta
211.	*Pennisetum americanum* (L.)R.Br.	Poaceae	Seeds	Retention of placenta, health
212.	*Pergularia daemia* (Forssk.) Chiov.	Asclepiadaceae	Leaf, latex	To relieve post- natal pains, for muscular pains in cattle, corneal opacity, ringworm
213.	*Persicaria chinensis* (L.) H. Gross.	Polygonaceae	Roots	Diarrhea
214.	*Phaseolus mungo* (L.) Hepper	Leguminosae	Seeds	Swellings, wounds
215.	*Phyllanthus amarus* L. (Figure 2a)	Euphorbiaceae	Leaf, root, fruit	Cough, fever
216.	*Phyllanthus virgatus* G. Forst.	Euphorbiaceae	Whole plant	Bleeding
217.	*Physalis peruviana* L.	Solanaceae	Leaves and dried seeds	Jaundice and glaucoma

TABLE 7.2   *(Continued)*

No	Botanical Name	Family	Plant Parts	Uses
218.	*Pilea angulata* Blume	Urticaceae	Aerials parts	Fodder
219.	*Pilea wightii* Wedd.	Urticaceae	Aerials parts	Fodder
220.	*Piper betel* L.	Piperaceae	Leaf	Gastro intestinal disorders, poison bite, three day sickness
221.	*Piper brachystachyum* Wall. ex Hook. f.	Piperaceae	Stem and fruit	Tooth ache and dyspepsia
222.	*Piper longum* L. (Figure 2e)	Piperaceae	Fruit	Skin diseases
223.	*Piper nigrum* L. (Figure 2f)	Piperaceae	Leaves	Anthrax, throat cancer, three day sickness, black quarter, pyrexia, dysentery, acidic indigestion
224.	*Plumbago zeylanica* L.	Plumbaginaceae	Whole plant	Wounds
225.	*Polycarpon tetraphyllum* L.	Caryophyllaceae	Leaves	Rheumatism
226.	*Polygonum hydropiper* (L.) Delabre	Polygonaceae	Leaves	Fractures
227.	*Pongamia pinnata* (L) Pierre	Fabaceae	Fruits, leaves, seed oil	Healthy growth with higher vitality in horses, for successful conception, acid indigestion, wound healing, skin diseases
228.	*Pothos scandens* L.	Araceae	Whole plant	Increases lactation
229.	*Pouzolzia bennettiana* Wight	Urticaceae	Whole plant	Cuts and fracture, antiseptic
230.	*Psidium guajava* L.	Myrtaceae	Leaf, flower, fruit	Hemorrhagic enteritis, calf diarrhea, remove ectoparasites, pyrexia, maggot wounds
231.	*Pterolobium hexapetalum* (Roth) Sant. & Wagh	Fabaceae	Stem bark	Dyspepsia.
232.	*Pterospermum rubiginosum* Heyne ex Wight & Arn.	Sterculiaceae	Bark, leaves	Arthritis, odema, bone fracture

TABLE 7.2   *(Continued)*

No	Botanical Name	Family	Plant Parts	Uses
233.	*Raphanus sativus* L.	Brassicaceae	Leaf, rhizome	Fever
234.	*Ricinus communis* L.	Euphorbiaceae	Seeds, root, leaves	Three day sickness, eye disease, maggots wounds, opacity of cornea
235.	*Rosa alba* L.	Rosaceae	Flower	Fever and cough
236.	*Rubia cordifolia* L.	Rubiaceae	Leaves	Post-natal diseases, jaundice, cataract, FMD
237.	*Rubus racemosus* Roxb.	Rosaceae	Fruits	Edible
238.	*Ruta graveolens* L.	Rutaceae	Leaves	Fever and cough
239.	*Sansevieria roxburghiana* Schult. & Schult.f.	Agavaceae	Stem	Ear disease
240.	*Santalum album* L.	Santalaceae	Bark oil, leaf	Peptic ulcers, throat cancer, to induce milk secretion
241.	*Sapindus laurifolia* Vahl.	Sapindaceae	Fruit	Ectoparasites (ticks and mites)
242.	*Senecio candicans* DC.	Asteraceae	Leaves	Influenza, cold, throat infection
243.	*Sesamum indicum* L.	Pedaliaceae	Seeds, leaf	Retention of placenta, constipation
244.	*Sesbania grandiflora* (L.) Poir	Fabaceae	Seeds	Uterine bleeding, healthy growth with higher vitality in horses, to increase and maintain sexual vigor and strength
245.	*Shorea roxburghii* G. Don	Dipterocarpaceae	Leaves	Swelling
246.	*Smilax aspera* L.	Smilacaceae	Whole plant	Intestinal diseases
247.	*Smilax zeylanica* L.	Smilacaceae	Root	Antidote, venereal diseases and skin troubles
248.	*Solanum nigrum* L.	Solanaceae	Fruit, leaf	Ulcer, heat, health
249.	*Solanum surattense* Burm. f. Syn.: *Solanum xanthocarpum* Schrad. & Wendl.	Solanaceae	Fruit, root	Fever, cough

TABLE 7.2 *(Continued)*

No	Botanical Name	Family	Plant Parts	Uses
250.	*Solanum torvum* Sw.	Solanaceae	Leaf, fruit	Anorexia, digestion
251.	*Spergula arvensis*, L.	Caryophyllaceae	Aerial parts	Diuretic, antibacterial, antifungal
252.	*Sphaeranthus indicus* L.	Asteraceae	Fruit, whole plant	Stomach disorders, wounds, remove ectoparaites
253.	*Spondias mangifera* Willd.	Anacardiaceae	Bark	Dysentery
254.	*Stereospermum colais* (Buch.-Ham. ex Dillwyn.) Mabb	Bignoniaceae	Stem bark	Rabies in dogs
255.	*Strychnos nux-vomica* L.	Loganiaceae	Fruit, leaf	Ulcer, sprains
256.	*Strychnos potatorum* L. f.	Loganiaceae	Seeds	Eye infections
257.	*Syzygium aromaticum* (L.) Merrill & Perry	Myrtaceae	Flower bud	Three day sickness
258.	*Syzygium caryophyllatum* Gaertn.	Myrtaceae	Bark	Tympanites
259.	*Syzygium cumini* (L.) Skeels	Myrtaceae	Fruit	Fed to horses to attain maximum and healthy growth with higher vitality; foot and mouth disease of cattle
260.	*Syzygium caryophyllatum* Gaertn.	Myrtaceae	Bark	Tympanites
261.	*Tagetes erecta* L.	Asteraceae	Leaves	Endometritus, fly infestation
262.	*Tamarindus indica* L.	Fabaceae	Leaves, fruit	Foot disease, swelling, odema, reduce swelling in the mammary of lactating livestock
263.	*Terminalia bellirica* (Gaertn.) Roxb.	Combretaceae	Bark	Curing blood dysentery, intestinal worms (round worms), wounds, FMD, maggot wound, wound healing
264.	*Terminalia chebula* Retz.	Combretaceae	Fruit, bark	Urinary disorders
265.	*Tinospora cordifolia* (Willd.) Hook. f. & Thomson	Menispermaceae	Leaves	Fever, FMD, to treat back generator disease in cattle

TABLE 7.2 *(Continued)*

No	Botanical Name	Family	Plant Parts	Uses
266.	*Toddalia asiatica* (L.) Lam.var. *floribunda* Gamble	Rutaceae	Roots and fruits	Fever and rheumatism
267.	*Trichosanthes tricuspidata* Lour.	Cucurbitaceae	Tuber	Bloat/Tympany
268.	*Trigonella foenumgraecum* L.	Fabaceae	Seed	Ulcer, heat, stomach pain
269.	*Tylophora asthmatica* (Burm. f.) Merrill.	Asclepiadaceae	Leaf, root	Lungs disease
270.	*Tylophora indica* Wight & Arn.	Asclepiadaceae	Leaf, milk	Snake bite
271.	*Vallaris solanacea* (Roth) Kuntze	Apocynaceae	Leaves	Bone fracture
272.	*Vanilla walkeriae* Wight	Orchidaceae	Stem	Fever in cattle, nutritive supplement for cattle
273.	*Vitex negundo* L.	Verbenaceae	Leaves	Poisonous bites, antibacterial, insecticidal, discharge of worms from stomach, fever, body pain
274	*Withania somnifera* (L.) Dunal	Solanaceae	Entire plant	Health improvement
275.	*Woodfordia fruticosa* (L.) Kurz	(Lythraceae)	Flower	Dysentery
276.	*Xanthium strumarium* L.	Asteraceae	Whole Plant	Swelling of the glands in cattle.
277.	*Zanthoxylum alatum* Roxb.	Rutaceae	Bark	Corneal opacity
278.	*Zehneria mysorensis* Wight & Arn.	Cucurbitaceae	Fruits	Blood purifier
279.	*Zingiber officinale* Roscoe	Zingiberaceae	Rhizome	Anorexia, bloat, Three day sickness, acidic indigestion
280.	*Ziziphus jujuba* Mill.	Rhamnaceae	Fruit	Impaction, bloat

7.3   DISCUSSION

Presently traditional veterinary health techniques and practices, traditional management as well as herbal and holistic medicines are increasingly being accepted in western societies (Khan et al., 2006; Dangol et al., 2006). The past prejudice on the Ethnoveterinary medicine seems to be reducing and scientific evaluation and validation of the ethnoveterinary preparations has been initiated by considerable number of researchers (Garforth, 2006; Mathias, 2006). It is now realized that a complementary medical approach is crucial and necessary to boost livestock production at community level. Because of their holistic nature, traditional remedies offer efficacy combined with safety more often than single cosmopolitan/conventional drugs. There are no harmful effects in most cases with the use of traditional medicine. Many plant based drugs are being tested for the control of the insects, worms and microorganisms affecting the domesticated animals (Butter et al., 2000; Rios and Recio, 2005; Max et al., 2006). Southeast Asia has the rich biological, cultural diversity and the knowledge to use the traditional medicine. It also has excellent infrastructure and expertise to develop human and animal health. There is an upsurge interest in the herbal/alternate medicine all over the world because widespread use of antimicrobials and other chemical drugs in livestock management led to the antibiotic and chemical drug residues in the milk and other animal products, and encourages spread of AMR. It is reported that Western Ghats houses several species used for animal health and need more documentation for further understanding the species now available. It is essential to conserve them otherwise several of the available species which are endangered now will extinct fast.

KEYWORDS

- **Dairy and Other Animal Products**
- **Ethno-Veterinary Practices**
- **Medicinal Plants**
- **Western Ghats**

REFERENCES

Alagesaboopathi, C. (2015). Medicinal plants used in the treatment of livestock diseases in Salem district, Tamilnadu, India. *J. Pharmaceu. Res., 7,* 829–836.

Anonymous (2004). Identification and evaluation of medicinal plants for control of parasitic diseases of livestock. *Technical Report. CIRG.* Makhdoom (Madhura).

Ashok, S.P., Sonawane, B.N. & Diwakar Reddy, P.G. (2012). Traditional Ethno-veterinary practices in Karanji Ghat areas of Pathardi Tasil in Ahmednagar District (MS) India. *Intern. J. Plants, Animals and Environ. Sci., 2,* 64–69.

Butter, N.L., Dawson J.M., Wakelin, D. & Buttery, J. (2000). Effect of dietary tannins and protein concentration on the nematodes infection (*Trichostrongylus colubriformis*) in Lambs. *J. Agric. Sci., 134,* 89–99.

Dangol, D.R., Teeling, C. & Grout, B.W.W. (2006). Ethnoveterinary medicine and traditional healers of Taru community in Citwan Terai of Nepal. Harvesting Knowledge, Pharming Opportunities. Ethnoveterinary Medicine conference. British Society of Animal Science, Writtle College, Chelmsford, UK.

Deokule, S.S. & Mokat D.N. (2004). Plants used as veterinary medicine in Ratnagiri District of Maharashtra. *Ethnobotany 16,* 131–135.

Garforth, C.J. (2006). Local knowledge as a resource in the developing Livestock systems. Harvesting Knowledge, Pharming Opportunities. Ethnoveterinary Medicine conference, British Society Animal Science, Writtle College, Chelmsford, UK.

Gopalakrishnan, S. (2013). Wound healing activity studies on some medicinal plants of Western Ghats of South India. *J. Bioanal. Biomed., 5(4),* http://dx.doi.org/10.4172/1948–593X.S1.011.

Harsha, V.H., Shripathi, V. & Hegde, G.R. (2005). Ethno-veterinary practices in Uttar Kannada district of Karnataka. *Indian J. Trad. Knowl. 4,* 253–258.

Hill, A.L. & McLaughlin, K. (2006). Antibacterial synergy between essential oils and its effect on antibiotic resistance bacteria. Harvesting Knowledge, pharming Opportunities, Ethnoveterinary Medicine conference, British Society Animal Science, Writtle College, Chelmsford, U.K.

Kalidass, C., Muthukumar, K., Mohan, V.R. & Manickam, V.S. (2009). Ethnoveterinary medicinal uses of plants from Agasthiamalai Biosphere Reserve (KMTR), Tirunelveli District, Tamil Nadu, India. *My Forest, 45(1),* 7–14.

Khan, M.A.S., Akbar, M.A. & Ahmed, T.U. (2006). Efficacy of Neem (*Azadirachta indica*) and pineapple (*Ananas comosus*) leaves as herbal dewormer and its effect on milk production of dairy cows under rural condition in Bangladesh. Harvesting Knowledge, Pharming Opportunities Ethnoveterinary Medicine conference, British society of animal Science, Writtle College, Chelmsford, UK. pp. 47–48.

Kholkunte, S.D. (2008). Database on ethnomedicinal plants of Western Ghats. *Final Report 05–07–2005 to 30–06–2008, ICMR* New Delhi.

Kiruba, S., Jeeva, S. & Dhass, S.S.M. (2006). Enumeration of ethnoveterinary plants of Cape Comorin, Tamil Nadu. *Indian J. Trad. Knowl. 5,* 576–578.

Krishna T.P.A., Ajeesh Krishna T.P., Chithra, N.D., Deepa, P.E., Darsana, U., Sreelekha, K.P., Juliet, S., Nair, S.N., Ravindran, R., Kumar, K.G.A. & Ghosh, S. (2014). Acaricidal activity of petroleum ether extract of leaves of *Tetrastigma leucostaphylum* (Dennst.) Alston against *Rhipicephalus* (*Boophilus*) *annulatus*. Hindawi Publishing Corporation. *Scientific World Journal*, Article ID 715481, 1–6 http://dx.doi.org/10.1155/2014/715481.

Mathias, E. (2006). Livestock keeper's knowledge verses test tubes and mice: Ethnoveterinary validation and use in animal health care projects. Harvesting Knowledge, Pharming Opportunities. Ethnoveterinary Medicine conference, British Society Animal Science, Writtle College, Chelmsford, UK. pp. 11.

Max, R.A., Buttery, P.J., Kassuku, A.A., Kimambo, A.E. & Mtenga, L.A. (2006). Plant based treatment of round worms: The potential of using tannins to reduce gastrointestinal nematode injection of tropical small rudiments. Harvesting Knowledge, pharming Opportunities. Ethnoveterinary Medicine conference, British Society of Animal Science, Writtle College, Chelmsford, UK. pp. 28–29.

Mini, V. & Sivadasan, M. (2007). Plants used in ethnoveterinary medicine by Kurichya tribe of Wyanada district in Kerala, India. *Ethnobot. 19,* 94–99.

Myers, N., Mittermeier, R., Mittermeier, C., Da Fonseca, G. & Kent, J. (2000). Biodiversity hot-spots for conservation priorities. *Nature 403,* 853–858.

Nair, M.N.B. (2005). Contemporary relevance of Ethno-veterinary medical traditions of India. *Proc Natl. Workshop.* October 17 and 18 at Kediyur Hotel, Udupi, Mangalore, Karnataka, India.

Nair, M.N.B. & Punniamurthy, N. (eds.) (2010). Ethno-veterinary practices. Mainstreaming traditional wisdom on livestock keeping and herbal medicine for sustainable rural livelihood, I-AIM, 74/2 Jarakabandekaval, Attur post, (via) Yelahanka, Bangalore 560106, India.

Nair, M.N.B. & Unnikrishnan P.M. (2010). Revitalizing ethnoveterinary medical tradition: A perspective from India (Chapter 5). In: David R. Katerere & Dibungi Luseba (Eds.) *Ethnoveterinary Botanical Medicine-Herbal Medicine for Animal Health.* CRC Press, Taylor & Francis group, USA. pp. 95–124.

Prasad, A.G., Benny, S.T. & Puttaswamy, R.M. (2014). Ethno -veterinary medicines used by tribes in Wayanad District, Kerala. *Int. J. Rec. Trends Sci & Tech, 10,* 331–337.

Prathipa, A., Senthilkumar, K., Gomathinayagam, S. & Jayathangaraj, M.G. (2014). Ethnoveterinary treatment of *Acariasis* infestation in rat snake (*Ptyas mucosa)* using herbal mixture. *Int. J. Vet. Sci., 3,* 61–64.

Rajagopalan, C.C.R. & Harinarayanan, M. (2003). Veterinary wisdom for the Western Ghats. *Amruth 2(2),* 15–17.

Rajesh, K., Rajesh N.V., Vasantha, S., Jeeva, S. & Rajashekaran, D. (2014 b). Anthelmentic efficacy of selected ferns in sheeps (*Ovis aries* Linn.). *Int. J. Ethnobiol., Ethnomed., 1,*1–14.

Rajesh, K., Rajesh N.V., Vasantha, S. & Geetha, V.S., (2015). Anti-parasitic action of *Actinopteris radiata, Acrostichum aureum* and *Hemionitis arifolia. Pteridology Res, 4,* 1–9.

Rajkumar, N. & Shivanna, M.B. (2012). Traditional veterinary healthcare practices in Shimoga district of Karnataka, India. *Indian J. Trad. Knowl. 2,* 283–287.

Ramdas, S.R. & Ghtoge, N.S. (2004). *Ethno-Veterinary Research in India: An Annotated Bibliography.* Anthra, Secunderabad, India.

Raveesha, H.R. & Sudhama, V.N. (2015). Ethno-veterinary Practices in Mallenahalli of Chikmagalur Taluk, Karnataka. *J. Med Plants Studies 3,* 37–41.

Renjini H., Thangapandian, V. & Binu T. (2015). Ethnomedicinal knowledge of tribe-Kattunayakans in Nilambur forests of Malappuram district, Kerala, India. *Int. J. Phytotherapy, 5,* 76–85.

Rios, J.L. & Recio, M.C. (2005). Medicinal plants and antimicrobial activity. *J. Ethnopharmacol. 100,* 80–84.

Somkuwar, S.R., Chaudhary, R.R. & Chaturvedi, A. (2015). Knowledge of Ethno-Veterinary Medicine in the Maharashtra State, India. *Int. J. Sci. Applied Res. 2,* 90–99.

Swarup, D. & Patra, R.C. (2005). Perspectives of ethno-veterinary medicine in veterinary Practice. Proceedings of the National conference on Contemporary relevance of Ethnoveterinary medical traditions of India, M.N.B. Nair (ed.). FRLHT, 74/2 Jarakabande Kavel, Attur Post, Yelehanka, Bangalore, 560064, Karnataka, India.

APPENDIX

PLATE 7.1

a. *Mimosa pudica*. b. *Acalypha indica*. c. *Eclipta alba*. d. *Ocimum basilicum*. e. *Cissus quadrangularis*. f. *Leucas aspera*.

PLATE 7.2

a. *Phyllanthus amarus*, b. *Andrographis paniculata*, c. *Oroxylum indicum*, d. *Withania somnifera*, e. *Piper longum*. f. *Piper nigrum*.

CHAPTER 8

MEDICINAL FLORA AND RELATED TRADITIONAL KNOWLEDGE OF WESTERN GHATS: A POTENTIAL SOURCE FOR COMMUNITY-BASED MALARIA MANAGEMENT THROUGH ENDOGENOUS APPROACH

B. N. PRAKASH,[1] P. M. UNNIKRISHNAN,[2] and
G. HARIRAMAMURTHI[1]

[1]Trans-Disciplinary University (TDU), School of Health Sciences, FRLHT, 74/2, Jarakabande Kaval, Attur post, Via Yelahanka, Bangalore – 560064, Karnataka, India. E-mail: vaidya_bnprakash@ yahoo.co.in; unnipm@gmail.com; hariram_01@yahoo.com

[2]United Nations University – Institute for the Advanced Study of Sustainability, Tokyo, Japan

CONTENTS

ABSTRACT

The ecosystem specific plants have been used by local communities since many centuries to take care of their primary health care problems. In malaria endemic tropical regions, more than 1200 plant species have been reported to be used for management of malaria. There are several strategies followed to develop new drugs for prevention and treatment of malaria. This chapter outlines a strategy based on an endogenous development approach with trans-disciplinary research methods that can be used to identify new botanicals for management of malaria. It also describes low-cost activities that can be carried out at the community level, with the aim of translating research directly into improvements in the local management of malaria. It also highlights the specific use of 31 plant species for malaria treatment in Western Ghats region.

8.1   INTRODUCTION

8.1.1   OVERVIEW: MALARIA: A PUBLIC HEALTH PROBLEM

Globally, an estimated 3.3 billion people are at risk of being infected with malaria and developing disease, and 1.2 billion are at high risk. According to the latest estimates, 198 million cases of malaria occurred globally in 2013 and the disease led to 584,000 deaths (WHO, 2014). India contributes to over one fifth (22.6%) of clinical episodes of *P. falciparum* and 42% of episodes of *P. vivax* globally (Guerra et al., 2010; Hay et al., 2010) and around 200,000 persons die annually due to malaria (Dhingra et al., 2010). Although, there are several strategies to control and treat malaria infections, mortality and morbidity due to malaria infection have not declined. This is because of the evolution and spread of drug resistance to all important classes of anti-malarials and resistance to almost all types of insecticides. Moreover, many of the anti-malarials are expensive and inaccessible to the needy populations. There are several sincere efforts to develop malaria vaccine. This could play a role in transmission reduction in the future by preventing the infection of mosquitoes and by preventing vaccinated individuals from becoming infected by mosquitoes. Although a malaria vaccine is technically feasible, making a vaccine to protect people against a parasite has never been done. Thus, there is no licensed vaccine against malaria. Therefore, the population in malaria endemic areas depends on easily accessible traditional plant-based solutions for prevention and treatment of malaria.

8.1.2 RELEVANCE OF TRADITIONAL MEDICINE

Medicinal plants have played a significant role in maintaining and improving the quality of human life for thousands of years. Herbal medicine is based on the principle that plants contain natural substances that can prevent disease, alleviate illness and promote health. The ecosystem specific plants continue to contribute to the health security of rural communities in India (Darshan and Unnikrishnan, 2004; Payyappallimana, 2010). Plant resources and the associated ethnobotanical knowledge of the communities are rich source of natural drug research and development (Albuquerque et al., 2014). In rural areas of developing countries, traditional medicines are often trusted, affordable and accessible, as they are made from locally available plants (Kitua and Malebo, 2004). A large proportion of the rural population continues to rely on traditional medical practitioners and local medicinal plants for primary health care, as a choice or when there is no access to other medicine (Phillipson, 2001).

In spite of the great advances observed in modern medicine in recent decades, around 75% of the population still depends on traditional medicine for treatment of malaria particularly in Asian and African countries (Willcox et al., 2004). In fact, nature is, as ever, an extremely rich source of potential anti-malarial agents. The usual reason for exploration in traditional medicine stems from the recognition that two of the major anti-malarial drugs, quinine and artemisinin trace their origin to traditional medicine (Hirt and M'Pia, 2008; Kayser et al., 2003). *Cinchona officinalis* L. was traditionally used to treat intermittent fever or malaria by South American healers. Chinese traditional medicine explains the use of *Artemisia annua* L. for malarial fevers. Today, Artemisinin based combination therapy is the drug of choice for the management of malaria in many chloroquine resistant areas like in African countries. Apart from these two main plants, earlier study has reported that 1277 plant species have been used for the management of malaria (Willcox et al., 2004). Communities in malaria endemic area find several advantages in use of local plant species as they are cost effective, highly accessible, acceptable, strong belief, empirical evidence for cure, relatively good knowledge in the people and long-term usage of traditional medicine.

8.1.3 ENDOGENOUS DEVELOPMENT METHODOLOGY AND TRANS-DISCIPLINARILY

In a situation where large-scale interventions and development approaches have either failed or are inaccessible to large percentage of the population, it is important to look at development strategies which are local resources based and are easily accessible to the communities. The approach proposed has its roots from the endogenous development practice which "is based on local communities' own criteria for change and their vision for well-being based on the material, social and spiritual aspects of their livelihoods but in a constant and dynamic interface with external actors and the world around them. Endogenous development seeks to overcome a western bias by making peoples' worldviews and livelihood strategies the starting point for development. Endogenous development moves beyond integrating traditional knowledge in mainstream development and seeks to build biocultural approaches that originate from local people's worldviews and their relationship with the earth. Organizations can support and strengthen the endogenous development that is already present within the communities, promoting the interface between tradition and modernity. In doing so, endogenous development emphasizes the cultural aspects within the development process, in addition to the ecological, social and economic aspects." (ETC, 2007). This is the guiding principle of the community outreach programs which mainly focuses on locally available resources, and related traditional knowledge of the communities. We believe that participatory, people-focused and local health knowledge driven knowledge and practices are relevant for primary health care interventions. This method is a judicious blending of locally available plant resources, knowledge, skills and expertise with appropriate elements of western scientific knowledge and modern technology to address specific health challenges (ETC, 2007). Most importantly this also requires a transdisciplinary and a multistakeholder approach.

As a methodology, trans-disciplinary research (TDR) refers to scientific inquiry that encompasses the various disciplines to address a 'real-world' problem. The basic concept of TDR is to transgress boundaries between scientific disciplines and between science and other societal fields and includes deliberation about facts, practices and values (Wiesmann et al., 2008). We adopt a participatory method which gives equal weightage to the strengths of every discipline and attempts to solve the problem holistically. While exploring traditional solutions for any health condition, this approach helps us in not only including the perspective of medicine related aspects (traditional medicine and modern bio-medicine), -but also other allied sciences as well as the social, economic and environmental dimensions in a community context.

The failure of national antimalarial programs over several decades, calls for developing such locally based solutions for the prevention as well as management of malaria. Following is such a methodology attempted by the Transdisciplinary University (at Jarakabande Kaval, Bangalore) for its malaria management interventions in communities of endemic regions. The method looks at the nature and patterns of the health condition, locally available plants and related traditional medical practice along with a strategic plan for local interventions. In the following section we elaborate the details of this methodology called DALHT (Documentation and Rapid Assessment of Local Health Traditions) with respect to management of malaria.

8.1.4 DOCUMENTATION AND ASSESSMENT OF LOCAL HEALTH TRADITIONS

Many local health traditions are sound, some are incomplete and a few may be distorted. One of the major challenges in advocating local health practices is to have clear documentation on the efficacy and safety of these practices. Finding out effective practices through elaborate pharmacological and clinical trials is a colossal task. For example, to validate a single practice would involve 5–8 years of laboratory research with a huge amount of capital investment.

In order to systematize and assess various malaria management practices in a rapid and cost effective way a participatory method called Documentation and Rapid Assessment of Local Health Traditions (RALHT) was applied (Figure 8.1). This method has been found very effective earlier in primary health care conditions in a number of locations in India. It is primarily community based and uses dialog and building consensus for assessing traditional health practices as their strategy. Community members who are knowledgeable in traditional practices, local healers, Ayurveda doctors, researchers, community members and other experts form the core group for dialog. To initiate the process in a community, nature and type of malaria, their incidences are documented along with comprehensive documentation of local knowledge on malaria management. Documentation is done in groups to facilitate maximum information sharing through cross referencing. Each group ideally consists of 5–7 community members, folk healers, Ayurveda doctor, a botanist and a documenter. Free, prior informed consent (FPIC) is taken before documentation. Detailed questionnaires have been prepared to document each of these aspects. Documentation includes details of the folk healer or community member sharing knowledge; details of the malarial fever as understood by the local healer or community member;

preventive or curative methods for malaria management or vector control; and medicinal plants used for each practice identified by botanist. Following documentation voucher specimens of medicinal plants used in these practices are identified and inventoried. Detailed literature research is carried out from codified traditional knowledge systems, such as Ayurveda and Siddha along with also data on phytochemistry and pharmacology. Desk research consists of correlation of the health condition as per Ayurveda and modern science and compiling detailed data from secondary literature. This exhaustive data based on global data bases are compiled systematically into dossiers which are sent to experts for review. This is followed by an assessment

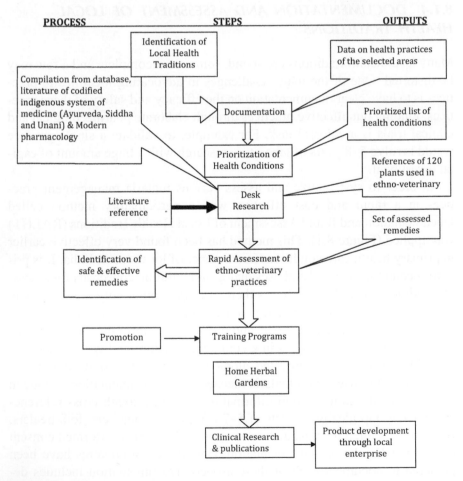

FIGURE 8.1 Process of rapid assessment of local health traditions (Adopted from Nair, 2006).

workshop. During this community-based workshop these comprehensive dossiers are used as the reference material for commenting on the safety and efficacy of each prioritized malaria management practice. The unique feature of the assessment workshop is that it is a pluralistic platform for dialog between local experts and scientific experts on each practice. A typical workshop involves folk healers, Ayurvedic doctors, botanists, field data collectors, clinical pharmacologists or modern medical doctors and community members. These subject experts comment on a specific health practice based on the dossier information available with them from the respective knowledge streams. Each practice with their details is presented to the group. Community's positive or negative experience about a practice and supporting evidence or precautions or additional information by any system of medicines is used for assessing a practice. Negative remarks from any system of knowledge are also documented and ambiguous practices are put aside for further detailed research. Positively assessed practices are then put through community based, participatory clinical studies especially for malaria prophylaxis. Once there is positive evidence from such clinical studies they are selected for wide promotion.

8.2 EPIDEMIOLOGY OF MALARIA IN AREAS OF WESTERN GHATS

Vector borne disease is a major health challenge in several districts of Western Ghats (Dhiman et al., 2011). *Anopheles fluviatilis*, *Anopheles stephensi* and *Anopheles culicifacies* are commonly seen vector species in Western Ghats which are responsible for transmission of malaria. The districts like Surat (Gujarath), Nandurbar (Maharashtra), Dakshina Kannada (Karnataka), Thrissur and Pathanamthitta (Kerala), Tirunelveli and Kanyakumari (Tamilnadu) are major malaria endemic districts in the region (http://www.malariasite.com/malaria-india/; http://www.nrhmtn.gov.in/vbdc.html; http://dhs.kerala.gov.in/). A recent study on projection of malaria by 2030, based on temperature and relative humidity at district level, indicates no major change in the transmission window, thus predicting the transmission of malaria to continue in similar trends in Western Ghats (Dhiman et al., 2011). As an illustration of current situation, Figure 8.2 shows the existing seasonal occurrence of malaria cases in a representative district of the Western Ghats, Dakshina Kannada (Karnataka).

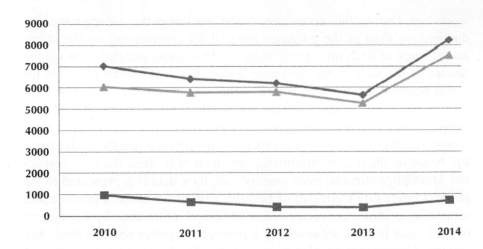

FIGURE 8.2   Incidence of malaria in Dakshina Kannada district, Karnataka.

8.3   BOTANICAL LISTING OF TRADITIONAL ANTI-MALARIALS FROM WESTERN GHATS

In Western Ghats, there are around 4,500 plant species in which 35 percent of them are endemic. As malaria is widespread health condition in the region, population in these endemic areas have also reported to be using diverse local flora as a means to prevent and treat malaria infection. A brief literature survey showed 104 plant species which are used for malaria, fever and intermittent fevers. From this list, we shortlisted 31 plant species which are specifically used for malaria treatment in Western Ghats region based on existing literature both traditional as well as contemporary (FRLHT database, Ganeshaiah, 2013). Table 8.1 shows the important plants species reported to be used by the communities. The plant species used in single or in different combinations, belong to 23 families. Among 31 species, majority were herbs (13 species) and shrubs (10 species). Leaf was the most commonly used plant part (13 species). The principal mode of remedy preparations was decoction (Table 8.1).

TABLE 8.1   Important Medicinal Plants Used for Malaria Management in Western Ghats

Sl. No.	List of plants	Family	Habit	Part used	Preparation
1	*Ageratum conyzoides*	Compositae	Herb	Whole plant	Decoction
2	*Adina cordifolia*	Rubiaceae	Tree	Bark	—
3	*Albizia amara*	Leguminosae	Tree	Fruit	—
4	*Alpinia galanga*	Zingiberaceae	Tree	Rhizome	—
5	*Alstonia scholaris*	Apocynaceae	Tree	Bark	—
6	*Alternanthera sessilis*	Amaranthaceae	Herb	Leaves	—
7	*Amaranthus spinosus*	Amaranthaceae	Herb	Leaves	—
8	*Argemone mexicana*	Papaveraceae	Herb	Aerial parts	Juice
9	*Aristolochia tagala*	Aristolochiaceae	Shrub	Leaves	Paste
10	*Artemisia nilagirica*	Compositae	Shrub	Seed	Paste
11	*Bauhinia racemosa*	Leguminosae	Tree	Leaves	Decoction
12	*Callicarpa tomentosa*	Lamiaceae	Shrub	Seeds	—
13	*Carica papaya*	Caricaceae	Tree	Fruit rind/ pulp	—
14	*Citrus sinensis*	Rutaceae	Tree	Fruit rind	—
15	*Clematis gouriana*	Ranunculaceae	Climber	Leaves	—
16	*Cyclea peltata*	Menispermaceae	Shrub	Root	Decoction
17	*Dendrocnide sinuata*	Urticaceae	Tree	Leaves	Decoction
18	*Duranta repens*	Verbenaceae	Climber	Leaves	—
19	*Erythrina variegata*	Fabaceae	Tree	Leaves	Decoction
20	*Hedyotis herbacea*	Rubiaceae	Herb	Whole plant	—
21	*Helminthostachys zeylanica*	Ophioglossaceae	Herb	Rhizome	—
22	*Indoneesiella echioides*	Acanthaceae	Herb	Leaves	Paste
23	*Leonotis nepetifolia*	Lamiaceae	Herb	Leaves	Decoction
24	*Momordica dioica*	Cucurbitaceae	Climber	Root	—
25	*Ocimum canum*	Lamiaceae	Herb	Leaves	—
26	*Oxalis corniculata*	Oxalidaceae	Herb	Leaves	Decoction
27	*Physalis peruviana*	Solanaceae	Herb	Whole plant	—

TABLE 8.1   *(Continued)*

Sl. No.	List of plants	Family	Habit	Part used	Preparation
28	*Piper longum*	Piperaceae	Herb	Fruit	—
29	*Pongamia pinnata*	Leguminosae	Tree	Leaves	Decoction
30	*Solanum myriacanthum*	Solanaceae	Shrub	Root	Decoction
31	*Swertia chirata*	Gentinaceae	Herb	Stem, leaves	—

8.4   ETHNO-PHARMACOLOGY OF TRADITIONAL ANTI-MALARIALS

The next step in a documentation and assessment methodology is to look at the existing scientific data on the local practices. In this section we outline the existing pharmacological data on the above selected plants.

Earlier studies have shown that there were great efforts to study the potential anti-malarial activity of Western Ghats plant species (Chenniappan and Kadarkarai, 2010; Samy and Kadarkari, 2011; Kaushik et al., 2015). The advantage in screening the plants which are widely used by traditional healers for treatment of malaria are significantly more active *in vitro* and/or *in vivo* against *Plasmodium* than plants which are not widely used, or not used at all, for the treatment of malaria (Leaman et al., 1995; Vigneron et al., 2005). Table 8.2 shows the list of plants which are studied for *antiplasmodial* and *antipyretic* activity using *in vitro* or *in vivo* models of *Plasmodium* species. However none of them have gone to stage of clinical trial and product development.

TABLE 8.2   Cross-References in Published Literature on *Antiplasmodial* and *Antipyretic* Activities of Plants Used in the Western Ghats Region in India

Sl. No.	Species	Selected antiplasmodial and antipyretic* activity
1	*Ageratum conyzoides*	Ukwe et al. (2010)
2	*Adina cordifolia*	—
3	*Albizia amara*	—
4	*Alpinia galanga*	Adhroey et al. (2010)
5	*Alstonia scholaris*	Keawpradub et al. (1999)

TABLE 8.2 *(Continued)*

Sl. No.	Species	Selected antiplasmodial and antipyretic* activity
6	*Alternanthera sessilis*	Nayak (2010)*
7	*Amaranthus spinosus*	Hilou et al. (2006)
8	*Argemone mexicana*	Simoes-Pires et al. (2014)
9	*Aristolochia tagala*	Chenniappan and Kadarkarai (2010)
10	*Artemisia nilagirica*	Bamunuarachchi et al. (2013)
11	*Bauhinia racemosa*	Gupta et al. (2005)*
12	*Callicarpa tomentosa*	—
13	*Carica papaya*	Bhat and Surolia (2001)
14	*Citrus sinensis*	Bhat and Surolia (2001)
15	*Clematis gouriana*	—
16	*Cyclea peltata*	Hullatti and Sharada (2007)*
17	*Dendrocnide sinuata*	—
18	*Duranta repens*	Ijaz et al. (2010)
19	*Erythrina variegata*	Herlina et al. (2010)
20	*Hedyotis herbacea*	Pravin et al. (2015)
21	*Helminthostachys zeylanica*	—
22	*Indoneesiella echioides*	—
23	*Leonotis nepetifolia*	—
24	*Momordica dioica*	Mishra et al. (1991)
25	*Ocimum canum*	Inbaneson et al. (2012)
26	*Oxalis corniculata*	Santhosh et al. (2015)*
27	*Physalis peruviana*	Khan et al. (2009)*
28	*Piper longum*	Sireeratawong et al. (2012)*
29	*Pongamia pinnata*	Simonsen et al. (2001)
30	*Solanum myriacanthum*	—
31	*Swertia chirata*	Bhat and Surolia (2001)

—, No information is available on the way the plant is used.

8.5   AN EXAMPLE OF A REMEDY AND ITS ASSESSMENT: AYURVEDA AND PHARMACOLOGICAL EVIDENCE

Before stepping into *in vitro* or *in vivo* activity of traditional anti-malarials, it is important to prioritize the plants on the basis of available Ayurvedic information and pharmacological information through RALHT approach. Here, we are showing an example of application of rapid assessment methodology on a single plant namely, *Alstonia scholaris*. Similar methodology can be adopted to prioritize other plants (Box-1 & 2).

8.6   COMMUNITY-BASED PARTICIPATORY STUDY

In the process of rapid assessment, we can identify an effective traditional remedy which provides an appropriate solution for the local problem. In addition to this, a community based, participatory research can be organized to test the efficacy of the traditional remedy. We implemented such a research approach through organizing health interventions called Decoction (*Kashaya camp*) camp in Odisha, India to study the prophylactic efficacy of a traditional herbal malaria prophylactic remedy. The *Kashaya camp* is a unique approach where selected healthy human volunteers were administered a decoction (twice every week) made from the local plants traditionally used for prevention of malaria. The camps were spread over a three months period during the monsoon, e.g., malaria endemic season. The outcome of the study showed a significant reduction in the malaria incidence rate in the population who consumed decoction (Prakash and Unnikrishnan, 2011). This is also a prelude for the 'reverse pharmacology' approach where evidence for drug efficacy is generated first from field to laboratory instead of conventional approach, e.g., laboratory to field.

8.7   CONCLUSION

Vector borne diseases continue to be a major health problem in the Western Ghats region today. Malaria epidemiological data shows there is high transmission rate in some of the districts in the Western Ghats region with a sudden increase in incidence rate in certain districts. Based on a study of 30 districts in the region it is predicted that the situation is expected to continue in a similar way for next 15 years.

Usage of large number of plants from the region show that the communities in the region have devised adaptation methods through local plant based innovations. Documenting these practices comprehensively and finding out most efficacious substances would be a pragmatic approach for malaria management. This has to follow a participatory, documentation and rapid assessment as a primary step to be followed by appropriate clinical studies to scientifically validate these practices.

When nation-wide interventions fail to tackle the problem or when conventional practices are not easily accessible for large section of the population such an endogenous, multi-stakeholder approach would be required for addressing such local development challenges.

ACKNOWLEDGEMENTS

We express sincere gratitude to the support extended by Dr. S.N. Venugopalan Nair, and Ms. Tabassum Ishrath Fathima, School of Conservation, TDU, Bangalore. Authors also thank Mrs. D. Tharini, School of Health Sciences, TDU for her support.

KEYWORDS

- **Endogenous Knowledge**
- **Herbal Anti-Malarial**
- **Malaria**
- **Traditional Medicine**
- **Western Ghats**

REFERENCES

Al-Adhroey, A.H., Nor, Z.M., Al-Mekhlafi, H.M. & Mahmud, R. (2010). Median lethal dose, antimalarial activity, phytochemical screening and radical scavenging of methanolic *Languas galanga* rhizome extract. *Mol. Basel Switz. 15,* 8366–8376.

Albuquerque, U.P., Medeiros, P.M. de, Ramos, M.A., Júnior, W.S.F., Nascimento, A.L.B., Avilez, W.M.T. & Melo, J.G. de (2014). Are ethnopharmacological surveys useful for the discovery and development of drugs from medicinal plants? *Rev. Bras. Farmacogn. 24,* 110–115.

Bamunuarachchi, G.S., Ratnasooriya, W.D., Premakumara, S. & Udagama, P.V. (2013). Antimalarial properties of *Artemisia vulgaris* L. ethanolic leaf extract in a *Plasmodium berghei* murine malaria model. *J. Vector Borne Dis. 50,* 278–284.

Bhat, G.P. & Surolia, N. (2001). *In vitro* antimalarial activity of extracts of three plants used in the traditional medicine of India. *Am. J. Trop. Med. Hyg. 65,* 304–308.

Chenniappan, K. & Kadarkarai, M. (2010). *In vitro* antimalarial activity of traditionally used Western Ghats plants from India and their interactions with chloroquine against chloroquine-resistant *Plasmodium falciparum. Parasitol. Res. 107,* 1351–1364.

Darshan, S. & Unnikrishnan, P. (2004). Challenging the Indian medical heritage. Foundation books, New Delhi.

Dhiman, R., Chavan, L., Manoj, P. & Sharmila, P., (2011). National and regional impacts of climate change on malaria by 2030. *Curr. Sci. 101,* 372–383.

Dhingra, N., Jha, P., Sharma, V.P., Cohen, A.A., Jotkar, R.M., Rodriguez, P.S., Bassani, D.G., Suraweera, W., Laxminarayan, R. & Peto, R. (2010). Million Death Study Collaborators, Adult and child malaria mortality in India: a nationally representative mortality survey. *Lancet 376,* 1768–1774.

ETC-Compas (2007). Learning endogenous development, Practical action publishing, United Kingdom.

FRLHT database; http://www.frlth.org (Accessed on 24th June, 2015).

Ganeshaiah, K.N. (2013). Plants of Western Ghats, vol. 1 & 2, UAS, Bangalore.

Guerra, C.A., Howes, R.E., Patil, A.P., Gething, P.W., Van Boeckel, T.P., Temperley, W.H., Kabaria, C.W., Tatem, A.J., Manh, B.H., Elyazar, I.R.F., Baird, J.K., Snow, R.W. & Hay, S.I. (2010). The international limits and population at risk of *Plasmodium vivax* transmission in 2009. *PLoS Negl. Trop. Dis. 4,* e774.

Gupta, M., Mazumder, U.K., Kumar, R.S., Gomathi, P., Rajeshwar, Y., Kakoti, B.B. & Selven, V.T. (2005). Anti-inflammatory, analgesic and antipyretic effects of methanol extract from *Bauhinia racemosa* stem bark in animal models. *J. Ethnopharmacol. 98,* 267–273.

Hay, S.I., Okiro, E.A., Gething, P.W., Patil, A.P., Tatem, A.J., Guerra, C.A. & Snow, R.W. (2010). Estimating the global clinical burden of *Plasmodium falciparum* malaria in 2007. *PLoS Med. 7,* e1000290.

Herlina, T., Supratman, U., Soedjanaatmadja, M.S., Subarnas, A., Sutardjo, S., Abdullah, N.R. & Hayashi, H. (2010). Anti-malarial compound from the stem bark of *Erythrina variegata. Indones. J. Chem. 9,* 308–311.

Hilou, A., Nacoulma, O.G. & Guiguemde, T.R. (2006). *In vivo* antimalarial activities of extracts from *Amaranthus spinosus* L. and *Boerhavia erecta* L. in mice. *J. Ethnopharmacol. 103,* 236–240.

Hirt, H. & M'Pia, B. (2008). Natural medicine in the tropics. Winnenden: Anamed.

Hullatti, K. & Sharada, M. (2007). Antipyretic activity of Patha: An Ayurvedic drug. *Pharmacogn. Mag. 3,* 173–176.

Ijaz, F., Ahmad, N., Ahmad, I., ul Haq, A. & Wang, F. (2010). Two new anti-plasmodial flavonoid glycosides from *Duranta repens. J. Enzyme Inhib. Med. Chem. 25,* 773–778.

Inbaneson, S.J., Sundaram, R. & Suganthi, P. (2012). *In vitro* antiplasmodial effect of ethanolic extracts of traditional medicinal plant *Ocimum* species against *Plasmodium falciparum. Asian Pac. J. Trop. Med. 5,* 103–106.

Jagetia, G.C. & Baliga, M.S. (2005). The effect of seasonal variation on the antineoplastic activity of *Alstonia scholaris* R. Br. in HeLa cells. *J. Ethnopharmacol. 96,* 37–42.

Kaushik, N.K., Bagavan, A., Rahuman, A.A., Zahir, A.A., Kamaraj, C., Elango, G., Jayaseelan, C., Kirthi, A.V., Santhoshkumar, T., Marimuthu, S., Rajakumar, G., Tiwari,

S.K. & Sahal, D. (2015). Evaluation of antiplasmodial activity of medicinal plants from North Indian Buchpora and South Indian Eastern Ghats. *Malar. J. 14,* 65. doi: 10.1186/s12936-015-0564-z.

Kayser, O., Kiderlen, A.F. & Croft, S.L., (2003). Natural products as antiparasitic drugs. *Parasitol. Res. 90 Suppl 2,* S55–62.

Keawpradub, N., Kirby, G.C., Steele, J.C.P. & Houghton, P.J. (1999). Antiplasmodial activity of extracts and alkaloids of three *Alstonia* species from Thailand. *Planta Medica. 65(8),* 690–694.

Khan, M.A., Khan, H., Khan, S., Mahmood, T., Khan, P.M. & Jabar, A. (2009). Anti-inflammatory, analgesic and antipyretic activities of *Physalis minima* Linn. *J. Enzyme Inhib. Med. Chem. 24,* 632–637.

Kirtikar, K.R. & Basu, B.D. (1999). Indian Medicinal Plants. Dehradun: International Book Distributors, vol. II.

Kitua, A. & Malebo, H. (2004). Malaria control in Africa and the role of traditional medicine, in: Traditional Medicinal Plants and Malaria. CRC Press, London, pp. 3–18.

Leaman, D.J., Arnason, J.T., Yusuf, R., Sangat-Roemantyo, H., Soedjito, H., Angerhofer, C.K. & Pezzuto, J.M. (1995). Malaria remedies of the Kenyah of the Apo Kayan, East Kalimantan, Indonesian Borneo: A quantitative assessment of local consensus as an indicator of biological efficacy. *J. Ethnopharmacol. 49,* 1–16.

Mishra, P., Pal, N., Guru, P., Katiyar, J. & Tandon, J. (1991). Antimalarial activity of traditional plants against erythrocytic stages of *Plasmodium berghei. Int. J. Pharmacogn. 29,* 19–23.

Murthy, Srikantha (2004). Vataadi varga, Bhavaprakasha of Bhavamishra. Vo. 1, p 306, Chowkhamba krishnadas academy, Varanasi.

Nair, M.N.B. (2006). Documentation and assessment of ethnoveterinary practices from an Ayurvedic viewpoint. In: Balasubramanian, A.V., Nirmala, D. (eds.). Traditional knowledge system of India and Sri Lanka. Centre for Indian Knowledge Systems, Chennai, p. 83.

Natesh, S. (2015). Remarkable Trees on NII Campus, New Delhi. http://www.nii.res.in/others/remarkable-trees-nii-campus (accessed 20th July, 2015).

Nayak, P. (2010). Pharmacological evaluation of ethanolic extracts of the plant *Alternanthera sessilis* against temperature regulation. *J. Pharm. Res. 3,* 1381.

Payyappallimana, U. (2010). Role of traditional medicine in primary health care: An overview of perspectives and challenges. *Yokohama J. Soc. Sci. 14,* 57–77.

Phillipson, J.D. (2001). Phytochemistry and medicinal plants. *Phytochemistry 56,* 237–243.

Prakash, B. & Unnikrishnan, P. (2011). Documentation and validation of traditional practices for management of malaria. In: Balasubramanian, A., Vijayalakshmi, K., Shylaja, R., Abarna, R. (Eds.), Testing and Validation of Indigenous Knowledge. Centre for Indian Knowledge Systems, Chennai.

Pravin, B., Nandkishor, J. & Shilpa., A. (2015). Evaluation of *in vitro* Antimalarial activity of *Hedyotis herbacea* by Schizont maturation inhibition assay. *Int. J. Pharm. Phytopharm. Res.*

Pratyush, K., Chandra, S.M., James, J., Lipin, D., Arun Kumar, T.V. & Thankamani, V. (2011). Ethnobotanical and Pharmacological Study of *Alstonia* (Apocynaceae) – A Review. *J. Pharm. Sci., Res 3,* 1394–1403.

Samy, K. & Kadarkari, M. (2011). Antimalarial activity of traditionally used Western Ghats plants from India and their interactions with chloroquine against chloroquine-tolerant *Plasmodium berghei. Vector Borne Zoonotic Dis. Larchmt. N 11,* 259–268. doi: 10.1089/vbz.2009.0208.

Santhosh, B., Bhanudas, S. & Sagar, B. (2015). Antipyretic Activity of β-Sitosterol Isolated from leaves of *Oxalis corniculata* Linn. *Am. J. Pharmtech Res. 5,* 214–222.

Simoes-Pires, C., Hostettmann, K., Haouala, A., Cuendet, M., Falquet, J., Graz, B. & Christen, P. (2014). Reverse pharmacology for developing an anti-malarial phytomedicine. The example of *Argemone mexicana*. Int. J. Parasitol. Drugs Drug Resist., Includes articles from two meetings: "Anthelmintics: From Discovery to Resistance," pp. 218–315, and "Global Challenges for New Drug Discovery Against Tropical Parasitic Diseases," pp. 316–357, 338–346.

Simonsen, H.T., Nordskjold, J.B., Smitt, U.W., Nyman, U., Palpu, P., Joshi, P. & Varughese, G. (2001). *In vitro* screening of Indian medicinal plants for antiplasmodial activity. *J. Ethnopharmacol. 74,* 195–204.

Sireeratawong, S., Itharat, A., Lerdvuthisopon, N., Piyabhan, P., Khonsung, P., Boonraeng, S. & Jaijoy, K. (2012). Anti-Inflammatory, Analgesic, and Antipyretic activities of the ethanol extract of *Piper interruptum* Opiz. and *Piper chaba* Linn. *Int. Sch. Res. Not.* e480265.

Ukwe, V., Epueke, E., Ekwunife, O., Okoye, T., Akudor, G. & Ubaka, C. (2010). Antimalarial activity of aqueous extract and fractions of leaves of *Ageratum conyzoides* in mice infected with *Plasmodium berghei*. *Int. J. Pharm. Sci. 2,* 33–38.

Vigneron, M., Deparis, X., Deharo, E. & Bourdy, G. (2005). Antimalarial remedies in French Guiana: a knowledge attitudes and practices study. *J. Ethnopharmacol. 98,* 351–360. doi: 10.1016/j.jep.2005.01.049.

WHO (2014). World Malaria Report, 2014. World Health Organization, Geneva.

Wiesmann, U., Biber-Klemm, S., Grossenbacher-Mansuy, W., Hadorn, G.H., Hoffmann-Riem, H., Joye, D., Pohl, C. & Zemp, E. (2008). Enhancing transdisciplinary research: A synthesis in fifteen propositions, In: Hadorn, G.H., Hoffmann-Riem, H., Biber-Klemm, S., Grossenbacher-Mansuy, W., Joye, D., Pohl, C., Wiesmann, U. & Zemp, E. (Eds.), Handbook of Transdisciplinary Research. Springer: Netherlands, pp. 433–441.

Willcox, M., Bodeker, G. & Rasoanaivo, P. (2004). Traditional Medicinal Plants and Malaria. CRC Press, London.

CHAPTER 9

PLANT-BASED ETHNIC KNOWLEDGE ON FOOD AND NUTRITION IN THE WESTERN GHATS

KANDIKERE R. SRIDHAR and NAMERA C. KARUN

Department of Biosciences, Mangalore University, Mangalagangotri, Mangalore – 574199, Karnataka, India, E-mail: kandikere@gmail.com

CONTENTS

ABSTRACT

The present chapter embodies a brief account on the ethnic plant-based tribal and traditional knowledge prevalent in the Western Ghats fulfilling basic human needs especially food and nutraceuticals. There are several major challenges deriving optimum benefits from the wild plants, which can be augmented by traditional knowledge or experience of tribals and aboriginals. Majority of the past studies in the Western Ghats dealt with floristics in relation to food and medicinal values. A few studies showed multiple applications of wild plant species. The present study has been divided into conventional (fruits and vegetables) nutraceuticals and non-conventional (roots/rhizomes/tubers) nutritional sources of wild plant species. Five important steps like documentation, vulnerability, conservation, product development and welfare of tribes have been suggested as very important to ethnobotanical research in the Western Ghats. The gaps in our knowledge on ethnic resource of the Western Ghats have been identified and approaches necessary to investigate indigenous knowledge/strategies of conservation of wild plant species are also discussed.

9.1   INTRODUCTION

International mega conventions were responsible for understanding the significance of biodiversity, which mitigated exploration and conservation of bioresources throughout the world (CBD, 1992; Millennium Ecosystem Assessment, 2005; Finlayson et al., 2011). Major parts of human nutrition and lifestyle requirements are fulfilled by wild or cultivated plant species. We are intimately dependent on plants to meet several demands especially nutrition, medicine, cosmetics, utensils, implements, musical instruments, sports equipments, handicrafts, furniture, boat/ships, jetties, bridges and shelter. The Western Ghats of India being a hotspot of biodiversity, it serves as a storehouse to meet several human needs. There is a major focus of attraction of the world due to its wealth of endemic biota and a wide variety of wildlife. Western Ghats of India is endowed with different vegetation types like moist evergreen, deciduous, semi-evergreen, Shola, grassland, freshwater swamps, bamboo brakes and scrub jungles supporting a wide variety of life forms (Shetty et al., 2002). In recognition of versatility in biodiversity, the UNESCO has recently recognized the Western Ghats as one of the important World Heritage Sites (UNESCO, 2012).

Tribals and village dwellers/rural-folk being intimately associated with forests and wildlife are gifted with the inherent knowledge on the use of forest product for their basic needs. Unlike modern society, they are not amenable for immediate nutritional needs (e.g., food scarcity), health care (e.g., first-aid) and sophisticated shelter. Thus, their inherent or acquired knowledge and skills instantaneously useful to meet their basic requirements like food, medicine and other day to day needs. Tribals also face several threats like scarcity of food, drought, floods and diseases and developed strategies to overcome or to manage such exigencies. The indigenous knowledge of using wild plant species for a specific purpose stems out of experience and interaction with nature as traditional, folklore and tribal. The Western Ghats being rich in tribal communities, tribal knowledge will go long way in recognizing indigenous plant sources, processes and products.

There are several note-worthy publications pertain to ethnobotanical research in the Western Ghats. Karun et al. (2014) inventoried up to 40 villages of Kodagu District of Karnataka to collect information on the use of traditional knowledge in identifying edible fruits from 45 plant species consisting of trees, shrubs, herbs and creepers. It has been proposed to develop sustainable development of fruit-yielding plant species in agroforests in Kodagu. Fruits of *Artocarpus hirsutus* (monkey jack) possess medicinal principles besides possessing nutritional value as Sarala and Krishnamurthy (2014) projected importance of dried powder of these fruits as indigenous dishes. Wild edible plants from Karnala bird sanctuary in Maharashtra state were enumerated by Datar and Vartak (1975). Gunjakar and Vartak (1982) enumerated wild edible legumes from Pune district of Maharashtra state.

The initial steps necessary to find out usefulness of plant resource in and around us include inventory, identification and documentation. Several studies have been performed in the Western Ghats especially to understand the nutritional and medicinal properties of plant species. Wild fruits provide a change in taste from that used by daily routine. Emphasis has been made on wild edible fruits, which can be immediately used as alternate food sources or as food supplements. Several wild fruits serve as appetizing agents, squashes and digesters. Some of them also serve as medicines (nutraceuticals), fish-poisons and anti-leech agents. Wild fruits are useful to develop several industrial food products like jam, jelly, juice, salted-fruit, wine and soap manufacture. Fruit powders are useful in preparation of a number of dishes either as a whole material or it may serve as ingredient in alternate food preparation. Besides wild fruit, a variety of whole wild plants, leaves stem, flowers, seeds, roots and tubers are used as vegetable. Some of such sources are also consumed directly as raw or cooked or after appropriate processing.

Likewise, whole range of plants are in use for medicinal purposes either as a whole plant or after processing. Many wild plants or their parts are used individually or in combination with other ingredients. Besides, some are worth upgrade for production in industrial scale owing to their versatile properties and novel applications. The major focus of this chapter is to consolidate and discuss various aspects of nutritional (fruit, vegetable and non-conventional sources) and nutraceutical perspectives of wild plant species in the Western Ghats based on the ethnic knowledge.

9.2 FRUITS

Wild fruits serve as valuable source of nutrients (minerals, proteins and vitamins) and nutraceuticals (antioxidants and edible vaccines). They are also useful in the preparation of wines and squashes. Usually fruits are consumed raw and some needs simple processing before consumption. From the Western Ghats parts of Maharashtra, Karnataka, Kerala and Tamil Nadu a variety of wild fruits are identified and those which are used by the tribal/ethnic communities/rural-folk (Table 9.1; Figure 9.1). Kulkarni and Kumbhojkar (1992) gave an account of wild edible fruits consumed by Mahadeo-Koli tribe in Western Maharashtra. Patil and Patil (2000) reported 36 wild edible angiospermic species used by tribals of Nasik district, Maharashtra. The aboriginies include Bhils, Thakur, Katkari, Warli, Kunbi-kokana and Mahadeo-Koli. Among these unripe or ripe fruits of 10 species are eaten raw. Less known 12 wild edible fruits and seeds of Uttar Kannada district was given by Hebber et al. (2010). Sasi et al. (2011) described 50 wild edible plant diversity of Kotagiri hills – a part of Nilgiri Biosphere reserve, of these 20 are fruits. From the Kolhapur region of Maharashtra, tribals and cattle breeders consume nearly 30 wild fruits (from 3 herbs, 10 shrubs and 17 trees) in raw, and many of them in unripe stage are useful as vegetable and pickle preparation (Valvi et al., 2011). Deshpande et al. (2015) recognized up to 28 wild plant species yielding edible fruits and those are consumed by the Rajgond tribe in Maharashtra. As early as 1990s, Uthaiah (1994) and his associates based on extensive survey in the Western Ghats of Karnataka recognized up to 50 edible fruit-yielding tree species. It was followed by Karun et al. (2014), who recently explored 40 villages to identify 45 plant species (32 trees, 7 shrubs, 3 herbs and 3 creepers) yielding edible fruits and gave commentary on their uses other than consumption as raw. Although these fruits are consumed in raw stage as source of nutrients, many of them have medicinal properties and provide immunity to many diseases of rural-folk. These wild fruits are also popular among tourists of Kodagu region as appetizers, digesters and squashes. Unconventional wild fruits and processing in tribal area of Jawahar, Thane district was discussed by Chothe et al. (2014). Recently, Greeshma and Sridhar (2015) reported nine edible fruits consumed by the villagers and tribes of Kodagu region. Some of them can be consumed as whole fruits, in some rind, pulp, juice and seeds can also be used either for consumption or to prepare several dishes without or with processing. Three tribes (Irula, Kurumba and Muduga) live in Attapadi Hills (Kerala) consume 21 wild fruits, seven seeds and one pod as food source (Nadanakunjidam, 2003). Narayanan et al. (2011) also documented 53 fruits and 9 seeds of wild plant species selectively consumed by three tribes (Kattunaikka, Kuruma and Paniya) in Wayanad. Shareef and Nazarudeen (2015) reported 13 edible raw fruits of tree species belonging to the family Euphorbiaceae distributed in entire Kerala. Interestingly one or the other fruits out of 13 species are available throughout the year for consumption. Fruits of *Aporosa cardiosperma*, *Baccaurea courtallensis* and *Phyllanthus emblica* are the most promising edible fruits. *Aporosa bourdillonii*, *A. indoacuminata* and *B. courtallensis* are endemic to the Western Ghats. Thirty eight species of wild edible fruits belonging to 25 genera and 17 families used by Muthuvans of Idukki district, Kerala were recorded by Ajesh et al. (2012a).

A versatile plant *Trichopus zeylanicus* was discovered from the Agasthia Hills of Kerala in the Western Ghats (https://en.wikipedia.org/wiki/Trichopus_zeylanicus). Fruits of this plant is used by the Kani tribe to boost their energy as they have anti-fatigue property. Scientists of Tropical Botanical Garden and Research Institute (TBGRI), Kerala successfully formulated and standardized a health drink called 'Jeevani' (means, 'life-giver'), which has been released for commercial production in 1995 by the Arya Vaidya Pharmacy (Mashelkar, 2001). This health drink (or tonic) has been considered equivalent to 'Ginseng' (a Korean energy drink) prepared out of roots of *Panax ginseng*.

Many reports are available on the tribal knowledge on fruits from the Western Ghats of Tamil Nadu. Arinathan et al. (2003b) reported 10 wild fruits (5 edible in unripe stage and 5 edible as raw) consumed by the Palliyar tribe in Tamil Nadu. Up to 20 wild fruits are eaten almost raw and these served as a nutritional source for the tribals of Anaimalai Hills (Sivakumar and Murugesan, 2005). Arinathan et al. (2007) documented 41 wild edible fruit-yielding plant species used by the Palliyar tribe living in the district of Virudhnagar. Many wild fruits are edible in raw stage (*Coccinia grandis*, *Gardenia resinifera*, *Opuntia stricta*, *Phyllanthus acidus*, *Syzygium cumini* and *S. jambos*) and some for culinary purposes (*Capsicum frutescens*). Under-ripened fruits are used as vegetable (*Carica papaya* and *Coccinia grandis*), for preparation of pickles (*Carissa carandas, Citrus aurantifolia* and *Phyllanthus acidus*) and juice/squash (*Citrus aurantifolia*) by the Paliyar tribe (Ayyanar et al., 2010). According to Jeyaprakash et al. (2011) tribes who live in Theni District use five wild fruits extensively for edible purpose. From Nilgiri Biosphere Reserve, fruits of 70 less known plant species (27 trees, 24 shrubs, 11 herbs and 8 climbers) used by the six tribes (Irula, Kattunayaka, Kota, Kurumba, Paniya and Toda) were documented by Sasi and Rajendran (2012). These fruits are eaten raw and also serve as potential source for local breweries as tribals trade them. Interestingly, these fruits also serve as nutraceuticals especially in reducing the risks of diseases like cancer, cardiovascular ailments and cataracts. Another report on wild fruits of 30 plant species (11 shrubs, 11 trees, 5 herbs, 2 climbers and 1 parasitic herb) meet nutritional needs of Badaga tribe in the Nilgiri Hills (Sathyavathi and Janardhanan, 2014). As Badaga tribe is aware of these wild plant species, plants are preserved and cultivated wherever possible to improve their economic status. In addition to nutritional properties, therapeutic uses of these fruits are also documented. Many fruits serve in first-aid treatments (e.g., diarrhea, vomiting, headache, toothache and cuts). Likewise, another tribe Irula in Maruthamalai Hills depends on 25 wild plant species as source of fruits (Sarvalingam et al., 2014). The wild jack (*Artocarpus hirsutus*) has many nutritional properties and can be used for preparation of a wide variety of dishes indigenously on drying and pulverizing and thus may lead to a future potential cottage industry (Sarala and Krishnamurthy, 2014).

9.3   VEGETABLES

The wild plant species are also major source of vegetable for the tribals and rural-folk (Table 9.1; Figure 9.1). Studies on wild vegetable comes from the states of Maharashtra, Karnataka, Kerala and Tamil Nadu. The Rajgond tribe in Maharashtra consumes nearly 76 wild plant species as vegetables (28 fruits, 22 greens, 9 tubers, 6 flowers, 4 seeds, 3 whole plants, 2 stem and 2 thalamus) either in monsoon season or rest of the year as and when such produce is available (Deshpande et al., 2015). Patil and Patil (2000) reported 36 wild edible angiospermic species used by tribals of Nasik district, Maharashtra. Among them leaves of 14 species, tubers/rhizomes of 4 species, 2 fruits and 1 young shoots are used as vegetable. Three tribals Irula, Kurumba and Muduga residing in Attapady Hills of Kerala use leaves of 23 wild plant species as vegetable (Nadanakunjidam, 2003). Leaves of 72 wild plant species (57 herbs, 11 shrubs and 4 trees) are used by the Paniya tribe in Wayanad (Kerala) as vegetable (Narayanan et al., 2003). Three ethnic communities (Paniya, Kattunaikka and Kuruma) and one migrant community in the Wayanad use leaves of 102 wild plant species as vegetable (Narayanan and Kumar, 2007). Tribals live in Parambikulam (Kerala) utilize leaves of 30 wild plant species and stem/shoot of 6 wild plant species as vegetable (Yesodharan and Sujana, 2007). Anamalai hills, Western Ghats, Coimbatore district were surveyed by Ramachandran (2007) to list out the edible plants utilized by the tribal communities, such as Kadars, Pulaiyars, Malasars, Malaimalasarss and Mudhuvars. About 75 plant species including 25 leafy vegetables, 4 fruit yielding and 45 fruit/seed yielding varieties have been identified. The local tribal communities for their dietary requirements, since a long time have utilized these forest produce. Leaves of 84 wild plant species are selectively useful to three tribes (Kattunaika, Kuruma and Paniya) of Wayanad (Narayanan et al., 2011). Sasi et al. (2011) described 50 wild edible plant diversity of Kotagiri hills – a part of Nilgiri Biosphere reserve. Of these 17 leafy vegetables and 20 tubers/rhizomes are used as vegetable. Greeshma and Sridhar (2015) have reported 10 wild plant species used as vegetables in Kodagu region of Karnataka with indigenous preparation methods.

Edible leaves, edible stem and edible seeds of five wild plant species each consumed by the Palliar tribe in Tamil Nadu have been reported by Arinathan et al. (2003b). Different ethnic groups of Anaimalai region of Tamil Nadu use 53 wild edible plant species as vegetable (Sivakumar and Murugesan, 2005). Of these, leaves of 25 species, wild fruits (as raw) of 20 species and different parts of 8 species (tubers, seeds and roots) are consumed. Ayyanar et al. (2010) carried out an ethnobotanical survey of wild plant species consumed by Paliyar Tribe of Tamil Nadu and reported that 5 plant species are extensively used as vegetable. Besides, these tribals also use pith, flower buds, aril, and bulbils of wild plant species as source of nutrition.

An extensive survey yielded information on leaves (54 species), seeds (45 species), pods (41 species) roots (19 species), pith/apical meristem (12 species) and flowers (10 species) those are nutritionally valuable to the Palliyar tribe in Theni District of Tamil Nadu (Arinathan et al., 2007). Among them, 7 species (*Atylosia scarabaeoides, Canavalia gladiata, Entada rheedi, Mucuna atropurpurea, Vigna bournaea, V. radiata* and *V. trilobata*) are endemic and except for two plant species (*V. radiata* and *V. trilobata*), the rest are threatened in that region. Deshpande et al. (2015) reported use of whole plants in tender stage as vegetable (e.g., *Amaranthus cruentus*, *Andrographis paniculata* and *Portulaca oleracea*) by Rajgond tribe in Vidharba of Maharashtra, out of which the last one is a weed in agricultural land. Mulay and Sharma (2014) described 85 weed species which are used as vegetables. Among them leaves of 46 species, fruits of 20 species, seeds of 7 species, underground parts of 4 species, flowers of 3 species and tender stems of 4 species are used as vegetables. Details of wild edible species of Amaranthaceae and Araceae used by Kuruma and Paniya tribes in Wayanad district in Kerala were given by Hema et al. (2006). Ajesh et al. (2012b) gave an account on the utilization of wild vegetables used by Muthuvan tribes of Idukki district, Kerala. Around 40 plant species were recorded. Among them 70% species contribute to vegetables by their leaf and stem, 18% by fruit, 8% by tubers, 2% by corm and 2% by calyx. Ashok and Reddy (2012) carried out extensive field surveys in Garbhagiri hills in Ahmednagar district in Maharashtra and reported that 35 species are being used as vegetables. Leaves of 17 plants, fruits of 11 plants, seeds of 3 plants, receptacles, bulbils and flower buds of one plant are being used as vegetables.

9.4   NUTRACEUTICALS

Besides fruits and vegetable (given in Sections 9.2 and 9.3), a variety of indigenous plant species worth employing as nutraceutical agents to overcome several diseases are hidden in traditional knowledge. Such knowledge has been spread in Western Ghats especially in Maharashtra, Karnataka, Kerala and Tamil Nadu (Table 9.1; Figure 9.1). Many simple and easily accessible wild plant species besides serving as nutritional source also possess nutraceutical potential. For example, eating raw leaves and fruits of *Oxalis comiculata* is helpful in preventing stomachache and helps in easy digestion (Lingaraju et al., 2013). Epiphyte *Remusatia vivipara* (also known as elephant-ear plant) has several edible and medicinal potentials (Asha et al., 2013). This epiphyte grows luxuriously on tree canopies during monsoon and serves as delicacy in Kodagu region. Its leaves and tuber has nutraceuticals as they possess several value-added pharmaceutical properties. Leaves and stem of *Justicia wynaadensis* are also traditionally used for medicinal purpose in Kodagu region and possess several bioactive principles (phenolics, flavonoids and antioxidants) including natural dye for food coloring and other industrial products (Medapa et al., 2011; Nigudkar et al., 2014) (Figure 9.1). Muthuvans are the prominent tribal group in Kerala following unique culture and ethnobotanical practices to meet the nutraceuticals (Ajesh and

Kumuthaklavalli, 2013). Ayyanar and Ignacimuthu (2013) documented 46 wild plant species used as food, 13 wild plants as nutraceuticals by the Kani tribe in Kalakad (Tamil Nadu) and unripe fruits (*Artocarpus heterophyllus*) and tubers (*Manihot esculenta*) are the favorite sources. Balakrishnan (2014) has reported ethnomedicinal information of about 35 tuberous plant species used by the tribes Malasar and Muthuvan in Coimbatore and also documented endangered nature of some of these plant species. Fruits of *Ziziphus rugosa* are edible and flowers are extensively used in treatment of hemorrhage and menorrhea and its bark also useful as astringent and anti-diarrheal (Kekuda et al., 2011). Madappa and Bopaiah (2012) reported various bioactive components of edible fruits of *Garcinia gummi-gutta* available in the Western Ghats. Recently, detailed studies on nutritional and bioactive component of fruits of *Cucumis dipsaceus* have also been performed (Chandran et al., 2013; Nivedhini et al., 2014). *Arenga wightii*, a palm, is a unique source of starch and beverage for Muthuvan tribe of Idukki district in Kerala (Manithottam and Francis, 2007a).

An endangered shurb, *Decalepis hamiltonii* distributed in the Western Ghats region of Kerala (Kothur Reserve Forest and Palode; also found in Andhra Pradesh) is a potential source of indigenous health drink (Vedavathy, 2004). Roots of this wild plant is called 'brown gold' used by the Yanadi tribe in Chittoor (Andhra Pradesh) as a main source of income. These bitter roots consists of 2-hydroxy-4-methoxybenzaldehyde (isomer of vanillin) useful as flavoring agent in ice-creams, chocolates and drinks (Thornell et al., 2000). Drinks prepared out of roots serve as an appetizer as well as health tonic as blood purifier and nutraceutical in preventing gastric and intestinal disorders (Vedavathy, 2004). Roots are also used in preparation of chutney as well as pickles with lemon juice. *Decalepis hamiltonii* is an endangered species and deserves utmost attention for cultivation and utilization for value-added food and nutraceutical products.

9.5   NON-CONVENTIONAL SOURCES

One of the popular traditional nutritional sources in the Western Ghats is bamboo shoots. Its use as nutritional source has been restricted owing to high toxic cyanogenic glycosides. Knowledge on removal of such toxins by tribals and rural-folk needs systematic approach for effective marketing and propagation of bamboo germsplasm (Nongdam and Tikendra, 2014). Underutilized velvet bean *Mucuna pruriens* is nutraceutically valuable as potent source of protein and essential amino acids (Sridhar and Bhat, 2007a;

Gurumoorthi et al., 2013). Various non-conventional sources (roots/tubers/rhizome) used for food have been listed in Table 9.1.

TABLE 9.1   Tribes and Ethnic Groups of the Western Ghats Using Wild Plant Species as Food and Nutraceuticals

Tribe/ ethnic group	State	Number of plant species	Product	Purpose	Reference	Remarks
Tribals and cattle grazers	Kolhapur, Maharashtra	30	Fruits	Edible	Valvi et al. (2011)	–
Rajgond	Kolhapur, Maharashtra	28	Fruits	Edible	Deshpande et al. (2015)	–
Tribals	Ahmednagar, Maharashtra	9	Fruits	Edible	Khyade et al. (2009)	–
Rural-folk	Kodagu, Karnataka	50	Fruits	Edible	Uthaiah (1994)	–
Rural-folk	Kodagu, Karnataka	45	Fruits	Edible	Karun et al. (2014)	11 species endangered
Tribals and rural-folk	Kodagu, Karnataka	7	Fruits	Edible	Greeshma and Sridhar (2015)	–
Paniya	Wayanad, Kerala	48	Fruits, nuts and seeds	Edible	Narayanan et al. (2003)	Some in riverside
Irula, Kurumba and Muduga	Attapadi Hills, Kerala	30	21 Fruits and 7 seeds	Edible	Nadanakunji-dam (2003)	–
Tribals	Parambi-kulam, Kerala	47	Fruits and seeds	Edible	Yasodharan and Sujana (2007)	–
Kattunai-kka, Kuruma and Paniya	Wayanad, Kerala	15–50	Fruits and seeds	Edible	Narayanan et al. (2011)	–
Muthuvans	Idukki, Kerala	38	Fruits	Edible	Ajesh et al. (2012a)	—

TABLE 9.1   *(Continued)*

Tribe/ ethnic group	State	Number of plant species	Product	Purpose	Reference	Remarks
Rural-folk	Kerala	13	Fruits	Edible	Shareef and Nazarudeen (2015)	All belong to Euphorbiaceae; 3 species endemic
Palliyar	Srivilliputhur, Tamil Nadu	15	10 Fruits and 5 seeds	Edible	Arinathan et al. (2003b)	
Irular	Nilgiris, Tamil Nadu	50	20 fruits	Edible	Sasi et al. (2011)	
Tribals	Anaimalai, Tamil Nadu	15	Fruits	Edible	Sivakumar and Murugesan (2005)	–
Palliyar	Virudhunagar, Tamil Nadu	41	Fruits	Edible	Arinathan et al. (2007)	–
Paliyar	Theni, Tamil Nadu	6	Fruits	Edible	Ayyanar et al. (2010)	–
Tribals	Theni, Tamil Nadu	5	Fruits	Edible	Jeyaprakash et al. (2011)	–
Toda, Kota, Kurumba, Paniya, Irula and Kattunayaka	Nilgiri, Tamil Nadu	70	Fruits	Edible	Sasi and Rajendran (2012)	Possess nutraceutical value
Kani	Kalakad, Tamil Nadu	29	21 Fruits, 7 seeds and 1 pod	Edible	Ayyanar and Ignacimuthu (2013)	
Badaga	Nilgiri, Tamil Nadu	30	Fruits	Edible	Sathyavathi and Janardhanan (2014)	Possess nutraceutical value
Irula	Maruthamalai, Tamil Nadu	25	Fruits	Edible	Sarvalingam et al. (2014)	–

TABLE 9.1 *(Continued)*

Tribe/ ethnic group	State	Number of plant species	Product	Purpose	Reference	Remarks
Rajgond	Kolhapur, Maharashtra	48	22 Leaves, 9 tubers, 6 flowers, 4 seeds, 3 whole plants, 2 stem and 2 thalamus	Vegetable	Deshpande et al. (2015)	–
Tribals	Ahmednagar, Maharashtra	22	Leaves, tubers	Vegetable	Khyade et al. (2009)	
Rural-folk	Kodagu, Karnataka	10	Leaves	Vegetable	Greeshma and Sridhar (2015)	–
Paniya	Wayanad, Kerala	72	Leaves	Vegetable	Narayanan et al. (2003)	Some in riverside
Irula, Kurumba and Muduga	Attapadi Hills, Kerala	30	23 Leaves and 7 Rhizomes/ tubers	Vegetable	Nadanakunjidam (2003)	3 species famine food
Paniya	Wayanad, Kerala	19	Roots and tubers	Vegetable	Narayanan et al. (2003)	Some in riverside
Paniya and Kuruma	Wayanad, Kerala	26	Leaves, tubers and fruits	Vegetable	Garcća (2006); Hema et al. (2006)	Transfer of knowledge
Tribals	Parambikulam, Kerala	47	Leaves, stem/ shoot and rhizomes/ tubers/ corms	Edible	Yasodharan and Sujana (2007)	–
Kattunaikka, Kuruma and Paniya	Wayanad, Kerala	21–71	Leaves	Vegetable	Narayanan et al. (2011)	–
Kattunaikka, Kuruma and Paniya	Wayanad, Kerala	6–25	Roots and tubers	Vegetable	Narayanan et al. (2011)	–

TABLE 9.1   *(Continued)*

Tribe/ ethnic group	State	Number of plant species	Product	Purpose	Reference	Remarks
Muthuvans	Idukki, Kerala	40	Leaves, stems, fruits, tubers, corm, calyx	Vegetables	Ajesh et al. (2012b)	–
Madhuvan	Idukki, Kerala	22	Various parts	Vegetable	Ajesh and Kumuthaklavalli (2013)	–
Palliyar	Srivilliputhur, Tamil Nadu	15	5 Leaves, 5 stem and 5 rhizomes/ tubers	Vegetable	Arinathan et al. (2003b)	–
Tribals	Anaimalai, Tamil Nadu	25	Leaves	Vegetable	Sivakumar and Murugesan (2005)	–
Irular	Nilgiris, Tamil Nadu	50	17 leaves, 5 tubers	Vegetable	Sasi et al. (2011)	
Tribals	Anaimalai, Tamil Nadu	8	Seeds, roots and tubers	Vegetable	Sivakumar and Murugesan (2005)	–
Paliyar	Virudhunagar, Tamil Nadu	181	54 Leaves, 45 seeds, 41 pods, 19 roots, 10 flowers and 12 meristem	Vegetable	Arinathan et al. (2007)	7 species endemic
Paliyar	Theni, Tamil Nadu	5	Various parts	Vegetable	Ayyanar et al. (2010)	–
Kani	Kalakad, Tamil Nadu	25	14 Leaves, 5 stem and 6 tuber	Nutraceuticals	Ayyanar and Ignacimuthu (2013)	–

Tubers of four plant species are utilized as vegetable in Kodagu region of Karnataka with specific method of processing (Greeshma and Sridhar, 2015). Seven rhizomes/tubers constitute vegetable of three tribes (Irula,

Kurumba and Muduga) live in Attapadi Hills of Kerala (Nadanakunjidam, 2003). Rhizomes/tubers of *Dioscorea bulbifera*, *D. pentaphylla* and *D. oppositifolia* serve as famine food of these three tribes. Roots and tubers of 19 wild plant species constitute vegetable of Paniya tribe in Wayanad (Kerala) and many of them are confined to riverside marshy regions (Narayanan et al., 2003). Rhizomes/tubers/corms of 10 wild plant species are used as vegetable by tribals of Parambikulam (Kerala) (Yesodharan and Sujana, 2007). Roots and tubers of 6–25 wild plant species constitute selective source of nutrition of three major tribes of Wayanad (Kattunaikka, Kuruma and Paniya) (Narayanan et al., 2011).

Arinathan et al. (2003b) reported edible tubers/rhizomes of five wild plant species used as vegetables by the Palliar tribe in Tamil Nadu. A survey of tribal communities in the Western Ghat region of Tamil Nadu (Coimbatore: Irular, Malasar and Muthuvan) revealed occurrence of 35 tuberous edible plant species (Balakrishnan, 2014). Another important tuber elephant-foot yam (*Amorphophallus paeoniifolius*) has potential to serve in food industry as it can be easily grown and preserved under normal conditions for prolonged periods. Besides its nutritional value, it has several medicinal properties especially it's methanolic and hydro-alcoholic extracts possess excellent antioxidant potential (Nataraj et al., 2008). *Diplazium esculentum*, a riparian fern is commonly used as vegetable (fiddle heads and tender petioles) in Kodagu region of the Western Ghats (Akter et al., 2014; Karun et al., 2014; Greeshma and Sridhar, 2015) (Figure 9.1). Taro starch of *Colocasia esculenta* is highly digestible compared to conventional starches due to its small grain size (70–80%) (Ahmed and Khan, 2013) (Figure 9.1). This can be achieved with duel advantage as its tuber and leaves are useful as vegetable.

Manithottam and Francis (2007b) discussed ethnobotany of finger millet among Muthuvan tribes of Idukki district, Kerala. *Katty* is a special dish prepared from the powdered grains of *Eleusine* by these people. *Katty* is the unique pudding prepared by the *Muthuvans* from the grains of finger millet. The dried grains are winnowed, dehusked in an *Ural* (Ponder) and powdered in a millstone. For 1 kg of powder, 4 L of water is required. Then with constant stirring using two thin strong sticks, the powder is added into the boiling water. At this time, the fire has to be regulated to adjust the softness of the preparation. This is a skilled work and requires some experience. If the stirring is not uniform the hot water will not reach every portion in the powder and if the heat of water is more, pudding will lose its property. After cooling, it is cut into pieces and consumed with or without side dishes

9.6   DISCUSSION

Integration of indigenous scientific knowledge of traditional communities or tribals assumes utmost importance to document as well as conserve wild plant resources (Berkes, 2008). Following five important steps that are needed to utilize the natural wild plant resource of the Western Ghats for human welfare: The first step of understanding the bounty of benefits of wild plants needs systematic documentation with geographic indication; Step second needs to follow up the vulnerability of wild plant species to specific environmental conditions and their plasticity to natural calamities; Thirdly, conservation measures to be implemented to enhance specific wild plant species *in situ* and *ex situ*; Fourth approach is mode of enforcement of desired technological measures to derive and upgrade products of wild plants; Finally, enhancement of economic benefits and incentives to ethnic population (tribes and aborigines) to uplift their current status through sharing of benefits.

Documentation of traditional knowledge of tribes and aboriginals and geographic indication are the most important steps to overcome biopiracy and helpful in sharing benefits. Floristics of wild plant species needs involvement of trained botanists/technicians to identify the plant species accurately and documentation of availability of such resources consistently or inconsistently (seasonal/annual/perennial) to develop database and future strategies to utilize those resources more effectively in regional scale. Identification of pockets of wild plant resources and possibilities of cultivation in available area would help independence of tribes. Unlike urban population, rural-folk in many developing countries rely on several indigenous food and medicinal plant species to meet their nutritional and nutraceutical needs (Arinathan et al., 2003a; FAO, 2004; Bhat and Karim, 2009; Bharucha and Pretty, 2010). Are there any possibilities to identify a plant resource which can be equivalent to the staple foods like rice, wheat and maize through traditional knowledge of tribes and aboriginals? In Kodagu region of Karnataka for instance, wild fruits of different varieties are available throughout the year and such knowledge would lead to economic gains of tribes exploiting one or the other fruits as staple ones.

Identification and classification of plant species as abundant, endemic, endangered, near endangered, threatened, re-identified and red-listed would help in future research to focus on rehabilitation and regeneration in local geographic conditions. Understanding the regional abiotic/edaphic factors governing distribution and competition of such plant species would help in future domestication. Large scale product-wise mapping based on traditional knowledge and inventories will go long way for *in situ* and *ex situ* conservation.

The above assessments lead to employing conservation measures. Several plants are used either as whole or its parts. Those plants used as whole for nutritional and medicinal purposes need priority of conservation. For example, whole plants in tender stage used as vegetable (e.g., *Amaranthus cruentus*, *Andrographis paniculata* and *Portulaca oleracea*) should be conserved on a priority basis (Deshpande et al., 2015). *Justicia wynaadensis* is another versatile nutraceutical plant species of Kodagu region of the Western Ghats deserves conservation priority (Medapa et al., 2011; Nigudkar et al., 2014; Greeshma and Sridhar, 2015) (Figure 9.1).

FIGURE 9.1   Representative wild plant species used as source of food and nutraceuticals by the ethnic groups in Western Ghats. Near ripened fruits of *Lycopersicum esculentum* (a); ripened fruits of *Solanum rudepannum* (b); unripe and ripe fruits of *Solanum americanum* (c); riparian fern *Diplazium esculentum* (d); edible fiddle heads and tender petioles of *D. esculentum* (e); leaves (f), flower (g) and fruits (h) of *Oxalis esculenta* serve as nutraceuticals; leaves of *Solanum americanum* serve as vegetable (i); shoots of *Colocacia esculenta* emerging from tuber (j); proliferating tubers of *C. esculenta* (k); nutraceutically valued plant species *Justicia wynaadensis* (l).

Techniques to establish repositories (seed or germplasm) will be based on the type of wild plant species. Various methods and strategies are necessary to follow up cultivation and conservation of herbs, climbers, shrubs and tree species. Identification and cultivation in agricultural fields close to the vicinity as hedge plants meet dual purposes of conservation and enhancement of economic gains. Some wild plant species are of special interest as they produce industrially important products. For instance, starch (taro starch) derived from common vegetable plant species (*Colocasia esculenta*) has high value in food and pharmaceutical industries, which needs immediate attention (Ahmed and Khan, 2013) (Figure 9.1). Similarly, wild fruits of *Garcinia gummi-gutta* are useful in preparation of products like high value vinegar, wine and also hydroxy citric acid (anti-cholesterol) (Karun et al., 2014). Studies are also available on the use of seeds of itch-bean (*Mucuna pruriens*) and lotus (*Nelumbo nucifera*) for nutritional and nutraceutical purposes (Sridhar and Bhat, 2007b; Bhat et al., 2008). Technological progress are necessary to process and preserve the essence and bioactive principles of fruits and wild plant species for prolonged period to gain benefit of nutraceuticals.

Food and health security of tribals and aboriginals are important to help them to continue pass on traditional knowledge from one generation to another. If any wild plant products are patented, some economic gains should be earmarked for tribal rehabilitation and welfare (like education, economic security and improvement of living status). For example, García (2006) projected an interesting connection between mother and child nexus in three socio-cultural groups (Paniya, Kuruma tribes and non-tribals) using wild food plants in Wayanad (Kerala) and its importance. The outcome of this exercise suggested that 26 wild food plants (leaf, tuber and fruit) can be used as nutritional source. This approach serves in prevention of erosion of traditional knowledge and involvement of children paving way for transfer of knowledge as well as future cottage industries. There are several plant species that have the capacity to serve as source of non-conventional products related to nutrition, medicine and other products for sustenance of tribes and aboriginals. Studies on gum-/resin-/dye-yielding, biofuel, essence and flavor from wild plants needs special attention in the Western Ghats.

9.7 CONCLUSIONS

The present study projected the traditional knowledge on various wild plant species occurring in the Western Ghats useful in human nutrition and health. Some studies are confined to list the wild plant resources and some projected knowledge of tribes towards their nutrition and nutraceuticals. Although there are some confirmed results for use in human nutrition and health, there are gaps in our knowledge on authenticity of use of parts of wild plant species and processing method needs advanced *in vitro* and *in vivo* studies for future applications. Documentation of wealth of traditional knowledge opens up ample opportunities for further progress in nutrition and health. Does the traditional knowledge helps solving protein-energy malnutrition in rural-folks? The scientific validation of several claims needs authenticity on priority to contribute benefits from regional towards global community. There are several approaches to make scientific studies more meaningful. There is a need for integrated approach towards confirmed studies and directions based on the publications on traditional values. For instance, it is possible to focus studies like one nutrient vs. several plant species or one plant species vs. several nutrients, likewise, one disease vs. several plant species or one plant species vs. several diseases. Cultivation of essential wild plant species need to be initiated in plantations and agroforests as an immediate measure. In addition to the current traditional knowledge base, research should orient towards gaining more information regarding wild plant species useful as food, fiber, fodder, fuel and fertilizer as an integrated approach. To progress further in upgrading the wild plant-based perspectives, it is necessary to respect and research the traditional and ethnic knowledge at grass root level. Threats of invasion of alien plant species in forests is one of the important issues to be addressed immediately to save useful wild plant species for future.

ACKNOWLEDGEMENTS

KRS acknowledges the University Grants Commission, New Delhi for the award of UGC-BSR Faculty Fellowship. NCK acknowledges Mangalore University for partial financial support through fellowship under the Promotion of University Research and Scientific Excellence (PURSE), Department of Science Technology, New Delhi. Authors are grateful to referees for meticulous corrections of early version of the manuscript.

KEYWORDS

- **Aboriginals**
- **Ethnic Food Plants**
- **Fruits**
- **Nutraceuticals**
- **Vegetables**

REFERENCES

Ahmed, A. & Khan, F. (2013). Extraction of starch from taro (*Colocasia esculenta*) and evaluating it and further using taro starch as disintegrating agent in tablet formulation with overall evaluation. *Inventi Rapid: Novel Excipients, 15*.

Ajesh, T.P. & Kumuthakalavalli, R. (2013). Botanical ethnography of muthuvans from the Idukki District of Kerala. *Int. J. Pl. Anim. Environ. Sci., 3,* 67–75.

Ajesh, T.P., Naseef, S.A.A. & Kumuthakalavalli, R. (2012a). Ethnobotanical documentation of wild edible fruits used by Muthuvan tribes of Idukki, Kerala – India. *Int. J. Pharm. Bio Sci, 3(3),* 479–487.

Ajesh, T.P., Naseef, S.A.A. & Kumuthakalavalli, R. (2012b). Preliminary study on the utilization of wild vegetables by Muthuvan tribes of Idukki district of Kerala, India. *Int. J. Appl. Biol. Pharmaceut. Tech. 3,* 193–199.

Akter, S., Hossain, M.M., Aral, I. & Akhtar, P. (2014). Investigation of *in vitro* antioxidant, antimicrobial and cytotoxic activity of *Diplazium esculentum* (Retz.). Sw. *Int. J. Adv. Pharm. Biol. Chem., 3,*723–733.

Arinathan, V., Mohan, V.R. & De Britto, A.J. (2003a). Chemical composition of certain tribal pulses in South India. *Int. J. Food Sci., 54,* 209–217.

Arinathan, V., Mohan, V.R., De Britto A.J. & Chelladurai, V. (2003b). Studies on food and medicinal plants of Western Ghats. *J. Econ. Taxon. Bot., 27,* 750–753.

Arinathan, V., Mohan, V.R., De Britto, A.J. & Murugan, C. (2007). Wild edibles used by Palliyars of the Western Ghats, Tamil Nadu. *Indian J. Trad. Know., 6,* 163–168.

Asha, D., Nalini, M.S. & Shylaja, M.D. (2013). Evaluation of phytochemicals and antioxidant activities of *Remusatia vivipara* (Roxb.) Schott., an edible genus of Araceae. *Der Pharmacia Lettre, 5,* 120–128.

Ashok, S. & Reddy, P.G. (2012). Some less known wild vegetables from the Garbhagiri hills in Ahmednagar district (M.S.) India. *J. Pharmaceut. Res. and Opinion, 2(1),* 9–11.

Ayyanar, M. & Ignacimuthu, S. (2013). Plants used as food and medicine: an ethnobotanical survey among Kanikaran community in South India. *Eur. J. Nutr. Food Safety 3,* 123–133.

Ayyanar, M., Sankarasivaraman, K., Ignacimuthu, S. & Sekar, T. (2010). Plant species with ethno botanical importance than medicinal in Theni district of Tamil Nadu, southern India. *Asian J. Exp. Biol. Sci., 1,* 765–771.

Balakrishnan, S.B. (2014). Ethnomedicinal importance of wild edible tuber plants from tribal areas of Western Ghats of Coimbatore, Tamil Nadu, India. *Life. Sci. Leaflets, 58,* doi: http://dx.doi.org/10.1234/lsl.v58i0.154

Berkes, F. (2008). Sacred ecology: Traditional ecological knowledge and management systems. Taylor and Frances, Philadelphia and London.

Bharucha, Z. & Pretty, J. (2010). The roles and values of wild foods in agricultural systems. *Phil. Trans. Royal Soc. B., 365,* 2913–2926.

Bhat, R. & Karim, A.A. (2009). Exploring the nutritional potential of wild and underutilized legumes. *Compare. Rev. Food Sci. Food Safety, 8,* 305–331.

Bhat, R., Sridhar, K.R., Young, C.-C., Arun, A.B. & Ganesh, S. (2008). Composition and functional properties of raw and electron beam irradiated *Mucuna pruriens* seeds. *Int. J. Food Sci. Technol., 43,* 1338–1351.

CBD (1992). Convention for Biological Diversity – Strategic Plan for biodiversity 2011–2020: http://www.cbd.int/decision/cop/?id=12268.

Chandran, R., Nivedhini, V. & Parimelazhagan, T. (2013). Nutritional composition and anti-oxidant properties of *Cucumis dipsaceus* Ehrenb. ex Spach leaf. *The Sci. World J.*, 1–9, doi: http://dx.doi.org/10.1155/2013/890451.

Chothe, A., Patil, S. & Kulkarni, D.K. (2014). Unconventional wild fruits and processing in tribal area of Jawahar, Thane district. *Bioscience Discovery 5(1),* 19–23.

Datar, R. & Vartak, V.D. (1975). Enumeration of wild edible plants from Karnala Bird Sanctuary, Maharashtra State. *Biovigyana., 1,* 123–129

Deshpande, S., Joshi, R. & Kulkarni, D.K. (2015). Nutritious wild food resources of Rajgond tribe, Vidarbha, Maharashtra State, India. *Indian J. Fund. Appl. Life Sci., 5,* 15–25.

FAO (2004). Annual Report: The state of food insecurity in the world, monitoring the progress towards the world food summit and millennium development goals. Food and Agricultural Organization, Rome.

Finlayson, C.M., Carbonell, M., Alarcó, T. & Masardule, O. (2011). Analysis of Ramsar's guidelines for establishing and strengthening local communities' and Indigenous peoples' participation in the management of wetlands (Resolution VII.8). In: Science and local communities – Strengthening partnerships for effective wetland management, Carbonell, M., Nathai-Gyan, N. & Finlayson C.M. (eds.). Ducks Unlimited Inc., Memphis, USA, pp. 51–56.

García, G.S.C. (2006). The mother – child nexus: knowledge and valuation of wild food plants in Wayanad, Western Ghats, India. *J. Ethnobiol. Ethnomed., 2,* 1–6.

Greeshma, A.A. & Sridhar, K.R. (2015). Ethnic plant-based nutraceutical values in Kodagu region of the Western Ghats. In: *Biodiversity in India*, Volume # 8, In: Pullaiah, T. & Rani, S. (eds.). Regency Publications, New Delhi, pp. 299–317.

Gunjakar, N. & Vartak, V.D. (1982). Enumeration of wild edible legumes from Pune district, Maharashtra state. *J. Econ. Taxon. Bot. 31,* 1–9.

Gurumoorthi, P., Janardhanan, K. & Kalavathy, G. (2013). Improving nutritional value of velvet bean, *Mucuna pruriens* (L.) DC. var. *utilis* (Wall. ex. Wight) L.H. Bailey, an under-utilized pulse, using microwave technology. *Indian J.Trad. Knowl., 12,* 677–681.

Hebbar, S.S., Hegde, G. & Hegde, G.R. (2010). Less known wild edible fruits and seeds of Uttar Kannada district of Karnataka. *Indian Forester, 136(9),* 1218–1222.

Hema, E.S., Sivadasan, M. & Anilkumar, N. (2006). Studies on edible species of Amaranthaceae and Araceae used by Kuruma and Paniya tribes in Wayanad district, Kerala, India. *Ethnobotany 18(1),* 122–126.

Jeyaprakash, K., Ayyanar, M., Geetha, K.N. & Sekar, T. (2011). Traditional uses of medicinal plants among the tribal people in Theni District (Western Ghats), Southern India. *Asian Pac. J. Trop. Biomed.*, S20–S25.

Karun, N.C., Vaast, P. & Kushalappa, C.G. (2014). Bioinventory and documentation of traditional ecological knowledge of wild edible fruits of Kodagu-Western Ghats, India. *J. For. Res., 3,* 717–721.

Kekuda, P.T.R., Raghavendra, H.L. & Vinayaka, K.S. (2011). Evaluation of pericarp of seed extract of *Zizypus rugosa* Lam. for cytotoxic activity. *Int. J. Pharmaceut. Biol. Arch., 2,* 887–890.

Khyade, M.S., Kolhe S.R. & Deshmukh, B.S. (2009). Wild Edible Plants used by the Tribes of Akole Tahasil of Ahmednagar District (MS), India. *Ethnobotanical Leaflets 13,* 1328–1336.

Kulkarni, D.K. & Kumbhojkar, M.S. (1992). Ethnobotanical studies on Mahadeo Koli tribe in western Maharashtra Part III Non-conventional wild Edible fruits. *J. Econ. Taxon. Bot. 10,* 151–158.

Lingaraju, D.P., Sudarshana, M.S. & Rajashekar, N. (2013). Ethnopharmacological survey of traditional medicinal plants in tribal areas of Kodagu district, Karnataka, India. *J. Pharm. Res. 5*, 284297.

Madappa, M.B. & Bopaiah, A.K. (2012). Preliminary phytochemical analysis of leaf of *Garcinia gummi-gutta* from Western Ghats. *IOSR J. Pharm. Biol. Sci., 4*, 17–27.

Manithottam, J. & Francis, M.S. (2007a). *Arenga wightii* Griff. – A unique source of starch and beverage for Muthuvan tribe of Idukki district, Kerala. *Indian J. Trad. Knowl. 6(1)*, 195–198.

Manithottam, J. & Francis, M.S. (2007b). Ethnobotany of finger millet among Muthuvan tribes of Idukki district, Kerala. *Indian J. Trad. Knowl. 6(1)*, 160–162.

Mashelkar, R.A. (2001). Intellectual property rights and the Third World. *Curr. Sci, 51*, 955–965.

Medapa, S., Singh, G.R.J. & Ravikumar, V. (2011). The phytochemical and antioxidant screening for *Justicia wynaadensis*. *Afr. J. Plant Sci., 5*, 489–492.

Millennium Ecosystem Assessment (2005). Ecosystems and human wellbeing: Synthesis. Island Press, Washington, DC.

Mulay, J.R. & Sharma, P.P. (2014). Some underutilized plant resources as a source of food from Ahmednagar District, Maharashtra, India. *Discovery, 9(23)*, 58–64.

Nadanakunjidam, N. (2003). Some less known wild food plants of Attapadi Hills, Western Ghats. *J. Econ. Taxon. Bot., 27*, 741–745.

Narayanan, M.K.R. & Kumar, N.A. (2007). Gendered knowledge and changing trends in utilization of wild edible greens in Western Ghats, India. *Indian J. Trad. Knowl., 6*, 204–216.

Narayanan, M.K.R., Kumar, N.A. & Balakrishnan, V. (2003). Uses of wild edibles among the Paniya tribe in Kerala, India, In: Conservation and Sustainable Use of Agricultural Biodiversity: A Source Book. CIP-UPWARD, Philippines, pp. 100–108.

Narayanan, M.K.R., Kumar, N.A., Balakrishnan, V., Sivadasan, M., Alfarhan, H.A. & Alatar, A.A. (2011). Wild edible plants used by the Kattunaikka, Paniya and Kuruma tribes of Wayanad District, Kerala, India. *J. Med. Pl. Res., 5*, 3520–3529.

Nataraj, H.N., Murthy, R.L.N. & Shetty, S.R. (2008). *In vitro* antioxidant and free radical scavenging potential of *Amorphophallus peoniifolius*. *Orient. J. Chem. 24*, 895–902.

Nigudkar, M., Patil, N., Sane, R. & Datar, A. (2014). Preliminary phytochemical screening and HPTLC fingerprinting analysis of *Justicia wynaadensis* (Nees). *Int. J. Pharma Sci., 4*, 601605.

Nivedhini, V., Chandran, R. & Parimelazhagan, T. (2014). Chemical composition and antioxidant activity of *Cucumis dipsaceus* Ehreng. ex Spach fruit. *Int. Food Res. J. 21*, 1465–1472.

Nongdam, P. & Tikendra, L. (2014). The nutritional facts of bamboo shoots and their usage as important traditional foods of Northeast India. *Int. Schol. Res. Not.*, 1–17; doi: http://dx.doi.org/10.1155/2014/679073

Patil, M.V. & Patil, D.A. (2000). Some more wild edible plants of Nasik district (Maharashtra). *Ancient Science of Life, 19(3&4)*, 102–104.

Ramachandran, V.S. (2007). Wild edible plants of the Anamalais, Coimbatore district, Western Ghats, Tamil Nadu. *Indian J. Trad. Knowl. 6(1)*, 173–176.

Sarala, P. & Krishnamurthy, S.R. (2014). Monkey jack: underutilized edible medicinal plant, nutritional attributes and traditional foods of Western Ghats, Karnataka, India. *Indian J. Trad. Knowl., 13*, 508–518.

Sarvalingam, A., Rajendran, A. & Sivalingam, R. (2014). Wild edible plant resources used by the Irulas of the Maruthamalai Hills, Southern Western Ghats, Coimbatore, Tamil Nadu. *Indian J. Nat. Prod. Res. 5*, 198–201.

Sasi, R. & Rajendran, A. (2012). Diversity of wild fruits in Nilgiri hills of the southern Western Ghats – ethnobotanical aspects. *Int. J. Appl. Biol. Pharmaceut. Technol., 3*, 82–87.

Sasi, R., Rajendran, A. & Maharajan, M. (2011). Wild edible plant Diversity of Kotagiri Hills – a Part of Nilgiri Biosphere Reserve, Southern India. *J. Research in Biol., 2*, 80–87.

Sathyavathi, R. & Janardhanan, K. (2014). Wild edible fruits used by Badagas of Nilgiri District, Western Ghats, Tamilnadu, India. *J. Med. Pl. Res., 8*, 128–132.

Shareef, S.M. & Nazarudeen, A. (2015). Utilization of potential of Euphorbiaceae wild edible fruits of Kerala, a case study. *J. Non-Timber For. Prod., 22*, 2529.

Shetty, B.V., Kaveriappa, K.M. & Bhat, K.G. (2002). *Plant resources of Western Ghats and lowlands of Dakshina Kannada and Udupi Districts.* Pilikula Nisarga Dhama Society, Mangalore.

Sivakumar, A. & Murugesan, M. (2005). Ethnobotanical studies on the wild edible plants used by the tribals of Anaimalai hills, the Western Ghats. *Ancient Sci. Life 25*, 1–4.

Sridhar, K.R. & Bhat, R. (2007a). Agrobotanical, nutritional and bioactive potential of un-conventional legume – *Mucuna. Livest. Res. Rural Develop., 19*, Article # 126: http://www.cipav.org.co/lrrd/lrrd19/9/srid19126.htm

Sridhar, K.R. & Bhat, R. (2007b). Lotus – A potential nutraceutical source. *J. Agric. Technol., 3*, 143–155.

Thornell, K., Vedavathy, S., Mulhooll, D.A. & Crouch, N.R. (2000). Parallel usage pattern of African and Indian periplocoids corroborate phenolic root chemistry. *South Afr. Enthnobot., 2*, 17–22.

UNESCO (2012). United Nations Educational, Scientific, and Cultural Organization. World Heritage Center: www.unesco.org/en/list/1342

Uthaiah, B.C. (1994). Wild edible fruits of Western Ghats – A survey. *Higher plants of Indian subcontinent.* Additional series of *Indian Journal of Forestry, 3,* 87–98.

Valvi, S.R., Deshmukh, S.R. & Rathod, V.S. (2011). Ethnobotanical survey of wild edible fruits in Kolhapur District. *Int. J. Appl. Biol. Pharmaceut. Technol., 2*, 194–197.

Vedavathy, S. (2004). *Decalepis hamiltonii* Wight & Arn. – an endangered source of indigenous health drink. *Nat. Prod. Rad., 3*, 22–23.

Yesodharan, K. & Sujana, K.A. (2007). Wild edible plants traditionally used by the tribes in the Parambikulam Wildlife Sanctuary, Kerala, India. *Nat. Prod. Rad., 6*, 74–80.

CHAPTER 10

USEFUL PLANTS OF WESTERN GHATS

S. KARUPPUSAMY[1] and T. PULLAIAH[2]

[1]Department of Botany, The Madura College (Autonomous), Madurai–625011, Tamilnadu, India.
E-mail: ksamytaxonomy@gmail.com

[2]Department of Botany, Sri Krishnadevaraya University, Anantapur–515003, Andhra Pradesh, India.
E-mail: pullaiah.thammineni@gmail.com

CONTENTS

ABSTRACT

Western Ghats of southern India represented rich and repository of useful plants with high anthropogenic diversity. The present review gives an account of fiber yielding plants, dye yielding plants, plants used as brooms, forage plants, wild ornamentals, sacred plants, resin and gum yielding plants from Western Ghats, which are used by traditional communities residing in the area. About 200 plant species are used for other than medicinal purposes and many of them are endemic to the region. The plant resources provide the quality life and socioeconomic support to the local communities. Hence the documentation and sustainable utilization are entrusted for long term uses and conservation of these useful plant resources.

10.1   INTRODUCTION

The ancient human started his nomadic life by using plant materials directly for covering and protecting his body, thatched leaf for shelter and huts, mats for household, coloring materials for his ornaments, arts and cloths, fibers and cordage materials for weaving and collecting vessels, gums and resins for adhesives, plants and plant products for keeping away evil spirit and other day to day activities. Gradually fast mobility and advancement in lifestyle led him to search for lighter, more durable and sophisticated looking material for routine use. There began an era of developing different plant sources for textiles, papers, basketries, dyes, woven cloths, mats, hats, ropes, cordage, fuel wood, timber wood, ornamental material for various uses.

The Western Ghats is concentrated with varied kind of vegetation and unimaginable topographical features. Biogeographically, the Western Ghats constitutes the Malabar province of the Oriental realm, running parallel to the west coast of India from 8° N to 21° N latitudes, 73° E to 77° E longitudes for around 1600 km. The average width of this mountain range is about 100 kms. This bioregion is hot spot diversity and under constant threat due to human pressure, and is considered one of the 34-biodiversity hot spots of the world. The tropical climate complimented by heavy precipitation from southwest monsoon and favorable edaphic factors provide an ideal condition for the luxuriant growth of plant life, which can be seen only in few parts of the world. With its rainfall regime, the western slopes of the Ghats have a natural cover of evergreen forest, which changes to moist and then dry deciduous type as one comes to the eastern slopes. The vegetation reaches its highest development towards the southern tip in Kerala with rich tropical rain forests. Many number of ethnic people settled all over the range of Ghats and their cultural diversity also varied. The Western Ghats presents a whole range of gradients, both altitudinal as well as latitudinal in climatic factors, such as total annual rainfall, maximum temperatures and anthropogenic diversity yielded the varieties of traditional systems with natural components. The present study revealed the useful plants of Western Ghats other than medicinal importance.

10.2   PLANT FIBERS

Fiber yielding plants have been of great importance to man and they ranked second only food plants in their usefulness. The fibers may be classified based on their application and uses that are textile fibers, brush fibers, rough weaving fibers, filling fibers, natural fibers and paper making fibers. All these fibers are of plant origin. Majority of the fiber yielding plants are exploited from the wild or semi-cultivated state. The plant fibers are extracted from different parts, such as stem, leaf, petioles, roots, fruits and seeds. A few species are the source of fibers from roots and stems both namely *Cissus quadrangularis* and *Homalocladus platycladus*. The fibers are mainly used in textile industries, paper manufacture, filling, making ropes, fishing nets and cordage, thatch, hats and other weaving materials and for brush making. For items like gunny bags, ropes, cordage, fishing nets, bast fibers of commercially exploited species, such as *Corchorus* spp., *Hibiscus* spp. have often been used (Pandey and Gupta, 2003).

10.3   DYE YIELDING PLANTS

Indigenous knowledge system has been practiced over the years in the past, the use of natural dyes has diminished over generations due to lack of documentation and with the arrival of synthetic dyes. Though there is large plant sources base available in Western Ghats, little has been exploited so far. Due to lack of availability of precise technical knowledge on the extracting and dyeing technique, it has not commercially succeeded so far. Many natural dyestuff and stains are obtained mainly from higher plants and dominate sources of natural dyes, producing different colors. Almost all parts of the plant like root, bark, leaf, fruit, wood, seed, flower, etc. produce dyes. It is interesting to note that over 2000 pigments are synthesized by various parts of plants, of which only about 150 have been exploited by human being for their coloring purposes (Siva, 2007).

10.4 PLANTS USED AS BROOMS

Cleaning of houses and courtyards is a daily practice in most of the Indian households and is ritualistically followed in many countries (Rasingam and Jeeva, 2013). The brooms are traditionally made by plant parts as a general practice. It has been used for a centuries to sweep caves, cabins and castles. Tree branches and young twigs of herbaceous plants were often used to sweep the floor and clean the ashes from the fire places. Sometimes crude brooms like straw, hay, fine twigs and infloresecence were also used by tying with thread or plant fiber for easy handling. However, in some places, where technology is unavailable or deficient and the bioresources are easily available, the traditional methods are still widely used.

10.5 FORAGE PLANTS

Forest inhabiting people maintain cattle, goat, sheep and farm animals for various purposes. The fodder and forages consumed by these animals should be easily digestible and should also yields more energy. The chemical composition and nutritive value of a fodder crop is very important. Presence of large amount of lignin and tannin and low amount of phosphorus affects the utility of fodder crops. Nair et al. (2014) reported two new fodder crop for Kerala region of Western Ghats.

10.6 WILD ORNAMENTAL PLANTS

The wild ornamental flowers have been more attractive and long prized for the beauty and planted in gardens around mankind dwelling places. The ornamental plants play an important role in environmental planning of urban and rural areas for abatement of pollution, social and rural friendly, wasteland development, afforestation and landscaping of outdoor and indoor space (Sarvalingam and Rajendran, 2014). These plants are grown usually for the purpose of beauty and esthetics and for their fascinating foliage, flowers and their pleasant fragrance. Maruthamalai hills of Western Ghats are reported to have 40 species of wild ornamental climbers belonging to 25 genera and 12 families. Most of the plants belonging to Convolvulaceae and Fabaceae, such as *Abrus precatorius, Argyreia pomacea, Canavalia mollis, C. virosa, Ipomoea pes-tigridis, I. quinata, I. cairica, I. obscura, I. quamoclit, I. staphylina, I. wightii, Mucuna monosperma* and *Rivea hypocrateriformis*. Many of the *Jasminum* species namely *J. auriculatum, J. angustifolium, J. azoricum, J. grandiflorum, J. malabaricum* and *J. sessiliflorum* are having ornamental potential. Madukkarai hills of southern Western Ghat alone have been recorded about 137 wild ornamental plants spread over 99 genera and 42 families (Palanisamy and Arumugam, 2014). Several species of them are endemic to Western Ghats, such as *Acalypha alnifolia, Asparagus fysonii, Barleria acuminata, Caralluma indica, Christisonia bicolor, Indigofera trita, I. uniflora, Moringa concanensis, Pterolobium hexapetalum, Sesamum laciniatum* and *Striga desniflora*.

10.7 SACRED PLANTS

The traditional worship practices show the symbiotic relation of human beings and conserved its valuable biodiversity (Sahu et al., 2013). Tribal pockets residing in the forest areas follow ancestral traditional worship in the form of deity worship, with central focus of worship on forest patches which signify sacred groves. Most of the sacred groves are open and do not have well demarcated boundaries. Well preserved sacred groves are store house of valuable medicinal and other useful plants (Ayyanar and Ignacimuthu, 2010). The sacred grove system was well explained for conservation of biodiversity in Western Ghats by Gadgil and Vartak (1975, 1976). Several useful trees, such as *Mangifera indica, Artocarpus integrifolia, Garcinia gummi-gutta, Caryota urens, Piper nigrum, Cinnamomum malabathrum, Canarium strictum, Dipterocarpus indicus, Vateria indica, Myristica fatua, Pinanga dicksonii, Semecarpus acuminata* and *Gymnacranthera canarica* are reported in many sacred groves of Western Ghats. Some rare and endangered plant species are recovered only from the sacred groves of Western Ghats, such as *Kuntsleria keralensis* (Mohanan and Nair, 1981), *Blepharistemma membranifolia, Buchanania lanceolata* and *Syzygium travancoricum*. Aesthetic values of selected floral elements of Khatana and Waghai forests of Dangs, Western Ghats were given by Kumar et al. (2005).

10.8 RESIN AND GUM YIELDING PLANTS

Gums and resins perhaps the most widely used and traded non-wood forest products other than items consumed directly as food, fodder and medicine. Human kinds have been using gums and resins in various forms for ages. Use of gums and resins for domestic consumption and for sale to earn some cash is very popular among the forest dwelling communities, particularly tribals in Western Ghats (Upathayay, 2013). Several woody species are sources of gums and resins in this region which include *Shorea robusta* (Sal), *Vateria indica* (Vallapine), *Pinus roxburghii* (Chair pine), *Dipterocarpus turbinatus* (Gurjan) and *Garcinia morella* (Gambog). Most of the gum and resin yielding trees belong to the families Pinaceae, Fabaceae, Burseraceae and Dipterocarpaceae in Western Ghats. Other gum and resin yielding trees are *Sterculia urens, Anogeissus latifolia, Canarium strictum, Boswellia serrata, Commiphora caudata* and *Gardenia gummifera* which are exploited by the tribals of Western Ghats for various types of gums and resins.

10.9 OTHER USES

Ajesh and Kumuthakalavalli (2013) reported that Muthuvans of Idukki district in Kerala are using 21 plants for construction of huts, 16 for domestic articles, 15 for cultural and traditional purposes, 12 for clothing and cosmetics and 20 for tools and weapons.

10.10 CONCLUSIONS

Plant genetic resources constitute the major biological basis of the world for livelihood security. By all means they support the livelihood of every form of life on planet earth. Biodiversity is the store house and acts as a cushion against potentially dangerous environmental changes and also economic reforms. Plant resources are only the ultimate support for all lives on the earth planet, hence their documentation and conservation are essentially needed for sustainable utilization of humankind.

List of plants yielding above products is given in Table 10.1.

TABLE 10.1   Useful Plants of Western Ghats Other Than Medicinal and Food Plants

S. No.	Plant species	Family	Uses	Reference
1	*Acacia catechu* L.	Fabaceae	Wood for agriculture implements	Kamble et al. (2015)
2	*Acacia nilotica* (L.) Del.	Fabaceae	Wood as fuel	Packiaraj et al. (2014)
3	*Acacia pennata* (L.) Willd.	Fabaceae	Fish stupefying	Kulkarni et al. (1990)
4	*Acalypha fruticosa* Forssk.	Euphorbiaceae	Whole plant used as brooms	Rasingam and Jeeva (2013)
5	*Achyranthes aspera* L.	Amaranthaceae	Whole plant used as brooms	Rasingam and Jeeva (2013)
6	*Acronychia pedunculata* (L.) Miq.	Rutaceae	Construction of huts	Ajesh and Kumuthakalavalli (2013)
7	*Aegle marmelos* L.	Rutaceae	Sacred tree worshipped as Lord Shiva	Sahu et al. (2013)
8	*Aerva lanata* (L.) Juss.	Amaranthaceae	Plants are kept or tied in front of houses during festivals	Packiaraj et al. (2014)
9	*Ailanthus malabarica* DC.	Simaroubaceae	Gum	Dev (1983)
10	*Albizia chinensis* (Osbeck) Merr.	Fabaceae	Timber, fire wood.	
11	*Albizia lebbeck* (L.) Benth.	Fabaceae	Wood as fuel	Packiaraj et al. (2014)
12	*Albizia procera* (Roxb.) Benth.	Fabaceae	Fish stupefying	Kulkarni et al. (1990)
13	*Anamirta cocculus* (L.) Wight & Arn. (Figure 1a)	Menispermaceae	Fiber used in house construction	Ramachandran et al. (2009)
14	*Anisomeles malabarica* (L.) R.Br. ex Sims.	Lamiaceae	Twigs are used to ward off evil spirits	Packiaraj et al. (2014)
15	*Anodendron manubrianum* Merr.	Apocynaceae	Strong wiry stems are used as a rope to collect honey from steep rock	Packiaraj et al. (2014)
16	*Anogeissus latifolia* (Roxb. ex DC.) Wall. ex Bedd. (Figure 1b)	Combretaceae	Wood used for house and hut construction	Ayyanar et al. (2010); Ayyanar and Ignacimuthu (2010); Packiaraj et al. (2014)

TABLE 10.1   *(Continued)*

S. No.	Plant species	Family	Uses	Reference
17	*Antiaris toxicaria* (Pers.) Lesch.	Moraceae	Bark for cloth preparation	Manithottam and Francis (2008)
18	*Ardisia solanacea* L.	Myrsinaceae	Flower buds yield dye	Siva (2007)
19	*Areca catechu* (L.) Willd.	Arecaceae	Midrib of leaves used as brooms	Rasingam and Jeeva (2013)
20	*Argyreia pilosa* Arn.	Convolvulaceae	Stem used in house construction	Ramachandran et al. (2009)
21	*Artemisia japonica* Thunb.	Asteraceae	Branches for brooms making	Patil et al. (2014)
22	*Artemisia nilagirica* (C.B. Clarke) Pamp.	Asteraceae	Branches used as brooms	Rasingam and Jeeva (2013)
23	*Artocarpus heterophyllus* L.	Moraceae	Wood for house construction	Ayyanar and Ignacimuthu (2010)
24	*Artocarpus heterophyllus* L.	Moraceae	Wood for house construction	Ayyanar and Ignacimuthu (2010)
25	*Azadirachta indica* A. Juss.	Meliaceae	Fresh leafy twigs are used for spraying cow's urine inside the hut to keep away insects and mites	Packiaraj et al. (2014)
26	*Bambusa arundinacea* (Retz.) Willd.	Poaceae	Stems used for making fences, cross bars, poles, baskets, bows, for brooms, hut construction and musical instruments Leaves are used for thatching	Rasingam and Jeeva (2013); Ayyanar and Ignacimuthu (2010)
27	*Bambusa bamboos* (L.) Voss	Poaceae	Hut construction	Ajesh and Kumuthakalavalli (2013)
28	*Barleria prionitis* Nees (Figure 1c)	Acanthaceae	Root yields dye	Siva (2007)
29	*Bauhinia malabarica* Roxb.	Fabaceae	Stem fiber	Ramachandran et al. (2009)
30	*Bauhinia tomentosa* L.	Fabaceae	Stem for making net and ropes	Ayyanar and Ignacimuthu (2010)
31	*Bauhinia vahlii* Wight & Arn.	Fabaceae	Leaves for packing material	Chauhan and Saklani (2013)

TABLE 10.1   *(Continued)*

S. No.	Plant species	Family	Uses	Reference
32	*Benkara malabarica* (Lam.) Tirv. (Figure 1d)	Rubiaceae	Fruits used for fish poison	Ramachandran et al. (2009)
33	*Borassus flabellifer* L.	Arecaceae	Leaves for making mats	Ayyanar and Ignacimuthu (2010)
34	*Boswellia serrata* Roxb. ex Colebr.	Burseraceae	Resins	Dev (1983)
35	*Brassica juncea* (L.) Czern.	Brassicaceae	Branches used as brooms	Rasingam and Jeeva (2013)
36	*Bridelia stipularis* (L.) Blume (Figure 1e)	Euphorbiaceae	Fruit yields blue dye	Siva (2007)
37	*Butea monosperma* (Lam.) Taub.	Fabaceae	Flowers used to extract dye (Brutin)	Siva (2007)
38	*Caesalpinia sappan* L.	Fabaceae	Bark dye	Menon (2002)
39	*Cajanus lineatus* (Wight & Arn.) Maesen	Fabaceae	Branches for brooms making	Patil et al. (2014)
40	*Calamus hookerianus* Becc.	Arecaceae	Hut construction	Ajesh and Kumuthakalavalli (2013)
41	*Calamus rotang* L. (Figure 1f)	Arecaceae	Stem for basket making	Ayyanar and Ignacimuthu (2010)
42	*Calotropis gigantea* (L.) R.Br.	Asclepiadaceae	The twigs are hung up in front of their huts to ward off evil spirits	Packiaraj et al. (2014)
43	*Canarium strictum* Roxb.	Burseraceae	Resin	Dev (1983)
44	*Capparis decidua* (Forssk.) Edgew.	Capparaceae	Fish stupefying	Kulkarni et al. (1990)
45	*Carica papaya* L.	Caricaceae	Few pieces of ripe fruits are taken orally for abortion at the time of first trimester	Packiaraj et al. (2014)
46	*Caryota urens* L.	Arecaceae	Hut construction	Ajesh and Kumuthakalavalli (2013)
47	*Casearia graveolens* Dalzell.	Salicaceae	Fish stupefying	Kulkarni et al. (1990)

TABLE 10.1   *(Continued)*

S. No.	Plant species	Family	Uses	Reference
48	*Casearia wynaden-sis* Bedd.	Salicaceae	Leaves – fish stupefying	
49	*Cassia auriculata* L.	Fabaceae	Petals yield yellow dye	Siva (2007)
50	*Cassia fistula* L.	Fabaceae	Stem bark for fish stupefying	Packiaraj et al. (2014)
51	*Cassytha filiformis* L.	Lauraceae	Stem yields yellow dye	Siva (2007)
52	*Catunaregum spinosa* (Thunb.) Tirv. [*Randia dumetorum* (Retz.) Poir.; *Xeromphis spinosa* (Thunb.) Keay]	Rubiaceae	Fruit used as fish poison	Ramachandran et al. (2009)
53	*Cayratia pedata* Juss.	Vitaceae	Stem for making ropes and nets	Ayyanar and Ignacimuthu (2010)
54	*Ceriscoides turgida* (Roxb.) Tirveng.	Rubiaceae	Fruit used for fish poison	Ramachandran et al. (2009)
55	*Chionanthuss linocieroides* (Wight) Bennet ex Raiz.	Oleaceae	Hut construction	Ajesh and Kumuthakalavalli (2013)
56	*Cocos nucifera* L.	Arecaceae	Leaves for making mats; midrib of leaflets used as brooms	Ayyanar and Ignacimuthu (2010); Rasingam and Jeeva (2013)
57	*Commiphora caudata* (Wight & Arn.) Engl.	Burseraceae	Fuel wood	Ayyanar et al. (2010); Packiaraj et al. (2014)
58	*Corypha umbraculifera* L.	Arecaceae	Leaves used for umbrella and basket making, hut construction	Chandran (1996); Ajesh and Kumuthakalavalli (2013)
59	*Coscinium fenestratum* (Goetgh.) Colebr.	Menispermaceae	Stem yields yellow dye (Berberine)	Siva (2007)
60	*Costus speciosus* (Kon. ex Roxb.) J.E. Smith	Zingiberaceae	Fish stupefying	Kulkarni et al. (1990)
61	*Crotalaria calycina* Schrank	Fabaceae	Whole plant for brooms making	Patil et al. (2014)
62	*Cupressus lusitanica* Mill.	Cupressaceae	Young branches used as brooms	Rasingam and Jeeva (2013)

TABLE 10.1   *(Continued)*

S. No.	Plant species	Family	Uses	Reference
63	*Curcuma longa* L.	Zingiberaceae	Rhizome for yellow dye (Curcumin)	Siva (2007)
64	*Cymbopogon caesius* (Nees ex Hook.f. & Arn.) Stapf	Poaceae	Hut construction	Ajesh and Kumuth-akalavalli (2013)
65	*Cymbopogon flexuosus* (Nees ex Steud.) Wats.	Poaceae	Thatching material	Ramachandran et al. (2009)
66	*Cynodon dactylon* (L.) Pers.	Poaceae	Sacred plant	Ayyanar and Ignaci-muthu (2010)
67	*Dalbergia latifolia* Roxb.	Fabaceae	Wood for house construction, agricultural implements, furniture,	Packiaraj et al. (2014)
68	*Dendrocalamus strictus* (Roxb.) Nees	Poaceae	Branches for building material, ladder making and thatching	Kamble et al. (2015)
69	*Dioscorea bulbifera* L.	Dioscoreaceae	Tuber yields dye	Siva (2007)
70	*Diospyros ebenum* Roxb.	Ebenaceae	Wood for house construction, furniture, agricultural implements	Ayyanar and Ignaci-muthu (2010); Packi-araj et al. (2014)
71	*Diospyros montana* Roxb.	Ebenaceae	Bark used as fish poison	Ramachandran et al. (2009)
72	*Diospyros cordifolia* Roxb.	Ebenaceae	Bark fish poison	
73	*Diospyros montana* Roxb.	Ebenaceae	Fish stupefying	Kulkarni et al. (1990)
74	*Diospyros ebenum* J. Koenig ex Retz.	Ebenaceae	Wood is used as agriculture implements	Ayyanar et al. (2010)
75	*Dodonaea angusti-folia* L.f.	Sapindaceae	Branches used as brooms	Rasingam and Jeeva (2013)
76	*Dodonaea viscosa* Jacq.	Sapindaceae	Stem for construc-tion purposes	Ayyanar and Ignaci-muthu (2010)
77	*Drypetes venusta* (Wight) Pax & Hoffm.	Euphorbiaceae	Hut construction	Ajesh and Kumuth-akalavalli (2013)

TABLE 10.1 *(Continued)*

S. No.	Plant species	Family	Uses	Reference
78	*Euphorbia tirucalli* L.	Euphorbiaceae	Fish stupefying	Kulkarni et al. (1990)
79	*Filicium decipiens* Thw. (Figure 2a)	Sapindaceae	Flowers used for worship	Ayyanar et al. (2010)
80	*Flueggea leucopyrus* Willd.	Euphorbiaceae	Branches used as brooms	Rasingam and Jeeva (2013)
81	*Galium rotundifolium* L.	Rubiaceae	Root yields yellow dye	Siva (2007)
82	*Gmeliina arborea* L.	Lamiaceae	Wood for making musical instruments	Ayyanar and Ignacimuthu (2010)
83	*Gnidia glauca* (Presen.) Gilg	Thymelaeaceae	Branches for brooms making; fish stupefying	Patil et al. (2014); Kulkarni et al. (1990)
84	*Grevillea robusta* A. Cunn.	Proteaceae	Branches and leaves used as brooms	Rasingam and Jeeva (2013)
85	*Grewia robusta* Drum.	Malvaceae	Bark for making ropes	Kamble et al. (2015)
86	*Gyrocarpus americanus* Jacq.	Hernandiaceae	Fuel wood	Packiaraj et al. (2014)
87	*Gyrocarpus asiaticus* Jacq.	Hernandiaceae	Fuel wood	Ayyanar et al. (2010)
88	*Haldina cordifolia* (Roxb.) Ridsdale	Rubiaceae	Stem used to plow	Ayyanar and Ignacimuthu (2010)
89	*Hardwickia binata* Roxb.	Fabaceae	Gums	Dev (1983)
90	*Helicteres isora* Wight & Arn. Figure 2 b	Malvaceae	Stem bark for making net and ropes	Ayyanar and Ignacimuthu (2010)
91	*Hibiscus cannabinus* L.	Malvaceae	Stem bark used to make rope and nets	Ayyanar and Ignacimuthu (2010)
92	*Holoptelia integrifolia* Planch.	Ulmaceae	Fish stupefying	Kulkarni et al. (1990)
93	*Ichnocarpus frutescens* (L.) R.Br.	Apocynaceae	Strong wiry stems used as cordage	Packiaraj et al. (2014)
94	*Indigofera aspalathoides* DC.	Fabaceae	Root yields yellow dye	Siva (2007)
95	*Indigofera tinctoria* L.	Fabaceae	Root dye-Indgotin	Menon (2002); Siva (2007)
96	*Jasminum malabaricum* Wight	Oleaceae	Branches for brooms making	Patil et al. (2014)

TABLE 10.1 *(Continued)*

S. No.	Plant species	Family	Uses	Reference
97	*Justicia adhatoda* (Nees) And.	Acanthaceae	Leaves used to extract dyes (Aha-todic acid)	Siva (2007)
98	*Lantana camara* L.	Verbenaceae	Stem for making baskets	Ayyanar and Ignacimuthu (2010)
99	*Lantana indica* Roxb.	Verbenaceae	Branches used as brooms	Rasingam and Jeeva (2013)
100	*Lawsonia inermis* L.	Lythraceae	Leaves – dye (Lawsone), Branches for brooms making	Siva (2007); Patil et al. (2014)
101	*Leucas stelligera* Wall. ex Benth.	Lamiaceae	Fish stupefying	Kulkarni et al. (1990)
102	*Ligustrum perrottetti* DC.	Oleaceae	Branches used as brooms	Rasingam and Jeeva (2013)
103	*Maclura spinosa* C.C. Berg.	Moraceae	Stems used as brooms	Rasingam and Jeeva (2013)
104	*Madhuca longifolia* (J. Konig. ex L.) J.F. Macbr.	Sapotaceae	Fish stupefying	Kulkarni et al. (1990)
105	*Maesa indica* (Roxb.) A. DC	Primulaceae	Fish stupefying	Kulkarni et al. (1990)
106	*Mallotus philippensis* Mull. Arg.	Euphorbiaceae	Fruit epicarp used to prepare dye (Rottlerin)	Siva (2007); Kamble et al. (2015)
107	*Mangifera indica* L.	Anacardiaceae	Wood for house construction	Ayyanar and Ignacimuthu (2010)
108	*Melastoma malabathricum* L.	Melastomataceae	Flower yields purple dye	Siva (2007)
109	*Mimosa instia* L.	Fabaceae	Stem bark used for stupefying fish	Ayyanar et al. (2010)
110	*Mimusops elengi* L.	Sapotaceae	Wood yields purple dye; Sacred tree	Siva (2007); Ayyanar and Ignacimuthu (2010)
111	*Morinda tinctoria* L.	Rubiaceae	Heartwood used to prepare yellow dye (Morindone)	Siva (2007)
112	*Morus alba* L.	Moraceae	Wood for furniture	Kamble et al. (2015)
113	*Mundulea sericea* (Willd.) A. Chev.	Fabaceae	Fish stupefying	Kulkarni et al. (1990)
114	*Naregamia alata* Wight & Arn. (Figure 2c)	Meliaceae	Leaves yields black dye	Siva (2007)

TABLE 10.1   *(Continued)*

S. No.	Plant species	Family	Uses	Reference
115	*Ochlandra travancorica* Benth.	Poaceae	Wiry stem used for basket making	Ayyanar et al. (2010)
116	*Oldenlandia umbellata* L.	Rubiaceae	Roots used to prepare yellow dye (Rubichloric acid)	Siva (2007)
117	*Olea dioica* Roxb.	Oleaceae	Hut construction	Ajesh and Kumuth-akalavalli (2013)
118	*Osbeckia chinensis* L.	Melastomata-ceae	Leaf extract used to removal of stains on cloths	Ayyanar et al. (2010)
119	*Ougenia oojeinensis* (Roxb.) Hochr.	Fabaceae	Fish stupefying	Kulkarni et al. (1990)
120	*Oxalis corniculata* L.	Oxalidaceae	Leaves yields yellow dye	Siva (2007)
121	*Pachygone ovata* (Poir.) Hook.f.	Menispermaceae	Stem used for cordage	Ayyanar et al. (2010); Packiaraj et al. (2014)
122	*Pandanus odoratis-simus* L.	Pandanaceae	Leaves for making mats	Ayyanar and Ignaci-muthu (2010)
123	*Parthenium hys-terophorus* L.	Asteraceae	Plant used as brooms	Rasingam and Jeeva (2013)
124	*Passiflora edulis* Sims.	Passifloraceae	Ornamental plant	Sarvalingam and Rajendran (2014)
125	*Pavetta indica* L.	Rubiaceae	Stem fiber used to basket making, cordage	Ayyanar et al. (2010); Packiaraj et al. (2014)
126	*Pheoenix loureiroi* Kunth. var. *humilis* S.C. Barrow	Arecaceae	Leaves used as brooms	Rasingam and Jeeva (2013)
127	*Phoenix sylvestris* (L.) Roxb.	Arecaceae	Leaves for making brooms and mats	Patil et al. (2014); Ayyanar and Ignaci-muthu (2010)
128	*Phyllanthus reticu-latus* Poir.	Phyllanthaceae	Fruit yields blue dye	Siva (2007)
129	*Pinus roxburghii* Sarg.	Pinaceae	Resins	Dev (1983)
130	*Piper betle* L.	Piperaceae	Leaves chewed with areca nut	Packiaraj et al. (2014)
131	*Polyalthia fragrans* (Dalz.) Bedd.	Annonaceae	Hut construction	Ajesh and Kumuth-akalavalli (2013)
132	*Pongamia pinnata* (L.) Pierre	Fabaceae	Fish stupefying	Kulkarni et al. (1990)

TABLE 10.1   *(Continued)*

S. No.	Plant species	Family	Uses	Reference
133	*Pterocarpus marsu-pium* Roxb.	Fabaceae	Stem exudates used to keep off evil spirit	Ayyanar et al. (2010); Packiaraj et al. (2014)
134	*Punica granatum* L.	Punicaceae	Fruit is used to extract dye (Petargonidin)	Siva (2007)
135	*Rubia cordifolia* L.	Rubiaceae	Root is used to extract yellow dye (Purpurin)	Siva (2007)
136	*Saccharum sponta-neum* L.	Poaceae	Thatching	Ajesh and Kumuth-akalavalli (2013)
137	*Sansevieria rox-burghiana* Schult. & Schult. f.	Liliaceae	Leaves for making ropes and nets	Ayyanar and Ignaci-muthu (2010)
138	*Santalum album* L	Santalaceae	Sacred tree	Packiaraj et al. (2014)
139	*Sapindus trifolia-tus* L.	Sapindaceae	Fruits for cleaning cloths	Meena Devi et al. (2012)
140	*Sapium insigne* (Royle) Trimen	Euphorbiaceae	Fish stupefying	Kulkarni et al. (1990)
141	*Semecarpus ana-cardium* L.f. (Figure 2d)	Anacardiaceae	Fruits used to ex-tract Bhilawanol	Siva (2007)
142	*Shorea robusta* Gaertn. (Figure 2e)	Dipterocar-paceae	Gums	Dev (1983)
143	*Sida acuta* Burm.f.	Malvaceae	Twigs used as brooms	Ramachandran et al. (2009); Rasingam and Jeeva (2013)
144	*Sporobolus wallichii* Munro	Poaceae	Fodder crop	Nair et al. (2014)
145	*Swietenia mahagoni* (L.) Jacq.		Fruits hung to drive away evil spirits	Packiaraj et al. (2014)
146	*Syzygium arnottia-num* Walp.	Myrtaceae	Hut construction	Ajesh and Kumuth-akalavalli (2013)
147	*Syzygium cumini* (L.) Skeels	Myrtaceae	Sacred tree; hut construction; fish stupefying	Ayyanar and Ignaci-muthu (2010); Ajesh and Kumuthakalavalli (2013); Kulkarni et al. (1990)
148	*Syzygium gardneri* Thwaites	Myrtaceae	Hut construction	Ajesh and Kumuth-akalavalli (2013)

TABLE 10.1 *(Continued)*

S. No.	Plant species	Family	Uses	Reference
149	*Terminalia catappa* L.	Combretaceae	Wood for construction purposes	Ayyanar and Ignacimuthu (2010)
150	*Terminalia chebula* Retz.	Combretaceae	Fruit dye; wood for construction purposes	Menon (2002); Ayyanar and Ignacimuthu (2010)
151	*Themeda cymbaria* (Roxb.) Hackel	Poaceae	Straw used for thatching	Ayyanar et al. (2010); Ajesh and Kumuthakalavalli (2013); Packiaraj et al. (2014)
152	*Themeda triandra* Forssk	Poaceae	Thatching	Ajesh and Kumuthakalavalli (2013)
153	*Toddalia asiatica* (L.) Lam.	Rutaceae	Leaves used to extract dye (Toddalin)	Siva (2007)
154	*Trema orientialis*		The fiber separated from bark is used as ropes.	
155	*Urochloa brizantha* Webster	Poaceae	Fodder crop	Nair et al. (2014)
156	*Vateria indica* C.F. Gaertn.	Dipterocarpaceae	Gums	Dev (1983)
157	*Vitex negundo* L.	Verbenaceae	Stem for making baskets and brooms	Ayyanar and Ignacimuthu (2010); Patil et al. (2014)
158	*Wattakaka volubilis* (L.f.) Stapf	Asclepiadaceae	Ropes	Ajesh and Kumuthakalavalli (2013)
159	*Woodfordia fruticosa* Kurz.	Lythraceae	Branches for fuel wood	Kamble et al. (2015)
160	*Wrightia tinctoria* R.Br. (Figure 2f)	Apocynaceae	Leaves yield dye (Beta amyrine)	Siva (2007)
161	*Xylopia parvifolia* (Wight) Hook f. & Thoms.	Annonaceae	Hut construction	Ajesh and Kumuthakalavalli (2013)
162	*Ziziphus jujuba* Mill.	Rhamnaceae	Used as hedge plant	Ayyanar and Ignacimuthu (2010)
163	*Ziziphus rugosa* Lam.	Rhamnaceae	Used as hedge plant	Ayyanar and Ignacimuthu (2010)

KEYWORDS

- **Brooms**
- **Dyes**
- **Fibers**
- **Forage**
- **Gums**
- **Ornamental**
- **Sacred Plants**

REFERENCES

Ajesh, T.P. & Kumuthakalavalli, R. (2013). Botanical ethnography of muthuvans from the Idukki District of Kerala. *Int. J. Pl. Anim. Environ. Sci.*, *3*, 67–75.

Ayyanar, M. & Ignacimuthu, S. (2010). Plants used for non-medicinal purposes by the tribal people in Kalakad Mundanthurai Tiger Reserve, Southern India. *Indian J. Trad. Knowl. 9(3)*, 515–518.

Ayyanar, M., Sankarasivaraman, K., Ignacimuthu, S. & Sekar, T. (2010). Plant species with ethnobotanical importance other than medicinal in Theni district of Tamilnadu, southern India. *Asian J. Exp. Biol. Sci. 1(4)*, 765–771.

Chandran, S. (1996). Talipot: A forgotten palm of the Western Ghats – Plea for its conservation. *Resonance*. 69–75.

Chauhan, R. & Saklani, S. (2013). *Bauhinia vahlii*: plant to be explored. *International Research J. Pharmacy 4(8)*, 5–9.

Dev, S. (1983). Chemistry of resins exudates of some Indian trees. *Proc. Indian Natn. Sci. Acad. 49A(3)*, 359–365.

Gadgil, M. & Vartak, V.D. (1975). Sacred groves in India; a plea for continued conservation. *J. Bombay Nat. Hist. Soc. 72*, 314–320.

Gadgil, M. & Vartak, V.D. (1976). The sacred groves of Western Ghats in India. *Economic Bot. 30*, 152–160.

Kamble, S.R., Deokar, R.R., Mane, S.R. & Patil, S.R. (2015). Ethnobotanical studies of some useful herbs, shrubs and trees of Shirala Tahsil of Sangli district, M.S., India. *Intern. J. Innovative Research Sci. 4(5)*, 3002–3008.

Kulkarni, D.K., Kumbhojkar, M.S. & Nipunage, D.S. (1990). Note on fish stupefying plants from western Maharashtra. *Indian Forester 116(4)*, 331–333.

Kumar, J.I.N., Soni, H. & Kumar, R.N. (2005). Aesthetic values of selected floral elements of Khatana and Waghai forests of Dangs, Western Ghats. *Indian J. Trad. Knowl. 4(3)*, 275–286.

Manithottam, J. & Francis, M.S. (2008). Preparation of Maravuri from *Antiaris toxicaria* (Pers.) Lesch. by Muthuvans of Kerala. *Indian J. Trad. Knowl. 7(1)*, 74–76.

Meena Devi, V.N., Rajakohila, M., Syndia, L.A.M., Prasad, P.N. & Ariharan, V.N. (2012). Multifacetious uses of soap nut tree – a mini review. *Research J. Pharmaceutical, Biological and Chemical Sci. 3(1),* 420–424.

Menon, P. (2002). Checklist and approximate quantity of non-wood forest produce (NWFP) collected from Peppara Wildlife Sanctuary (www.mtnforum.org).

Mohanan, C.N. & Nair, N.C. (1981). *Kunstleria* Prain – a new genus record for India and a new species in the genus. *Proceedings of the Indian Academy of Sciences* B40: 207–210.

Nair, G.G., Dileep, P. & Meera Raj, R. (2014). *Sporobolus wallichii* and *Urochloa brizantha* (Poaceae) – two new records for Kerala. *Indian J. Plant Sci. 3(1),* 111–114.

Packiaraj, P., Suresh, K. & Venkadeswaran, P. (2014). Plant species with ethno botanical importance other than medicinal in Paliyars community in Virudhunagar District, Tamil Nadu, India. *Int. J. Applied Bioresearch, 20,* 6–9.

Palanisamy, J. & Arumugam, R. (2014). Exploration of wild ornamental flora of Madukkarai hills of southern Western Ghats, Tamilnadu. *Biolife. 2(3),* 834–841.

Pandey, A. & Gupta, R. (2003). Fiber yielding plants of India. Genetic resources, perspective for collection and utilization. *Nat. Prod. Rad. 2(4),* 194–204.

Patil, P.V., Taware, S. & Kulkarni, D. (2014). Traditional knowledge of broom preparation from Bhor and Mahad region of Western Maharashtra, India. *Bioscience Discovery 5(2),* 218–220.

Ramachandran, V.S., Joseph, S. & Aruna, R. (2009). Ethnobotanical studies from Amaravathy range of Indira Gandhi Wildlife Sanctuary, Western Ghats, Coimbatore District, Southern India. *Ethnobotanical Leaflets 13,* 1069–1087.

Rasingam, L. & Jeeva, S. (2013). Indigenous brooms used by the aboriginal inhabitant of Nilgiri Biosphere Reserve, Western Ghats, India. *Indian J. Trad. Knowl. 4(8),* 312–316.

Sahu, P.K., Kumari, A., Sao, A., Singh, M. & Pandey, P. (2013). Sacred plants and their ethno-botanical importance in Central India: A mini review. *International J. Pharmacy & Life Sci. 4(8),* 2910–2914.

Sarvalingam, A. & Rajendran, A. (2014). Wild ornamental climbing plants of Maruthamalai hills in southern Western Ghats, Tamilnadu state, India. *World J. Agriculture Sci. 10(5),* 204–209.

Siva, R. (2007). Status of natural dyes and dye yielding plants in India. *Curr. Sci. 92(7),* 916–925.

APPENDIX

PLATE 10.1

a. *Anamirta cocculus*

b. *Anogeissus latifolia*

C. *Barleria prionitis*

d. *Benkara malabarica*

e. *Bridelia stipulacea*

f. *Calamus rotang*

PLATE 10.2

a.   *Filicium decipiens*   b. *Helicteres isora*

b.   *Naregamia alata*   d. *Semecarpus anacardium*

e. *Shorea robusta*   f. *Wrightia tinctoria*

CHAPTER 11

ETHNOBOTANY OF MANGROVES WITH PARTICULAR REFERENCE TO WEST COAST OF PENINSULAR INDIA

T. PULLAIAH,[1] BIR BAHADUR,[2] and K. V. KRISHNAMURTHY[3]

[1]Department of Botany, Sri Krishnadevaraya University, Anantapur-515003, AP, India.
E-mail: pullaiah.thammineni@gmail.com

[2]Department of Botany, Kakatiya University, Warangal 506009, Telangana, India. E-mail: birbahadur5april@gmail.com

[3]Consultant, R&D, Sami Labs, Peenya Industrial Area, Bangalore – 560058, Karnataka, India. E-mail: kvkbdu@yahoo.co.in

CONTENTS

ABSTRACT

Mangroves constitute an important ecosystem. Mangrove forests in India cover an area of 4,461 sq.m./6700 km² which constitutes *ca* 7% of the world's mangroves. 82 species belonging to 52 genera belonging to 36 families are distributed in mangroves of India. In the present chapter ethnobotany of mangroves is described. Plant names, part used, usage and references are given.

11.1   INTRODUCTION

Mangroves are salt tolerant plant communities occurring in tidal and intertidal regions of the tropics and subtropics and are known as 'mangals' 'tidal forests,' 'coastal woodlands' or 'oceanic rain forests.' These wetlands with distinctive flora and fauna are unique, complex and with adaptations to perform unique functions. These forests, constitute *ca* 1,81,077 sq. km. in area worldwide (Spalding et al., 1997). The maximum diversity in mangroves occurs in the Asian coasts alone with over 40% of total mangrove forests. Australia, Bangladesh, Brazil, India, Indonesia and Nigeria are with the largest mangrove formations. Mangrove forests in India cover an area of 4,461 sq.m./6700 km² which constitutes *ca* 7% of the world's mangroves, that constitute 0.14% of its total geographical area. The largest stretch of mangroves in India occurs in Sundarbans (West Bengal) which covers an area of about 4200 km². Sundarbans has been designated as World Heritage site of which 80% of them are restricted to Sundarbans (West Bengal) and in Andaman & Nicobar Islands (Chowdhery and Murti, 2000). The remaining taxa are scattered in the coastal areas of Andhra Pradesh, Tamil Nadu, Orissa, Maharashtra, Gujarat, Goa and Karnataka. 82 species belonging to 52 genera belonging to 36 families are distributed in mangroves of India (Mandal and Naskar 2008; Debnath, 2004). Some of the dominant mangrove species are *Avicennia marina, A. officinalis, Bruguiera gymnorrhiza, B. parviflora, Ceriops tagal, Heritiera fomes, Lumnitzera* spp., *Rhizophora mucronata, R. apiculata, R. stylosa, Sonneratia* spp., *Xylocarpus* spp., etc. The shrubby *Aegialitis rotundifolia* and *Acanthus ilicifolius* occur commonly on poor saline plains. Herbaceous succulent halophytes are represented by *Aegieceras corniculata, Suaeda brachiata, Sesuvium portulacastrum* and *Salicornia brachiata* while *Nypa fruticans* and *Phoenix paludosa* are characteristic mangrove palms found in Sundarbans and Andaman and Nicobar Islands.

11.2 ETHNOBOTANY

Ravindran et al. (2005) described medicinal properties of 11 species of mangroves and halophytes used by local inhabitants. Prabhakaran and Kavitha (2012) gave an account of ethnomedicinal importance of mangroves of Pichavaram in Tamil Nadu. Ethnobotanical and medical aspects of Mangroves from southern Kokan (Maharashtra) was given by Sathe et al. (2014). Chowdhury et al. (2014) reported that a total of 31 species of mangrove flora were found to have different medicinal properties and other usage. Ethnomedicinally useful mangrove species occurring in the Diu and surroundings have been studied by Jadeja et al. (2009) to assess the potentiality of various mangroves for ethnobotanical uses. This work is based on interviews with local physicians practicing indigenous system of medicine, village head men, priests and tribal folks. Many tribes like Khaniya Koli Kharwa, Vaniya, Brahmin, Vanja, Khoja possess a good deal of information about properties and medicinal use of mangrove plants. Further these authors have provided the traditional mangrove uses prevalent amongst the aboriginals of Diu. Ethnobotany of mangroves and their management in Andhra Pradesh was discussed by Swain and Rama Rao (2008) while Venkatesan et al. (2005) gave ethnobotanical report of mangroves of Pichavaram in Tamil Nadu state. According to Govindasamy and Kannan (2012) several mangroves species are used to treat range of conditions from toothache and even diabetes. Dahodouh-Guebas et al. (2006) analyzed ethnobotanical and fishery-related importance of mangroves of the East Godavari delta in Andhra Pradesh for conservation and management purposes.

Selvam et al. (2004) in their publication have mentioned the uses and applications of wetland mangroves of India as under:

1. Minor timber, poles and posts and fire wood.
2. Non-wood products, such as fodder, honey, wax, tannins, dyes and plant mats for thatching
3. Aquaculture: Aquatic food like fishes, prawns, shrimps, crabs, molluscs, etc.
4. Wild life conservation particularly wide variety of birds and other small mammals.
5. Mitigates storms, cyclones, tsunamis, global warming, etc.

11.3   ETHNOBOTANICAL USES

The details of ethnobotanical uses of mangrove plants are given below.

1. ***Acanthus ilicifolius*** L. (Acanthaceae)—used for reducing the poisonous snake bite, curing skin diseases, kidney stone, smallpox, ulcer, asthma, cough, diabetes and rheumatism (Prabhakaran and Kavitha, 2012). Given as nerve tonic, for dressing wounds and boils, as aphrodisiac (Sathe et al., 2014). Leaves are traditionally used for treating Tiger bites. Roots are boiled and extract used to treat various diseases like asthma, paralysis, leucorrhoea and debility (Chowdhury et al., 2014). Crushed fruits are used for dressing snake bite. The plant is boiled in water and the patient drinks half of glass each time until the signs and symptoms of kidney stone disappear. The whole plant extract and paste is used for curing skin diseases, small pox, ulcer and for detoxification and health promotion (Ravindran et al., 2005).

2. ***Acanthus volubilis*** L. (Acanthaceae). Leaves are dried and taken as a remedy for stomach ulcer (Chowdhury et al., 2014).

3. ***Acrostichum aureum*** L. (Pteridaceae). Rhizome paste is used to treat boils and carbuncles.

4. ***Aegiceras corniculatum*** (L.) Blanco (Myrsinaceae). Cure for asthma, diabetes, rheumatism, and as a fish poison (Prabhakaran and Kavitha, 2012). Root, bark and stem used as fish poison (Sathe et al., 2014; Dahdouh-Guebas, 2006).

5. ***Atalantia correa*** Roem. (Rutaceae). Oil from the fruit is used for treatment of rheumatism.

6. ***Avicennia alba*** Bl. (Avicenniaceae). Resinous substance exuded and used for birth control purposes (Ravindran et al., 2005).

7. ***Avicennia marina*** (Forsk.) Vierh. (Avicenniaceae). Leaves used in the treatment of rheumatism, small pox and ulcers; Leaves are also used as fodder for livestock (Prabhakaran and Kavitha, 2012). Bitter aromatic fruit juice is used in a concoction to facilitate abortion. Mainly used by tribal population that settled in Sundarbans during British colonial period (Chowdhury et al., 2014).

8. ***Avicennia officinalis*** L. (Avicenniaceae). Leaves are used to treat smallpox, joint pain, urinary disorders, bronchial asthma, stomach disorders, as an aphrodisiac, diuretic, cure for hepatitis, leprosy (Prabhakaran and Kavitha, 2012). Root bark and seeds used in small pox, boils, abscesses, skin parasites and wounds (Sathe et al., 2014).

Seed bitter, but edible. Unripe fruit is used as a remedy to treat boils (Chowdhury et al., 2014). Leaves are used for the treatment of joints pain, urinary disorders, bronchial asthma, stomach disorders and detoxification (Ravindran et al., 2005).

9. ***Barringtonia racemosa*** (L.) Spr. (Barringtoniaceae). Bark, fruits and kernels used for curing asthma, cough jaundice, skin disorder (Sathe et al., 2014).

10. ***Bruguiera cylindrica*** (L.) Blume (Rhizophoraceae*).* Used in the treatment of hepatitis (Prabhakaran and Kavitha, 2012).

11. ***Bruguiera gymnorhiza*** L. (Rhizophoraceae). Bark and fruit as eye medicine, also used as astringent (Sathe et al., 2014). Bark is macerated and the extract is said to be useful in controlling diarrhea (Chowdhury et al., 2014). The whole plant boiled in water is given twice daily after meals to relieve constipation (Ravindran et al., 2005).

12. ***Bruguiera sexangula*** Poir (Rhizophoraceae). Bark is macerated and the extract is used to control diarrhea (Chowdhury et al., 2014).

13. **Cerbera odollam** Gaertn. (Apocynaceae). Bark and fruits used for hydrophobia, rheumatism, hemorrhage, ulcer (Sathe et al., 2014).

14. ***Ceriops decandra*** (Griff.) Ding Hou. (Rhizophoraceae): Cure for hepatitis, ulcers (Prabhakaran and Kavitha, 2012). The poles of the stem are used as fencing (Chowdhury et al., 2014). Bark used for dyeing fishing nets (Dahdouh-Guebas, 2006).

15. ***Ceriops tagal*** (Perr.) C.B. Rob. (Rhizophoraceae). Bark, shoot and fruit used to treat hemorrhage and ulcers (Sathe et al., 2014). Stem bark extract is used to stop hemorrhages. It is said that bark is useful for ailment that resembles peptic ulcers. The poles of the stem are used as fencing material (Chowdhury et al., 2014).

16. ***Clerodendron inerme*** Gaertn. (Verbenaceae). Leaves contain bitter extract that is used as febrifuge (Chowdhury et al., 2014). Leaves are used for removing pain and in jaundice. Sap of leaves is used for washing dishes. Leaf extract and paste are used in the treatment of malaria, infected wounds, inflammation and itching diseases (Ravindran et al., 2005).

17. ***Derris heterophylla*** Willd. (Fabaceae). Roots are used as fish poison and as larvicide (Sathe et al., 2014).

18. ***Derris trifoliata*** Roxb. (Fabaceae). Root is dried and powdered and used to treat person affected by chronic alcoholism, useful as stimulant and antispasmodic (Chowdhury et al., 2014).

19. *Dolichandrone spathacea* (L.) Schum. (Bignoniaceae). Seed powder is used as antiseptic and in enteric spasms (Chowdhury et al., 2014).

20. *Excoecaria agallocha* L. (Euphorbiaceae): Latex is used as medication for toothache. The wood smoke is used as anti-epileptic. The roots are used for anti-inflammation (Ravindran et al., 2005). Used as an uterotonic, as purgative, in the treatment of epilepsy, conjunctivitis, dermatitis, haematuria, leprosy, toothache, as a piscicide, dart poison, and a skin irritant, Swelling hands and feet; flatulence; epilepsy, antiinflammation (Prabhakaran and Kavitha, 2012). Root, branches and leaves are used for the treatment of epilepsy, ulcer and leprosy (Sathe et al., 2014). Latex is acrid and poisonous. In local myth, it is said to be blessed by snake god "Manasha" (Chowdhury et al., 2014).

21. *Heritiera fomes* Buch.-Ham. (Malvaceae). Seed is grounded and used to treat dysentery (Chowdhury et al., 2014).

22. *Hibiscus tiliaceus* L. (Malvaceae). Leaf extract is used as laxative. Bark mucilage is used to treat dysentery like symptoms. Root is used to prepare herbal tonic for treatment of rheumatism. Seed is used as an emetic (Chowdhury et al., 2014).

23. *Kandelia kandel* (L.) Druce (Rhizophoraceae). Bark is used to treat diabetes (Sathe et al., 2014). Medicinally useful in the treatment of problems related to frequent urination (Chowdhury et al., 2014).

24. *Lumnitzera racemosa* Willd. (Combretaceae): Used in antifertility, treatment of asthma, diabetes and snake bite (Prabhakaran and Kavitha, 2012). Bark used to treat asthma and in antifertility (Sathe et al., 2014). Fluid from the stem is used to treat rashes and itch (Chowdhury et al., 2014).

25. *Nypa fruticans* Wurmb. (Arecaceae). Alcohol production is done by fermenting the fruit pulp (Chowdhury et al., 2014).

26. *Phoenix paludosa* Roxb. (Arecaceae). The fruit pulp reduces inflammation and used during persistent fevers (Chowdhury et al., 2014).

27. *Rhizophora apiculata* Blume (Rhizophoraceae). Bark extract is used for diarrhea, nausea, vomiting and amoebiasis, as antiseptic and to stop bleeding (Ravindran et al., 2005). Astringent, used for curing diarrhea, treatment of nausea, vomiting, typhoid, hepatitis, an antiseptic, insecticide and Amoebiasis (Prabhakaran and Kavitha, 2012).

28. *Rhizophora lamarckii* Montrouz. (Rhizophoraceae). Leaves are used to cure hepatitis (Prabhakaran and Kavitha, 2012).

29. ***Rhizophora mucronata*** Lamk. (Rhizophoraceae*)*). Bark extract is used for controlling diarrhea, nausea and vomiting (Ravindran et al., 2005). Leaves used for the treatment of elephantiasis, haematoma, hepatitis, ulcers, and a febrifuge, Bark-powerful astringent useful in diabetics, hemorrhage (Prabhakaran and Kavitha, 2012). Bark is used to treat diabetes, leprosy, hemorrhage and dysentery (Sathe et al., 2014).

30. ***Salicornia brachiata*** Roxb. (Chenopodiaceae). Whole plant ash is applied to treat itches (Ravindran et al., 2005).

31. ***Scyphiphora hydrophyllacea*** Gaertn. (Rubiaceae). Shoot extract is warmed slightly and used for enteric diseases and also used to treat liver ailments (Chowdhury et al., 2014).

32. ***Sonneratia alba*** J. Sm. (Sonneratiaceae). Fruits used to treat hemorrhage and swellings (Sathe et al., 2014).

33. ***Sonneratia apetala*** Buch.-Ham. (Sonneratiaceae;). Fruit is used as a spice and to improve flavor of cooking (Chowdhury et al., 2014).

34. ***Sonneratia caseolaris*** Engler (Sonneratiaceae). Fruits edible and is used to prepare a local cuisine and is valued for it's sour taste. Fruit extract is used as an anthelmintic medicine (Chowdhury et al., 2014).

35. ***Sonneratia griffithii*** Kurz. (Sonneratiaceae). Fruit is used as spice and to enhance flavor of cooked food (Chowdhury et al., 2014).

36. ***Suaeda nudiflora*** Moq. (Chenopodiaceae). Leaves made into ointment used to treat wounds, used in ophthalmia and as emetic (Sathe et al., 2014).

37. ***Xylocarpus granatum*** Pierre. (Meliaceae). Bark extract cures dysentery. Seed oil extract is used as illuminant of hair. Bark decoction is used for curing diarrhea and cholera (Ravindran et al., 2005). Bark used to treat fevers, malaria, cholera (Prabhakaran and Kavitha, 2012). Leaves, seeds and bark are used for treating jaundice, cholera, dysentery, fever, cough in the new born babies, dysentery, tonic, astringent, for breast cancer, cholera, diarrhea (Sathe et al., 2014). Bark extract is used to treat dysentery. Wood is durable and is suitable for making furniture (Chowdhury et al., 2014).

38. ***Xylocarpus mekongensis*** Pierre (Meliaceae). Bark extract is used to treat dysentery. Wood is durable and is suitable for making furniture.

Like many ecohabitats, mangrove forests have been degraded and destroyed over the years by humans. Mangrove forests have iconic loss is a source of global concern. It may be relevant to point out that there is considerable loss of mangrove area in Indian coastal and tidal area due to indiscriminate exploitation and multiple uses like fodder, timber, building material, alcohol, paper, charcoal and medicines (Upadhya et al., 2002).

KEYWORDS

- **Ethnobotany**
- **Indian Mangroves**
- **West Coast of India**

REFERENCES

Chowdhury, A., Sanya, P. & Maiti, S.K. (2014). Ethnobotanical understanding of mangroves: An investigation from central part of Indian Sundarbans. *Int. J. Bot. Res. 4(1),* 29–34.

Dahdouh-Guebas, F., Collin, S., Lo Seen, D., Ronnback, P., Depommier, D., Ravishankar, T. & Koedam, N. (2006). Analyzing ethnobotanical and fishery-related importance of mangroves of East-Godavari Delta (Andhra Pradesh, India) for conservation and management purposes. *J. Ethnobiol. Ethnomed. 2,* 24. doi: 10.1186/1746-4269-2-24.

Debnath, H.S. (2004). Diversity and management of mangroves in India. In: Aggarwal, P.K., Gariola, S. & Rao, K.S. (eds.) *Proc. Natl. Workshop on Conservation, Restoration and Sustainable Management of Mangrove forest in India,* pp. 54–55.

Govindasamy, C. & Kannan, R. (2012). Pharmacognosy of mangrove plants in the system of Unani medicine. *Asian Pacific Journal of Tropical Disease;* S38–S51.

Jadeja, B.A., Modhvadia, A.R. & Jha, K. (2009). Some ethnobotanical useful mangroves in Diu (U.T.), India. *Plant Archives 9(1),* 221–225.

Mandal, R.N. & Naskar, K.R. (2008). Diversity and classification of Indian Mangroves: a review. *Tropical Ecol. 49(2),* 131–146.

Prabhakaran, J. & Kavitha, D. (2012). Ethnomedicinal importance of mangroves of Pichavaram. *Int. J. Res. Pharmaceutical and Biomedical Sci. 3(2),* 611–614.

Ravindran, K.C., Venkatesan, K., Balakrishnan, V. Chellappan, K.P. & Balasubramanian, T. (2005). Ethnomedicinal studies of Pichavaram mangroves of East Coast, Tamil Nadu. *Indian J. Trad. Knowl. 4(4),* 409–411.

Sathe, S.S., Lavate, R.A. & Patil, S.B. (2014). Ethnobotanical and medicinal aspects of mangroves from southern Konkan (Maharashtra). *Int. J. Emerging Trends in Pharmaceut. Sci. 3(4),* 12–17.

Selvam, V., Ravichandran, K.K., Karunagaran, V.M., Mani, K.G. & Beula, G.E.J. (2004). Joint Mangrove Management in Tamil Nadu: Processes, Experiences and Prospects: Part 1 to 4. M.S. Swaminathan Research Foundation, Chennai, India.

Spalding, M.D., Blasco, F. & Field, C.D. (Eds.) (1997). *World Mangrove Atlas.* The International Society for Mangrove Ecosystems, Okinawa, Japan.

Swain, P.K.& Rama Rao, N. (2008). Ethnobotany of mangroves and their management in Andhra Pradesh, India. In: Soloman Raju, A.J. (ed.) *Bioresources Conservation and Management.* Today and Tomorrow publications, New Delhi. pp. 65–70.

Upadhyay, V.P., Ranjan, R. & Singh, J.S. (2002). Human mangrove conflicts: The way out. *Curr. Sci. 83,* 1328–1336.

Venkatesan, K., Balakrishnan, V., Ravindran, K.C. & Devanathan, V. (2005). Ethnobotanical report from mangroves of Pichavaram, Tamil Nadu state, India. *Sida 21(4),* 2243–2248.

CHAPTER 12

SACRED GROVES OF WESTERN GHATS: AN ETHNO-BASED BIODIVERSITY CONSERVATION STRATEGY

K. V. KRISHNAMURTHY[1] and S. JOHN ADAMS[2]

[1]Consultant, R&D, Sami Labs, Peenya Industrial Area, Bangalore–560058, Karnataka, India

[2]Department of Pharmacognosy, R&D, The Himalaya Drug Company, Makali, Bangalore, India

CONTENTS

ABSTRACT

This chapter deals with sacred groves (SGs), patches of native vegetation associated with local deities/spirits/ancestors, established and maintained by the local ethnic communities of the western peninsular India. Established long back, the SGs are shown to have survived under the maintenance of the ethnic communities retaining their original structural composition when compared to adjacent vegetation. This chapter highlights how the traditional wisdom of local people in associating religious belief systems with vegetation on which they were dependent on for their sustainable resources can conserve them effectively.

12.1   INTRODUCTION

India is one of the top 12 megabiodiversity countries of the world. It is equally remarkable for its rich cultural diversity, with modes of subsistence ranging from hunting and gathering, nomadic herding, sustainable cultivation to intensive agriculture. Given the heterogeneity of cultures, the present day pressures of large human populations and the enormous resources demands, conserving India's biodiversity is a very daunting task today. Indian people belong to three major ecological categories of whom the ecosystem people form an important category; the majority of them live in villages, hamlets and forests and depend on living resources around them (Gadgil, 1991) for their survival. These people "rooted in a locality dependent on resources drawn from a limited area that they are personally familiar with" utilize these resources sustainably. They also practice prudence as they perceive that the resource base is finite and limited. Such prudent use of natural resources is of great survival value for traditional societies which are often in active territorial conflict with adjacent societies and which are in danger of cultural, if not, genetic, elimination if they exhaust their resource base (Gadgil and Berkes, 1991). Instead of following biodiversity conservation "molded by interests of those in power," these people follow oft-tested traditional conservation strategies for not only conserving individual species but also for a whole ecosystem. One of the most common strategies followed by the traditional ethnic people is the maintenance of sacred groves (SGs). Although the institution of SGs has been well-studied in India from various perspectives (see Ramakrishnan et al., 1998; Malhotra et al., 2001), it has not been analyzed from traditional conservation perspective, particularly in the western peninsular India. This chapter deals with the SGs of this region from an ethnoconservation perspective.

12.2 WHAT ARE SACRED GROVES?

SGs may be defined as "definite segments of landscape containing vegetation and other forms of life and geographical features that are delimited and protected by human societies under the belief that to keep them in a relatively undisturbed state is expressive of an important relationship of humans with the divine or with nature" (Hughes and Subash Chandran, 1998). While establishing SGs ethnic people have provided protection to patches of forested landscape that were dedicated by them to deities or ancestral spirits. Thus, traditional societies have inputted sacred qualities to these SGs and their various elements, such as plants (especially trees), animals, and even abiotic components and that these imputations go back to the dominant, pantheistic traditions before the evolution of major formal religions, such as Hinduism, Christianity, Buddhism, Jainism, and Islam. SGs, depending on their location, may consist of a multi-species, often multi-tier, primary virgin forests or clumps of vegetation. This dedication of forested areas happened when the early ethnic settlers were clearing certain forested areas, for their living and cultivation activities and allotting certain other forested areas without disturbing them either for meeting their basic minimal requirements in a sustainable manner or for keeping as reservoirs of their requirements. Thus, initially SGs were common property institutions (see Berkes, 1989) that were serving for entire ethnic societies. Because of this, SGs, over the years, have served as refugia that were easily perceived and were most efficiently guarding against resource depletion (Gadgil and Vartak, 1976).

The different human cultures have perceived this relationship between humans and divine spirits/deities in diverse ways and have institutionalized the various rules of behavior with regard to these sacred spaces and their various biotic and abiotic elements. SGs are a very ancient and widespread institution in the Old World cultures, particularly in India. According to Kosambi (1962), SGs date back to the pre-agrarian hunting-gathering age and, thus, were conceptualized and established by the primitive hunter-gatherer societies themselves. SGs can belong to any vegetation type depending upon the type of forest that the place originally had before the SGs were established. There are as many SG types as there are forest types. Usually the SGs have retained the characteristics of their original vegetation type or may today look like climax vegetations with over mature trees, although their surrounding vegetations might have undergone drastic degradative changes. This has made, for example, Brandis and Grant (see Brandis, 1897) to wonder as follows: "Why should a certain locality be covered with evergreen, and another in its immediate vicinity with dry forest?" This also enables people to name SGs as "safety forests" (Malhotra, 1990). Many SGs, for various reasons, have even been totally eliminated. For example, the number of SGs got reduced from 755 to 346 between 1900 and 1992 in Coorg, with an area reduction of 60%.

12.3   SACRED GROVES OF WESTERN PENINSULAR INDIA

It is very difficult to make a guess regarding the total number of SGs, as well as their total area of occupation in India or in any of its regions. This is partly due to the lack of correct census efforts as well as to difficulties in defining the minimum size for assigning the SG status to a patch of natural vegetation associated with a deity. According to a conservative estimate there may be around 100 to 150 thousand SGs in India and at least nearly half of that number in western peninsular India. Some figures have been provided for various states of India (Malhotra et al., 2001). The six states which occupy western peninsular India are from (north to south) Gujarat, Maharashtra, Goa, Karnataka, Kerala, and Tamil Nadu. A review of the SGs of W. Ghats has been made by Subash Chandran et al. (1988). This review covers only Kerala, Karnataka and Maharashtra and deals with various aspects relating to SGs. A review of the SGs of Kerala has been made by Chandrashekara and Sankar (1998) and by Pushpangadan et al. (1998) and of the SGs of North Kerala by Unnikrishnan (1995). SGs of Maharashtra were reviewed by Deshmukh et al. (1998) and of the SGs of North Kerala by Unnikrishnan (1995). SGs of Maharashtra were reviewed by Deshmukh et al. (1998) and Gadgil and Vartak (1981). Different investigators have given different numbers of SGs in different regions as also to some of the same regions making it difficult to correctly count the number of SGs. The available information is given here to emphasize the above point. 12 SGs have been reported in Kanyakumari district, 31 in Nilgiris (the most important of which is Benagudi shola sacred grove with about 615 ha area), around 15 in Anamalais. 15,000 in former Travancore area according to census Report of 1891, 115 in Thiruvananthapuram district, around 2,000 with 500 hectares in the entire Kerala State (Rajendra Prasad, 1995), 17 *Myristica* swamp SGs in Northern Kannada, 1214 Kodagu in Karnataka State covering 6299.61 acres, and 873 in the same region covering 10,865 acres (Kalam, 1996),

1424 in Karnataka W. Ghats (Kalam, 1996; Gokhale, 2000), 54 in Siddapur taluk of North Kannada 953 (including 233 studied by Gadgil and Vartak 1981) in Maharashtra (233 to 1600 according others), nearly two-thirds of which occur in Western Maharashtra, and 1600 in Maharashtra (Deshmukh et al., 1998).

SGs have been variously called in different parts of western peninsular India: *Kavu, Kans, Devara Kadu. Koil Kudu, Devarabana, Huli Devarakadu, Nadabana, Bhutappanbana, Jatakappanbana, Chowdibana, Devarai, Devarahati, Devaragudi*, etc. The SGs vary very greatly in size from a few acres to several hectares. The smaller ones are often just around 0.02 ha or even smaller than this. 79% of the SGs of Kerala are reported to belong to the small category.

12.4 MAINTENANCE AND CONSERVATION OF SACRED GROVES BY TRADITIONAL COMMUNITIES

It was mentioned earlier that SGs present a case study of a resource conservation practice grounded in religious beliefs. This natural experiment suggests that pre-scientific traditional societies can and do adopt biodiversity conservation practices on the basis of their experience with nature and natural objects, both living and non-living. Thus, conservation practices were implemented through the medium of religious beliefs. This has also resulted in SGs and the deities/spirits resting in them worshiped for different religious functions and in playing an important role in social and cultural aspects of the ethnic societies that maintain and conserve them.

The divine entities associated with the SGs are invariably traditional Gods, Goddesses, Spirits and ancestors. They are either idolized or symbolized in natural objects like stones or specific tress. The gender of the presiding deity in SGs is quite variable. In Maharashtra SGs female deities are associated with 15 out of 21 SGs, male deities 5 out of 21 SGs and with Spirits in one SG. In Kerala/Karnataka some SGs have both male and female deities. These deities or Spirits/ancestors are known by different names for different SGs. The *vanadevada* or ancestral Spirits of some SGs of Kerala are known as *Madan* or *Yekshi*. During the course of history some of these traditional Gods/Goddesses are either replaced by some so-called 'modern' Gods/Goddesses or the latter are also introduced to co-exist with traditional Gods/Goddesses. For example, some SGs of Kerala are now dedicated to Lord Ayyappa or Bhagavati Amman. Temples are often erected to house these organized Hindu deities and the SGs has suffered in this process (Unnikrishnan, 1995). Most SGs as already stated, are associated with primitive religion/spirits. As characteristic of primitive religions/ ancestral Spirit worship, SGs serve as places of religious rituals and sacrifices. The latter are usually animals, such as country goats, chicken, buffaloes or oxen depending on the tribal communities associated with the SGs. There are totem (trees, rocks or other natural objects) worships, often associated with specific ceremonies like *Vasantha Vizha* (spring festival) in Tamil Nadu and *Theyyam* ritual/ceremony in Kerala (Unnikrishnan, 1990). Many ethnic societies conduct their festivals and marriages in their own SGs. In addition, community gatherings are often held in SGs. The SGs also serve for political functions. Tribal-controlled SGs assert group identity and solidarity. According Kosambi (1962) SGs are usually found along preagrarian trade routes and cross-roads where long-distance traders rested/bartered.

Smaller SGs are either fully or better protected than larger ones mainly due to the fact that most of former are under the control of local communities. For instance, the Mahadev Koli tribe of the W. Ghats of Maharashtra and the Kuntis of Kolhapur own a number of SGs (Roy Burman, 1992). The Benagudi Shola SG of Nilgiris is maintained by the *Irula* tribe. There are regional variations in terms of ethnic association of SGs and they are associated with different castes. SGS are often considered as symbols of ethnic identify. There is also a division of SGs into fully protected ones and restricted ones in which certain activities alone were allowed. Invariably the larger SGs are put under the control of the temple priests, who are also the village doctors associated with them, who use these SGs for collecting flowers, leaves, fruit or seeds used for worshipping the presiding deity/Spirit of the concerned SG. Sometimes the larger SGs are used for collecting dried and fallen twigs/wood pieces for fuel, non-destructive collection of useful non-timber produce and for tapping toddy from the palm, *Caryota urens* (Subash Chandran and Gadgil, 1993). Thus, these SGs serve as safety forests, as mentioned previously. The larger SGs are also reported to serve for ecological security by providing services much needed for a better life of the ethnic societies living near them.

12.5 IMPACT OF TRADITIONAL CONSERVATION OF SACRED GROVES

The impact of the traditional way of conservation through SGs by the local ethnic communities of western peninsular India is seen not only in the maintenance of structural integrity of SGs in the way they were originally established and developed towards climax vegetation but also in its biodiversity composition, especially when compared to those that were taken away from traditional ethnic control either by the government or governmental agencies. In the traditionally maintained/conserved SGs there is often a rich diversity at the gene and species (especially in taxic diversity and species richness) levels and presence of IUCN-recognized threat category taxa surviving only there, medicinal plants (Ayurvedic, Siddha, Tribal and folk medicines), new taxa, unreported taxa, endemic taxa, keystone species, land races, wild relatives of some cultivated taxa, etc. Pushpangadan et al. (1998) demonstrated that the biological spectrum of SGs in Kerala resembles the typical biological spectrum of a rain forest. They cite a SG of just 1.4 km^2 having 722 species of flowering plants compared to 960 angiosperms of a 90 km^2 area of silent valley forest. This diversity is largely due to the efforts of the local ethnic communities in protecting and conserving the SGs.

In the SGs of Thiruvananthapuram district (Kerala state) maintained by local communities there are 531 species, 5 subspecies and 20 varieties of plants of which five taxa are new to science, three were earlier unreported in India, seven were new to Kerala State and 117 were additions to the Flora of Thiruvananthapuram. A new plant genus and species (*Kunstleria keralensis* a climbing legume) was discovered in a SG of Kerala (Mohanan and Nair, 1981). Indeed the only surviving stand of *Dipterocarpus indicus* in the Uttar Kanda (North Karnataka) district is part of a SG of Karikanamma. The Village forest (SG) of Kallable managed by the local ethnic community supports a larger standing biomass and a greater variety of trees when compared to reserve forests in the vicinity (Gadgil, 1993). A SG of Kerala (Alappuzha district) contains *Cinnamomum quilonensis*, a rare plant taxon (Unnikrishnan, 1995) and an endemic taxon *Syzygium travancoricum*. The other rare species found in SGs of Kerala are *Blepharistemma membranifolia* and *Buchanania lanceolata* (Nair and Mohanan, 1981).

SGs not only have protected and conserved RET plants but also abiotic components. They hold water resource in the form of springs, ponds, lakes, streams or even as small rivers (Pushpangadam et al., 1998). Many of these have retained their perennial water-holding capacity. SGs are also crucial for soil conservation.

12.6   CONCLUSION

Sacred Groves of western peninsular India like those in any other region of India, have been established by the various traditional ethnic communities with the dual purpose of getting sustainable resources in a non-destructive manner and as a conservation institution, of course incidentally serving for the first purpose also. The conservation value of SGs is evident not only from its biotic and abiotic composition but also when we compare them with adjacent vegetation, if any, or with those forcibly taken over control from the traditional ethnic communities. Although may not have the minimum viability population size for its biological taxa, SGs still serve as one of the best traditional methods of ecosystem conservation.

KEYWORDS

- **Ethnic Communities**
- **Sacred Groves**
- **Safety Forests**
- **Traditional Conservation**
- **Western Ghats**

REFERENCES

Berkes, F. (Ed.). (1989). Common Property Resources. Belhaven, London.

Brandis, D. (1897). Indigenous Indian Forestry: Sacred Groves. pp. 12–13. In: Indian Forestry: Oriental institute, Working.

Chandrashekara, U.M. & Sankar, S. (1998). Structure and function of sacred groves: case studies in Kerala. pp. 323–336. In: Ramakrishnan, P.S., Saxena, K.G., Chandrashekara, U.M. (Eds.). Conserving the Sacred for Biodiversity Management. Oxford & IBH Publ. Co., New Delhi.

Deshmukh, S. Gogati, S.G. & Gupta, A.K. (1998). Sacred grove and biological diversity: Providing new dimensions to conservation Issues. pp. 397–414. In: Ramakrishnan, P.S., Saxena, K.G., Chandrashekara, U.M. (Eds.). Conserving the Sacred for Biodiversity Management. Oxford & IBH Publ. Co., New Delhi.

Gadgil, M. (1991). Environment update. Conserving India's biodiversity: the social context. *Evol. Trends Plants 5*, 3–8.

Gadgil, M. (1993). Biodiversity and India's degraded lands. *Ambio. 22,* 167–172.

Gadgil, M. & Berkes. (1991). Traditional Resource Management Systems. *Resource Management and Organization 8,* 127–141.

Gadgil, M. & Vartak, V.D. (1976). The Sacred groves of Western Ghats in India. *Econ. Bot. 30,* 152–160.

Gadgil, M. & Vartak, V.D. (1981). Sacred groves in Maharashtra—An inventory. pp. 279–294. In: Jain, S.K. (Ed.). Glimpses of Indian Ethnobotany. Oxford & IBH Pub. Ltd., New Delhi.

Gokhale, Y. (2000). Sacred conservation in India: An overview. Abstract Nat. Workshop on community Strategies on the Management of natural Resources. Bhopal, India.

Hughes, J.D. & Subash Chandran, M.D. (1998). Sacred groves around the earth: An overview. pp. 69–86. In: Ramakrishnan, P.S., Saxena, K.G. and Chandrashekara, U.M. (Eds.). Conserving the Sacred for Biodiversity Management. Oxford & IBH Publ. Co., New Delhi.

Kalam, M.A. (1996). Sacred groves in Kodagu district of Karnataka (South India): A socio-historical study. French Institute, Pondicherry.

Kosambi, D.D. (1962). Myth and Reality: Studies in the Formation of Indian Culture. Popular Press, Bombay.

Malhotra, K.C. (1990). Village supply and safety forest in Mizoram: a traditional practice of protecting ecosystem. pp. 439. In: abstracts of V international congress of Ecology.

Malhotra, K.C., Gokhale, Y. Chatterjee, S. & Srivastava, S. (2001). Cultural and Ecological Dimensions of Sacred Groves in India. Indian National Science Academy, New Delhi and Indira Gandhi Rashtriya Manav Sangrahalaya, Bhopal. India.

Mohanan, C.N. & Nair, N.C. (1981). *Kunstleria* Prain-a new genus record of India and a new species in the genus. *Proc. Indian Acad. Sci. B90*, 207–210.

Nair, N.C. & Mohaman, C.N. (1981). On the rediscovery of form threatened species from the sacred groves of Kerala. *J. Econ. Taxon. Bot. 2*, 233–235.

Pushpangadan, P., Rajendraprasad, M. & Krishnan, P.N. (1998). Sacred groves of Kerala: A synthesis on the state of art of knowledge. pp. 193–210. In: Ramakrishnan, P.S., Saxena, K.G. & Chandrashekara, U.M. (Eds.). Conserving the Sacred for Biodiversity Management. Oxford & IBH Publ. Co., New Delhi.

Rajendraprasad, M. (1995). The Floristic, Structural and Functional Analysis of sacred groves of Kerala. PhD Thesis, Univ. Kerala, Thiruvananthapuram.

Ramakrishnan, P.S., Saxena, K.G. & Chandrashekara, U.M. (Eds.). (1998). Conserving the Sacred for Biodiversity Management. Oxford & IBH Publ. Co., New Delhi.

Roy Burman, J.J. (1992). The institution of Sacred groves. *J. Indian Anthropol. Soc. 27*, 219–238.

Subash Chandran, M.D. & Gadgil, M. (1993). Sacred groves and sacred trees of Uttar Kannada (a pilot study) Mimeograph. Centre for Ecological Sciences, Indian Institute of Science, Bangalore.

Subash Chandran, M.D., Gadgil, M. & Hughes, J.D. (1998). Sacred groves of the Western Ghats. pp. 211–232. In: Ramakrishnan, P.S., Saxena, K.G. and Chandrashekara, U.M. (Eds.). Conserving the sacred for Biodiversity Management. Oxford and IBH Publ. Co., New Delhi.

Unnikrishnan, E. (1990). Part played by the sacred Groves local environment. Centre for Science and Environment.

Unnikrishnan, V. (1995). Sacred Groves of North Kerala: An Eco-Folklore Study (in Malayalam). Jeevarekha, Thrissur.

CHAPTER 13

ETHNOBRYOLOGY OF INDIA

AFROZ ALAM

Department of Bioscience and Biotechnology, Banasthali University, Rajasthan–304022, India. E-mail: afrozalamsafvi@gmail.com

CONTENTS

ABSTRACT

About 58 years ago the term "ethnobryology" was introduced in a paper regarding the bryophytes utilized by the Gosiute people of Utah (Flowers, 1957). Even if there are smaller number of text reports about human beings using bryophytes than those concerning higher plants, several indications about ethnobotanically important bryophytes are present. These examples of ethnobryological uses are more interesting because of their comparative scarcity, and for the information that they can provide about the relations between inhabitants of a place and miniature plants, occurring there, like bryophytes. This article presents a summary of conventional uses of bryophytes around India. A list of about 58 ethnobotanical species of bryophytes is given. The most common use of bryophytes is for medicinal purposes. However, genera like *Marchantia*, *Sphagnum* and *Polytrichum* whose species are the most commonly reported to have ethnobotanical uses.

13.1   INTRODUCTION

The understanding of ethnobotany is extremely ancient. It might have initially come into the reality possibly during "stone age," when early man had seen the animals like the monkeys and apes taking particular plants/plant parts as food. Occasionally, when required they used specific plants to cure their wounds and for relieving their of pains, distress and diseases. Thus, initially animals started to use plants and later on it was followed by human beings. Subsequently, the concepts of Ethnozoology and Ethnobotany were developed, which combined into a common term Ethnobiology. Hence, it can be said that ethnobotany is the learning of interrelationships between humans and plants in their environment.

The term Ethnobotany was first used at the end of the 19th century by J.W. Harshberger (1895) to specify the interrelationship of plants with indigenous people or tribal societies. According to him, "Ethnobotany is a specific discipline of plant sciences which deals with studies about total interrelationships between plants and prehistoric man." In a broader sense the word can apply to any ethnic group or society of any region or ecological area and hence can be used to explain the varied uses of plants by different ethnic (tribal) societies in the diverse parts of the world. Even though, this branch of plant science had been neglected in the past, in the last three or four decades with the rediscovery of some amazing uses of various plants for human society based on ethnic knowledge, it achieved great significance.

India is a country of great diversity. The inhabitants of India are closely associated with the plants since time immortal. References to use of plants by human beings are found in ancient Sanskrit and Tamil literature. Diverse usage of plants and plant based products had been documented, though in bits and pieces. The first ever book on ethnobotany was written by Faulks (1958), and after 23 years, the first ever book on Indian ethnobotany entitled 'Glimpses of Indian Ethnobotany' was presented by Jain (1981). He did extensive field work amongst the tribes and their surroundings, scrutinized of literature, herbaria and Musea and also studied archaeological remains. Raghavaiah (1956) provided the much needed guidance for approaching the tribals for collecting ethnobotanical data. Many workers became active and studied the ethnobotany of various regions of India, such as Mayurbhanj (Odisha), Madhya Pradesh, Kumaon Hills (Uttarakhand), Araku Valley, Ratanmahal Hills, Brahmaputra Valley, Saurasthra, Santal Pargana, Sind Valley, Andaman & Nicobar Islands (Bhargava, 1983), Andhra Pradesh (Banerjee, 1977), Arunachal Pradesh (Dam and Hazra, 1981), Assam (Baruah and Sarma, 1984), Bihar (Goel et al., 1984), Gujarat (Bedi, 1978), Haryana (Jain, 1984), Himachal Pradesh (Arora et al., 1980), Jammu & Kashmir (Dar et al., 1984), Karnataka (Razi and Subramaniam, 1978), Kerala (John, 1984), Madhya Pradesh (Jain, 1963), Maharashtra (Kamble and Pradhan, 1980), Meghalaya (Rao, 1981), Nagaland (Rao and Jamir, 1982), Odisha (Mudgal and Pal, 1980), Punjab (Lal and Lata, 1980), Rajasthan (Mishra and Billore, 1983), Uttar Pradesh (Dixit and Pandey, 1984), Tripura (Deb, 1968), Tamil Nadu (Ramachandran and Nair, 1981) and West Bengal (Chaudhuri and Pal, 1975).

Mostly the angiosperms were documented initially and the other plant groups were neglected. The work related to other plant groups was carried out initially by foreign workers, for example, Liano (1956) explored the ethnic uses of Lichens and Zaneveld (1959) worked on marine algae. Subsequently, in India, Deb et al. (1974) worked on edible algae and Dixit (1982) studied ethnic uses of pteridophytes. Bryophytes were invariably ignored in ethnobotanical study by most of the workers in India and abroad as well.

Despite its 58 year old existence, the term ethnobryology has not entered seriously into the ethnobiological literature. In his paper Harris (2008) considered two main factors for its limited ethnic uses: first, there are not much ethnobotanical uses of bryophytes, and second, there may be a prejudice on the part of ethnobotanical researchers to focus on large plants because of their easy availability and identification. As a result, the ethnobotanical uses of bryophytes are very scarcely documented and this led, Henry S. Conrad

(1956) to write as follows: "Perhaps no great group of plants has so few uses, commercial or economic, as the mosses."

Humans usually identify their environs at a firm spatial and temporal scale. Unlike other extremely large plants, most dimensions of morphological characteristics in bryophytes (Lilliputians/amphibians of plant kingdom) are in micrometers or millimeters, centimeters at the biggest. They also lie in less noticeable places and hence they are often unnoticed by humans. Therefore, bryophytes remain underexplored in many aspects including their medicinal value. However the conventional use of few taxa of bryophytes on the basis of 'doctrine of signatures' is an old practice among native tribes of India (Glime, 2007). Stone Age people living in ancient Germany once collected the moss *Neckera crispa* (Grosse-Brauckmann, 1979). Other sporadic evidences suggest a variety of uses by various cultures around the world (Pant and Tewari, 1990; Glime, 2007). Numerous secondary compounds make bryophytes inedible and their nutritional value is doubtful. They are now increasingly being used as new sources of pharmaceutical products (Kumar et al., 2000), antioxidants (Dey et al., 2013), medicines and even food (Glime, 2007). The various ethnobotanical uses of bryophytes in India are discussed in this chapter.

13.2 FOLK MEDICINE

Like most of the higher plants, bryophytes are also primarily used for medicinal purposes. The term 'liverwort' is a good example, where the thallus shape is an indicative of its use in liver-related problems. In the last few decades the medicinal value of bryophytes has received interest to a great extent and many chemicals and secondary metabolites have been isolated from many bryophytes. Recently, Harris (2008) compiled and documented a comprehensive account about the medicinal uses of bryophytes in diverse parts of the world and listed about 150 species. According to him the medicinal uses of bryophytes are mostly explored by Native North Americans (28%), followed by Chinese (27%). But in India the figure is not very encouraging. However, the first ever description of the traditional and cultural uses of bryophytes in the Indian region, along with a detailed description and illustration, originally appeared in *Hortus Malabaricus* (Harris, 2008).

Bryophytes are customarily used for burns, cuts, wounds and skin diseases by some tribes of Himalayan region of India, which suggests that they have some protective substances that protect the skin and open wounds from microbial pathogens (Pant and Tewari, 1989a, b; Saxena and Harinder,

2004). Kumar et al. (2000) also reported the use of *Plagiochasma appen-diculatum* by the people of Gaddi tribe of Himachal Pradesh, India, for the treatment of cuts, wounds and burns. Singh et al. (2006; 2007) studied and validated the antimicrobial, wound healing and antioxidant activity of the same. Later, they studied the antibacterial activity of four bryophytes viz., *Plagiochasma appendiculatum, Conocephalum conicum, Bryum argenteum, Mnium marginatum*, for the treatment of burn infections. All these plants are used by these tribes in skin related diseases. Earlier, use of *Philonotis, Bryum, Mnium* spp. were also reported in the treatment for pain of burns, to cover the bruises and wounds and as padding under splints in setting broken bones (Flowers, 1957). Joshi et al. (1994) in a new hypothesis, considered bryophytes as the source of the origin of *Shilagitu*. Sharma et al. (2015) have assessed antifungal and antioxidant profile of *Pellia endivaefolia* and *Plagiochasma appendiculatum*. These are used by indigenous tribes of district Reasi of North West Himalayas.

Unlike mosses, the plants of class Hepaticae (Liverwort) hold oil bodies in their cells, which can be easily extracted with organic solvents. The liverworts have remarkable medicinal value and are said to have definite biological activity and effect (Alam et al., 2015). A number of taxa are used to heal burns, external wounds, bruises, etc. In the Himalayas, tribals use a mixture of plant ashes with fat and honey as a soothing and healing ointment for cuts, burns, and wounds (Pant et al., 1986).Western Himalayan tribes use *Marchantia polymorpha* and *M. palmata* to treat boils and eruptions because the immature archegoniophore looks like a boil as it comes out from the thallus (Pant and Tewari, 1989a, b). *Riccia* spp. (Plate 13.2: Figure 4) are used by the tribes of Himalayan region as a cure for ringworm because of the semblance of the growth habit of this liverwort to the rings made by the worm. Latest research on *R. fluitans* however, indicated that there is nothing in it to inhibit bacterial growth (Glime, 2007) yet, the tribes believe in its potential as local medicine.

The ethnobotanical uses of bryophytes are primarily somewhat restricted to the Western Himalayas. However, few reports are also available from other bryorich regions of India. Recently, during the comprehensive ethnobotanical studies on the tribal groups of Palakkad district, two bryophytes, for example, *Targionia hypophylla* and *Frullania ericoides*, which were earlier not known to have any economic uses, were recorded from the Attappady Valley (close to Silent Valley National Park) of Western Ghats, Kerala. The three tribes namely, Irula, Kurumba and Muduga inhabit this valley. Except Kurumba, other two tribes have fair ethnomedicinal knowledge on bryophytes and they are using these taxa in their local medical system (Ramesh and Manju, 2009).

In Nilgiri hills (Tamil Nadu), Toda tribe also uses a few liverworts as ointment to cure skin infection. Species of *Marchantia, Plagiochasma, Riccia* and *Lunularia* are their common choice, however, they consider all these taxa as a single plant. The potentiality of many liverworts as antibacterial agents has been confirmed by many recent research publications (Banerjee, 2001; Singh et al., 2007; Vats and Alam, 2013). Liverworts also showed significant antifungal activity against many pathogenic fungal strains (Alam et al., 2012). The antimicrobial potential of ethnomedicinally important plants, namely *Marchantia polymorpha, Riccia fluitans, Targionia hypophylla, Plagiochasma rupestre, Atrichum undulatum* and *Entodon nepalensis* was validated by recent researches (Alam, 2012; Alam et al., 2015).

13.3   CLOTHING

Many bryophytes are also used in day to day clothing by the tribes of Himalayan regions. Stuff mosses viz. *Hylocomium, Hypnum, Trachypodopsis* are frequently used into cloth sacks to make head cushions (*sirona*) that also absorb leaking water as they carry water vessels by the women in the townships of Kumaun hills (Pant and Tewari, 1989a).

13.4   HOUSEHOLD USES

In the past, mosses were used as a substance for bedding, packing, plugging and stuffing due to their soft elastic texture, and also because of their insecticidal and antimicrobial potential. In India, *Sphagnum* spp., *Hypnum cupressiforme, Macrothamnium submacrocarpum, Neckera crenulata, Trachypodopsis crispatula* and *Thuidium tamariscellum* are used for packing of fruits in the Western Himalayas. *Sphagnum* especially, functions as an insulator, cushion, mattress, and furnishings filling, to keep the liquid cool or warm, to stuff into doormats to dirt free shoes, to weave reception mats, and also to support baby cradles, keeping the toddler hygienic, dried up, and temperate. The sturdiness and softness of mosses may well have contributed to stuffing balls and dolls with *Hypnum* (Pant et al., 1986).

In Kumaun Himalayas, mosses, such as *Anomodon, Entodon, Hypnum, Meteoriopsis* are also used for door covers (Pant, 1998). Kumaun Indians use meager mosses (*Entodon, Hypnum, Meteoriopsis*) and liverworts

(*Herbertus*, and *Scapania*) as a filter of handmade cigarette that is wrapped in a cone of *Rhododendron campanulatum* leaves. The tribals of this region also clean household utensils with a mixture of mosses and ashes (Pant and Tewari, 1989a).

13.5 PACKING MATERIALS

Wet *Sphagnum* spp. are most commonly used by the nurserymen in India for transportation of live plants from one place to another. The remarkable absorbent characteristics of *Sphagnum* make it the most utilized bryophytes till date, both in eastern and western Himalayas as well. Besides *Sphagnum* for packing apples and plums, *Brachythecium salebrosum, Cryptoleptodon flexuosus, Hypnum cupressiforme, Macrothamnium submacrocarpum, Neckera crenulata, Trachypodopsis crispatula*, and *Thuidium tamariscellum* are used as packing material (Pant and Tewari, 1989a, b).

13.6 FOOD SOURCE

Bryophytes are very rarely used as a food source by the tribals of India. However, in the Kumaun Himalayas, tribals use slender bryophytes, such as *Anomodon, Entodon, Hypnum, Meteoriopsis, Herbertus*, and *Scapania* as rare food source (Pant and Tewari, 1989a, b).

13.7 CONSERVATION OF SOIL

Bryophytes are excellent soil binders hence they keep the soil intact during rainy season. Alam et al. (2012) have observed the impact of bryophytes depletion on landslides in the Nilgiri hills (Tamil Nadu). Several moss genera viz. *Actinodontium, Aerobryum, Atrichum, Archidium, Barbella, Barbula, Bartramia, Bartramidula, Braunia, Campylodontium, Ctenidium, Dicranoloma, Distichophyllum, Entoshodon, Hedwigidium, Platydictya, Lyellia, Porotrichum, Philonotis, Garckea*, etc., are contributing in conservation of soil. The local Thoda tribe knows the importance of these miniature plants and contributing to the protection of these hills. Several latest researches also signified the usefulness of these plants in bringing up moisture and nutrients from the underground soil, which is valuable for the other flora of the region (Table 13.1).

TABLE 13.1   Bryophytes of Ethnic Importance and Their Distribution in India

S. No.	Plant species	Used as	Area	Tribes	Reference
1.	*Actinodontium* spp.	Soil conservation	Nilgiri Hills	Thodas, Badagas, Kurumbas, Irulas	Alam et al. (2012)
2.	*Aerobryum* spp.	Soil conservation	Nilgiri Hills	Thodas, Badagas, Kurumbas, Irulas	Alam et al. (2012)
3.	*Anomodon* spp.	food source	Kumaun Hills	Bhotia, Raji, Tharus, Boxas	Pant and Tewari (1989)
4.	*Archidium* spp.	Soil conservation	Nilgiri Hills	Thodas, Badagas, Kurumbas, Irulas	Alam et al. (2012)
5.	*Atrichum undulatum*	Soil conservation and skin treatment	Nilgiri Hills	Thodas, Badagas, Kurumbas, Irulas	Alam et al. (2012, 2015)
6.	*Barbella* spp.	Soil conservation	Nilgiri Hills	Thodas, Badagas, Kurumbas, Irulas	Alam et al. (2012)
7.	*Barbula* spp. (Plate 13.2: Figure 2)	Soil conservation	Nilgiri Hills	Thodas, Badagas, Kurumbas, Irulas	Alam et al. (2012)
8.	*Bartramia* spp.	Soil conservation	Nilgiri Hills	Thodas, Badagas, Kurumbas, Irulas	Alam et al. (2012)
9.	*Bartramidula* spp.	Soil conservation	Nilgiri Hills	Thodas, Badagas, Kurumbas, Irulas	Alam et al. (2012)
10.	*Bazzania* spp.	medicinal	Kumaun Hills	Bhotia, Raji, Tharus, Boxas	Pant et al. (1986)
11.	*Brachythecium salebrosum*	Packing material	Himalayas	Bhotia	Pant and Tewari (1989) a, b
12.	*Braunia* spp.	Soil conservation	Nilgiri Hills	Thodas, Badagas, Kurumbas, Irulas	Alam et al. (2012)
13.	*Bryum argenteum* (Plate 2: Figure 1)	medicinal	Kumaun Hills	Bhotia, Raji, Tharus, Boxas	Pant et al. (1986)
14.	*Campylodontium* spp.	Soil conservation	Nilgiri Hills	Thodas, Badagas, Kurumbas, Irulas	Alam et al. (2012)
15.	*Conocephalum conicum*	medicinal	Kumaun Hills	Bhotia, Raji, Tharus, Boxas	Pant et al. (1986)
16.	*Cryptoleptodon flexuosus*	Packing material	Himalayas	Gaddi, Boto, Chamgpa	Pant and Tewari (1989) a, b

TABLE 13.1 *(Continued)*

S. No.	Plant species	Used as	Area	Tribes	Reference
17.	*Ctenidium* spp.	Soil conservation	Nilgiri Hills	Thodas, Badagas, Kurumbas, Irulas	Alam et al. (2012)
18.	*Dicranoloma* spp.	Soil conservation	Nilgiri Hills	Thodas, Badagas, Kurumbas, Irulas	Alam et al. (2012)
19.	*Distichophyllum* spp.	Soil conservation	Nilgiri Hills	Thodas, Badagas, Kurumbas, Irulas	Alam et al. (2012)
20.	*Dumortiera hirsuta*	medicinal	Kumaun Hills	Bhotia, Raji, Tharus, Boxas	Pant et al. (1986)
21.	*Entodon* spp.	food source	Kumaun	Bhotia, Raji, Tharus, Boxas	Pant and Tewari (1989 a, b)
22.	*Entoshodon* spp.	Soil conservation	Nilgiri Hills	Thodas, Badagas, Kurumbas, Irulas	Alam et al. (2012)
23.	*Frullania ericoides*	Removal of lice and nourishment of hair.	Western Ghats: Kerala	Oosimooppan and Mudugars	Ramesh and Manju (2009)
24.	*Garckea* spp.	Soil conservation	Nilgiri Hills	Thodas, Badagas, Kurumbas, Irulas	Alam et al. (2012)
25.	*Hedwigidium* spp.	Soil conservation	Nilgiri Hills	Thodas, Badagas, Kurumbas, Irulas	Alam et al. (2012)
26.	*Herbertus* spp.	food source	Kumaun	Bhotia, Raji, Tharus, Boxas	Pant and Tewari (1989) a, b
27.	*Hylocomium* spp.	clothing	Kumaun	Bhotia, Raji, Tharus, Boxas	Pant and Tewari (1989) a, b
28.	*Hypnum cupressiforme*	Household uses, Packing material	Western Himala-yas	Bhotia, Raji, Tharus, Boxas, Dogri	Pant and Tewari (1989) a, b; Pant and Tewari (1990)
29.	*Hypnum* spp.	Clothing, food source	Kumaun	Bhotia, Raji, Tharus, Boxas	Pant and Tewari (1989) a, b
30.	*Lunularia cruciata* (Plate 13.1: Figure 1)	Biomonitoring	Nilgiri Hills	Thodas, Badagas, Kurumbas, Irulas	Alam and Sharma (2012)

TABLE 13.1   *(Continued)*

S. No.	Plant species	Used as	Area	Tribes	Reference
31.	*Lyellia* spp.	Soil conservation	Nilgiri Hills	Thodas, Badagas, Kurumbas, Irulas	Alam et al. (2012)
32.	*Macrothamnium submacrocarpum*	Household uses, Packing material	Western Himalayas	Bhotia, Raji, Tharus, Boxas	Pant and Tewari (1989a, b, 1990)
33.	*Marchantia furcata*	Medicinal	Kumaun Hills	Bhotia, Raji, Tharus, Boxas	Pant et al. (1986)
34.	*Marchantia paleacea* (Plate 13.1: Figure 3)	Biomonitoring	Nilgiri Hills	Thodas, Badagas, Kurumbas, Irulas	Alam and Srivastava (2007)
35.	*Marchantia palmata* (Plate 13.1: Figure 2)	medicinal	Kumaun Hills	Bhotia, Raji, Tharus, Boxas	Pant et al. (1986)
36.	*Marchantia polymorpha* (Plate 13.1: Figure 4)	medicinal	Kumaun Hills; Darjeeling	Bhotia, Raji, Tharus, Boxas	Pant et al. (1986); De et al. (2015)
37.	*Meteoriopsis* spp. (Plate 13.2: Figure 5)	Food source	Kumaun	Bhotia, Raji, Tharus, Boxas	Pant and Tewari (1989)
38.	*Mnium marginatum*	medicinal	Kumaun Hills	Bhotia, Raji, Tharus, Boxas	Pant et al. (1986)
39.	*Neckera crenulata*	Household uses, packing material	Western Himalayas	Raji, Boxas, Dogri	Pant and Tewari (1989, 1990)
40.	*Pellia endiviifolia*	medicinal	Kumaun Hills; Tribes of District Reasi; Darjeeling	Bhotia, Raji, Tharus, Boxas; Dongri	Biswas (1956); Pant et al. (1986); Sharma et al. (2015); Dey et al. (2015); Mukherjee et al. (2012)
41.	*Philonotis* spp.	Medicinal, soil conservation	Pithoragarh, Kumaun Hills	Bhotia, Raji, Tharus, Boxas, Thodas, Badagas, Kurumbas, Irulas	Pant et al. (1986); Alam et al. (2012)

TABLE 13.1   *(Continued)*

S. No.	Plant species	Used as	Area	Tribes	Reference
42.	*Plagiochasma appendiculatum*	medicinal	Himachal Pradesh; Tribes of District Reasi	Gaddi tribe	Kumar et al. (2000); Sharma et al. (2015)
43.	*Plagiochasma* sp. (Plate 13.1: Figure 5)		Himachal Pradesh	Gaddi	Kumar et al. (2000)
44.	*Plagiochasma rupestre*	Soil conservation and biomonitoring	Tamil Nadu, Rajasthan	Thodas, Badagas, Kurumbas, Irulas, Garasia, Damor	Alam (2014)
45.	*Plagiochila* spp. (Plate 13.2: Figure 6)	medicinal	North Western Himalayas	Bhotia, Raji, Tharus, Boxas	Pant et al. (1986)
46.	*Platydictya* spp.	Soil conservation	Nilgiri Hills	Thodas, Badagas, Kurumbas, Irulas	Alam et al. (2012)
47.	*Porella platyphylla*	medicinal	Kumaun Hills	Bhotia, Raji, Tharus, Boxas	Pant et al. (1986)
48.	*Porella vernicosa*	medicinal	Kumaun Hills	Bhotia, Raji, Tharus, Boxas	Pant et al. (1986)
49.	*Porotrichum* spp.	Soil conservation	Nilgiri Hills	Thodas, Badagas, Kurumbas, Irulas	Alam et al. (2012)
50.	*Radula* spp.	medicinal	Kumaun Hills	Bhotia, Raji, Tharus, Boxas	Pant et al. (1986)
51.	*Riccia aravalliensis*	Biomonitoring	Rajasthan	Bhil, Meena	Alam (2014)
52.	*Riccia billardieri* (Plate 13.1: Figure 6)	Biomonitoring	Rajasthan	Garasia, Damor, Bhil	Alam et al. (2014)
53.	*Scapania* spp.	food source	Kumaun	Bhotia, Raji, Tharus, Boxas	Pant and Tewari (1989)
54.	*Sphagnum* spp. (Plate 13.2: Figure 3)	Household uses, packing and fuel	Western Himalayas and eastern Himalayas	Tharus, Boxas	Pant and Tewari (1990)

TABLE 13.1　*(Continued)*

S. No.	Plant species	Used as	Area	Tribes	Reference
55.	*Targionia hypophylla*	Paste of plants used against scabies, itches and other skin diseases	Kerala	Kali and Jadiya of hamlet Bhuthu- vazhi from Irulars of Attappady	Ramesh and Manju (2009)
56.	*Thuidium tamariscellum*	Household uses, Packing material	Western Himala- yas	Boxas, Tharus	Pant and Tewari (1989, 1990)
57.	*Trachypodopsis crispatula*	Household uses, packing material	Western Himala- yas	Bhotia, Boxas, Dogri	Pant and Tewari (1989); (1990)
58.	*Trachypodopsis* spp.	clothing	Kumaun	Bhotia, Raji, Tha- rus, Boxas	Pant and Tewari (1989)

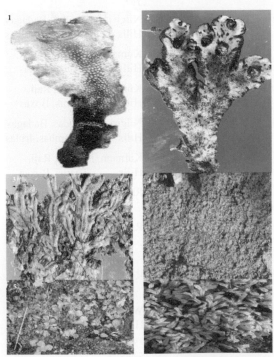

PLATE 13.1　1. *Lunularia cruciata*, 2. *Marchantia palmata*, 3. *Marchantia paleacea*, 4. *Marchantia polymorpha*, 5. *Plagiochasma* sp., 6. *Riccia* sp.

PLATE 13.2 1. *Bryum* sp., 2. *Barbula* sp., 3. *Sphagnum* sp., 4. *Riccia* sp., 5. *Meteoriopsis* sp., 6. *Plagiochila* sp.

13.8 CONCLUSION

As far as the ethnobotanical uses of the bryophytes in India is concerned, the documentation is sporadic (Pant et al., 1986; Pant, 1998; Pant and Tewari, 1989 a, b; Pant and Tewari, 1990; Glime and Saxena, 1991; Joshi et al., 1994; Banerjee, 2001; Kumar et al., 2000; Saxena and Harindar, 2004; Singh et al., 2006; Ramesh and Manju, 2009; Alam et al., 2015). Usually the ethnobotanical researches were also centralized in Western Himalayan region. Countrywide consolidated account regarding ethnobryology was lacking till date. India, being one of the main centers of bryo-diversity, is still far back in applied bryological research and therefore need some serious efforts. Through their long evolutionary history bryophytes have obtained many useful properties including an array of biochemical assets that may one day prove to be a substantial source of human beings. For that there is an urgent need to extract knowledge from local tribes of India regarding the various uses of this neglected plant group. The information obtained would be very useful in coming future, because concrete is always made from raw materials.

ACKNOWLEDGEMENTS

The author is grateful to Professor Aditya Shastri, Vice Chancellor and Professor Vinay Sharma, Dean, Faculty of Science and Technology, Banasthali University, Rajasthan, India, for their kind support for this work.

KEYWORDS

- **Bryophytes**
- **Ethnobotany**
- **India**
- **Medicinal Plants**
- **Tribes**

REFERENCES

Alam, A. (2012). Antifungal activity of *Plagiochasma rupestre* (Forst.) Steph. extracts. *Researcher 4(3)*, 61–64.

Alam, A. (2013). Bio-monitoring of metal deposition in Ranthambhore National Park (Rajasthan), India using *Plagiochasma rupestre* (G. Frost) Stephani. *Arch. Bryol. 186*, 1–10.

Alam, A. (2014). Bio-Monitoring of metal deposition in Ranthambhore national park (Rajasthan), India using *Riccia aravalliensis* Pande et Udar. *Elixir Bio Tech. 69*, 22838–22842

Alam, A. & Srivastava, S.C. (2009). *Marchantia paleacea* Bert. As an indicator of heavy metal pollution. *Indian J. Forest. 32(3)*, 465–470.

Alam, A. & Sharma, V. (2012). Seasonal variation in accumulation of heavy metals in *Lunularia cruciata* (Linn.) Dum. at Nilgiri hills, Western Ghats, *Int. Jour. Biol. Sci. Engg. 3 (2)*, 91–99.

Alam, A., Tripathi, A., Vats, S., Behera, K.K. & Sharma, V. (2011). *In vitro* antifungal efficacies of aqueous extract of *Dumortiera hirsuta* (Swaegr.) Nees against sporulation and growth of postharvest phytopathogenic fungi. *Arch. Bryol. 103*, 1–9.

Alam, A., Behera, K.K., Vats, S., Sharma, D. & Sharma, V. (2012). Impact of Bryo-Diversity depletion on Land Slides in Nilgiri Hills, Western Ghats (South India) *Arch. Bryol. 122*, 1–7.

Alam, A., Shrama, V., Rawat, K.K. & Verma, P.K. (2015). Bryophytes – The Ignored Medicinal Plants. *SMU Medical Journal, 2(1)*, 299–316.

Arora, R.K., Maheshwari, M.L., Chandel, K.P.S. & Gupta, R. (1980). Mano (*Inula racemosa*): Little known aromatic plants of Lahaul valley, India. *Econ. Bot. 34*, 175–180.

Banerjee, D.K. (1977). Observations on the ethnobotany of Araku Valley, Vishakhapatnam district, Andhra Pradesh. *J. Sci. Club. 31*, 14–21.

Banerjee, R.E. (2001). Anti microbial activity of bryophytes: a review. Pages 55–74. In: V. Nath & A.K. Asthana (eds.), *Perspectives in Indian Bryology*. Bishen Singh Mahendra Pal Singh, Dehra Dun, India.

Baruah, Parakutty & Sarma, G.C. (1984). Studies in medicinal uses of plants by the Boro tribals of Assam-II. *J. Econ. Tax. Bot. 5*, 599–604.

Bedi, S.J. (1978). Ethnobotany of Ratan Mahal Hills, Gujarat, India. *Econ. Bot. 32*, 278–284.

Bhargava, N. (1983). Ethnobotanical studies of the tribes of Andaman and Nicobar Islands, India. I Onge. *Econ. Bot. 37*, 110–119.

Chaudhuri, Rai H.N. &Pal, D.C. (1975). Notes on magico-religious belief about plants among Lodhas of Midnapur, W. Bengal. *Vanyajati 13*, 210–222.

Conard, H.S. (1956). How to Know the Mosses and Liverworts. Dubuque, Iowa W.C. Brown Co.

Dam, D.P. & Hajra, P.K. (1981). Observations on the ethnobotany of the Monpas of Kameng District, Arunachal Pradesh. In: Jain, S.K. (Ed.) Glimpses of Indian Ethnobotany, Oxford & IBH Publishing Co. New Delhi, pp. 107–114.

Dar, G.H., Vir Jee, Kachroo, P. & Buth, G.M. (1984). Ethnobotany of Kashmir-I. Sind valley. *J. Econ. Tax. Bot. 5*, 668–675.

Deb, D.B. (1968). Medicinal plants of Tripura state. *Ind. For. 94*, 753–765.

Deb, D.B., Krishna, B., Mukherjee, K., Bhatacharya, S., Chowdhury, A.N., Das, H.B. & Singh, Sh.T. (1974). An edible alga of Manipur (*Lemanea australis*): presence of silver. *Curr. Sci. 43*, 629.

Dey, A. & De, J.N. (2011). Antifungal Bryophytes: A Possible Role against Human Pathogens and in Plant Protection. *Res. Jour. Bot. 6*, 129–140.

Dey, A. & De, J.N. (2012). Antioxidative Potential of Bryophytes: Stress Tolerance and Commercial Perspectives: A Review. *Pharmacologia 3*, 151–159.

Dey, A., De, A., Ghosh, P. & Mukherjee, S. (2013). Altitude and tissue type influence antioxidant potential of *Pellia endiviifolia* from Darjeeling Himalaya. *Jour. Biol. Sci. 13*, 707–711.

Dixit, R.D. (1982). *Selaginella bryopteris* (L.) Box.—an ethnobotany study—IV. *J. Econ. Tax. Bot. 3*, 309–311.

Dixit, R.S. & Pandey, H.C. (1984). Plants used of folk-medicine in Jhansi and Lalitpur sections of Bundelkhand. Uttar Pradesh. *Int. J. Crude Drug. Res. 22*, 47–51.

Faulks, P.J. (1958). An introduction to Ethnobotany (Moredale Publication Ltd. London). pp. 3–5

Flowers, S. (1957). Ethnobryology of the Gosiute Indians of Utah. *Bryologist 60*, 11–14.

Glime, J.M. & Saxena, D. (1991). Uses of Bryophytes. New Delhi Today & Tomorrow's Printers & Publishers.

Glime, J.M. (2007). Economic and ethnic uses of Bryophytes. In: Flora of North America. Editorial committee, eds., 1993+, Flora of North America North of Mexico 15+ vols. New York and Oxford. *27*, 14–41.

Goel, A.K., Sahoo, A.K. & Mudgal, V. (1984). A contribution to the ethnobotany of Santal paragana, Botanical Survey of India, Hawrah.

Grosse-Brauckmann, G. (1979). Major plant remains of moor profiles from the area of a stone-age lakeshore settlement on Lake Duemmer, West Germany. *Phytocoenologia 6*: 106–117.

Harris, E.S.J. (2008). Ethnobryology: Traditional Uses and Folk Classification of Bryophytes. *Bryologist, 111 (2)*, 169–217.

Harshberger, J.W. (1896). The purpose of ethnobotany. *Bot. Gaz. 21*, 146–158.

Jain, S.K. (1963). Studies in Indian Ethnobotany-II, plants used in medicine by the tribals of Madhya Pradesh. *Bull. Reg. Res. Lab. Jammu 1*, 126–128.

Jain, S.K. (1981). Glimpses of Indian Ethnobotany. Oxford & IBH Publishing Co. New Delhi, 1–366.

Jain, S.P. (1984). Ethnobotany of Morni and Kabsar (district Ambala, Haryana). *J. Econ. Tax. Bot. 5*, 809–813.

John, D. (1984). One hundred useful drugs of the Kani tribes of Trivandrum forest division, Kerala, India. *J. Crude drug Res. 22*, 17–39.

Joshi, G.C., Tiwari, K.C., Pande, N.K. & Pandey, G. (1994). Bryophytes as the source of the origin of Shilagitu: a new hypothesis. *Bull. Medico-Ethno. Bot. Res. 15*, 106–111.

Kamble, S.Y. & Pradhan, S.G. (1980). Ethnobotany of the 'Korkus' in Maharashtra. *Bull. Bot. Surv., India. 22*, 201–202.

Kumar, K., Singh, K.K., Asthana, A.K. & Nath, V. (2000). Ethnotherapeutics of bryophyte *Plagiochasma appendiculatum* among the Gaddi Tribes of Kangra Valley, Himachal Pradesh, India. *Pharmaceutical Biology 38*, 353–356.

Lal, S.D. & Lata, K. (1980). Plants used by the Bhat community for regulation fertility. *Econ. Bot. 37*, 299–305.

Llano, G.A. (1956). Utilization of lichens in the Arctic and Subarctic. *Econ. Bot. 10*, 367–392.

Mishra, R. & Billore, K.V. (1983). Some Ethnobotanical loses from Banswara district. *Nagarjun 26*, 229–231.

Mudgal, V. & Pal, D.C. (1980). Medicinal plants used by tribals of Mayurbhanj (Orissa). *Bull. Bot. Surv. India 22*, 179–180.

Mukherjee, S., De, A., Ghosh, P. & Dey, A. (2012). *In vitro* antibacterial activity of various tissue types of *Dumortiera hirsuta* (Sw.) Nees from different altitudes of eastern Himalaya. *Asian Pac. J. Trop. Dis. 2*, S285–S290.

Pant, G. & Tiwari, S.D. (1990) Bryophytes and Mankind. *Ethnobotany 2*, 97–103.

Pant, G.P. (1998). Medicinal uses of bryophytes. Pages 112–124. In R.N. Chopra (ed.), Topics in Bryology. Allied Publishers Limited, New Delhi.

Pant, G.P. & Tewari, S.D. (1989a). Various human uses of bryophytes in Kumaun region of Northwest Himalaya. *Bryologist 92*, 120–122.

Pant, G.P. & Tewari, S.D. (1989b). Bryological activities in Northwest Himalaya III. A field excursion to the Pindari Glacier region of the District Almora (Kumaun Himalaya). *Bryol. Times 52*, 1–5.

Pant, G.P., Tewari, S.D., Pargaien, M.C. & Bisht, L.S. (1986). Bryological activities in Northwest Himalaya II. A bryophyte foray in the Askot region of district Pithoragarh (Kumaun Himalaya). *Bryol. Times 39*, 2–3.

Raghavaiah, V. (1956). How to approach the Adivasi. *Vanyajati 4*, 31–38.

Ramachandran, V.S. & Nair, N.C. (1981). Ethnobotanical observation on Irulars of Tamil Nadu, India. *J. Econ. Tax. Bot. 2*, 183–190.

Rao, R.R. (1981). Ethnobotany of Meghalay medicinal plants used by Khasia and Garo tribes. *Econ. Bot. 35*, 4–9.

Rao, R.R. & Jamir, N.S. (1982). Ethnobotanical studies in Nagaland, I. Medicinal Plants. *Econ. Bot. 36*, 176–181.

Razi, B.A. & Subramaniam, K. (1978). Collection, cultivation and conservation of Medicinal plants in Karnataka State. *Agri. Agro. Ind. II*, 9–16.

Remesh M. & Manju, C.N. (2009). Ethnobryological notes from Western Ghats, India. *Bryologist 112(3),* 532–537.

Rheede tot Drakestein, H.A. van (1693). Horti Malabarici: Pars Doudecima. Amsterdami.

Saxena, D.K. & Harindar (2004). Uses of bryophytes. *Resonance 9(6),* 56–65.

Sharma, A., Slathia, S., Gupta, D., Handa, N., Choudhary, S.P., Langer, A. & Bhardwaj, R. (2015). Antifungal and Antioxidant Profile of Ethnomedicinally Important Liverworts (*Pellia endivaefolia* and *Plagiochasma appendiculatum*) Used by Indigenous Tribes of District Reasi: North West Himalayas. *Proc. Natl. Acad. Sci. India B85,* 571–579.

Singh, M., Rawat, A.K.S. & Govindarajan, R. (2007). Antimicrobial activity of some Indian mosses. *Fitoterapia 78,* 156–158.

Singh, M., Govindarajan, R., Nath, V., Rawat, A.K.S. & Mehrotra, S. (2006). Antimicrobial, wound healing and antioxidant activity of *Plagiochasma appendiculatum* Lehm. et Lind. *J. Ethnopharmacol. 107,* 67–72.

Vats, S. & Alam, A. (2013). Antibacterial Activity of *Atrichum undulatum* (Hedw.) P. Beauv. Against some pathogenic Bacteria. *Jour. Biol. Sci. 13(5),* 427–431.

Zaneveld, J.S. (1959). The utilization of marine algae in tropical south and East Asia. *Econ. Bot. 13(2),* 89–131.

INDEX

Printed and bound by CPI Group (UK) Ltd, Croydon, CR0 4YY

23/10/2024

01777704-0004